THE FRACTAL PHYSICS OF POLYMER SYNTHESIS

THE FRACTAL PHYSICS OF POLYMER SYNTHESIS

G. V. Kozlov, DSc, A. K. Mikitaev, DSc,
and Gennady E. Zaikov, DSc

Apple Academic Press

TORONTO   NEW JERSEY

Apple Academic Press Inc. | Apple Academic Press Inc.
3333 Mistwell Crescent | 9 Spinnaker Way
Oakville, ON L6L 0A2 | Waretown, NJ 08758
Canada | USA

©2014 by Apple Academic Press, Inc.
Exclusive worldwide distribution by CRC Press, a member of Taylor & Francis Group

No claim to original U.S. Government works
Printed in the United States of America on acid-free paper

International Standard Book Number-13: 978-1-926895-63-5 (Hardcover)

Library of Congress Control Number: 2013951499

Library and Archives Canada Cataloguing in Publication

Kozlov, G. V., author
The fractal physics of polymer synthesis/G.V. Kozlov, DSc, A.K. Mikitaev, DSc, and Gennady E. Zaikov, DSc.

Includes bibliographical references and index.
ISBN 978-1-926895-63-5
1. Polymerization. 2. Polymers--Structure. 3. Fractal analysis. I. Mikitaev, Abdulakh K., author II. Zaikov, G. E. (Gennadi™i Efremovich), 1935-, author III. Title.

QD281.P6K69 2013   547'.7   C2013-907721-9

Apple Academic Press also publishes its books in a variety of electronic formats. Some content that appears in print may not be available in electronic format. For information about Apple Academic Press products, visit our website at **www.appleacademicpress.com** and the CRC Press website at **www.crcpress.com**

ABOUT THE AUTHORS

G. V. Kozlov, DSc

G. V. Kozlov, DSc, is a Senior Scientist at UNIID of Kabardino-Balkarian State University in Nal'chik, Russian Federation, during 1981–1994 and from 1997 until now. His scientific interests include the structural grounds of properties of polymeric materials of all classes and states: physics of polymers, polymer solutions and melts, and composites and nanocomposites. He proposed to consider polymers as natural nanocomposites. He is the author of more than 1500 scientific publications, including 30 books, published in the Russian Federation, Ukraine, Great Britain, German Federal Republic, Holland, and USA.

A. K. Mikitaev, DSc,

A. K. Mikitaev, DSc, is Professor and Head of the Chair of organic chemistry and high-molecular compounds at Kabardino-Balkarian State University in Nal'chik, Russia during 1971–1990 and from 2004 until now. He set up the Scientific-Educational Center "Polymer and Composites" in 2008. He developed and worked in industry on the production technology of poly(butylenes terephtalate) and nanocomposite materials and its basis, including nanocomposite thermoelastoplastics. His scientific interest in synthesis of polymers includes developing strategies for manufacture of polymers of different classes. He is the author of more than 1000 scientific works, including 14 books, and he has received 160 author's certificates and patents.

Gennady E. Zaikov, DSc

Gennady E. Zaikov, DSc, is Head of the Polymer Division at the N. M. Emanuel Institute of Biochemical Physics, Russian Academy of Sciences, Moscow, Russia, and Professor at Moscow State Academy of Fine Chemical Technology, Russia, as well as Professor at Kazan National Research Technological University, Kazan, Russia. He is also a prolific author, researcher, and lecturer. He has received several awards for his work, including the the Russian Federation Scholarship for Outstanding Scientists. He has been a member of many professional organizations and on the editorial boards of many international science journals.

CONTENTS

LIST OF ABBREVIATIONS

CATA	Chloroanhydride of terephtalic acid
DDM	Diaminodiphenylmethane
DLA	Diffusion limited aggregation
DMDAACh	Dimethyldiallylammoniumchloride
EAEA	Ethylallylethylacrylate
IGC	Inversed gas chromatography
MWD	Molecular weight distribution
PC	Polycarbonate
PMMA	Poly(methyl methacrylate)
PPQX	Polyphenylquinoxalines
PPX	Polyphenylxalines
PPX	Polyphenylxalines
PSA	Ammonium persulphate
PSF	Polysulfone
PVC	Poly(vinyl chloride)
SANS	Small angle neutron scattering

LIST OF SYMBOLS

$\dot{\gamma}$ = Shear rate

t_g^T = Theoretical magnitude

$\overline{D}_f$ = Fractal dimension

$\Sigma\sigma$ = Gammet's constant

$\vec{\delta}$ = Solubility parameter

T_{g_1} and T_{g_2} = Homopolymers glass transition temperatures

M_n^* and M_{n+1}^* = Threshold masses of atoms

Df = Macromolecular coil fractal dimension.

$K\eta$ = Constant

m0 = Monomer link

Q = Conversion degree

Rg = Gyration radius

Su = Specific surface of nanofiller particles

T   Temperature

$Z\infty$ = Parameter limiting value

GREEK SYMBOLS

α = Macromolecular coil swelling coefficient

βe = Fraction of macromolecular coil

δ = Feigenbaum's constant

δf = absolute value of components

η = intrinsic viscosity

ηred = reduced viscosity

ν = flory exponent

ρEuc = Euclidean object

ρp = polymer density

φ^* = threshold value of concentration

χs = empirical parameter

PREFACE

One of the principal objects of theoretical research in any field of knowledge is to find the point of view from which the subject appears in its greatest simplicity.

In spite of the enormous number of papers dealing with the influence of the medium on the rate of chemical reactions (including synthesis of polymers), no strict quantitative theory capable of "universal" application has been put forward up to now. And so it is possible to describe the relationship between the reaction rate constants and the equilibrium constants with the nature of the medium in which the reactions take place by means of a single equation.

The absence of general theory of the influence of the environment on the kinetics of chemical reactions can be explained by the fact that the change of solvent (or transition from liquid to solid phase) cannot only influence the process rate but also frequently results in complication of the reaction mechanism. The calculation of the individual contributions made by each factor is thus, in most cases, rather complicated and requires a deep and comprehensive study of the properties of the medium and of the reacting particles. This is because of a quantitative evaluation of all types of interactions between the reacting particles with medium can occur only by arriving ones on the basis of full knowledge of these properties.

Aristotle asserted that "among the unknown in the nature surrounding us, the most unknown thing is time, because nobody knows what time is and how it can be controlled". Since then definite positive changes have happened in this field. Particularly, the branch of science named chemical kinetics was established, which gave people the opportunity to predict the behavior of chemical reagents with time, and disclose the inner mechanism of the interaction between particles (molecular, ions, radicals and atoms) in various chemical processes.

The main aspects of the fractal analysis application for the description of the behavior of macromolecular coils in the diluted solution are also considered to emphasize the intercommunication of the classical and fractal (structural) characteristics of macromolecular coils. Developed in the physical chemistry of polymer solutions, the basic ideas are the basis of our understanding of the peculiar properties of polymers. Such an approach allows one to receive the direct correlations "structure-properties", which is the main task of any physical domain including the physical chemistry of polymer solutions and polymers synthesis.

Hence, the fractal analysis, that is a purely physical (structural) conception, and the irreversible aggregation models, closely connected with it, provide a simple quantitative description of both environment and time, whereas a reaction mechanism change also influences the reaction course of high-molecular systems. This is possible just owing to introduction of the polymer structure in its different states.

— **G. V. Kozlov, DSc, A. K. Mikitaev, DSc,**
and Gennady E. Zaikov, DSc

ABSTRACT

In the present monograph the description of the main reactions of high-molecular substances (polycondensation, radical polymerization, branched polymers synthesis, curing of cross-linked polymers, synthesis of polymer nanocomposites *in situ*, catalyzed reactions) is proposed within the frameworks of the fractal analysis and an irreversible aggregation models. The synergetics and percolation theory were also used. The notion of the polymer structure in solution (macromolecular coil) and its condensed state is introduced, and their intercommunication is established. This allows predicting the solid-state polymer characteristics that are already at the stage of synthesis. The influence of both environment and reaction duration and also of aggregation (synthesis) mechanism change effect can be taken into account within the frameworks of the mentioned conceptions. The polymers synthesis in a melt is also considered. It is significant that the fractal analysis, being purely physical conception, gives an exact and simple quantitative description of both indicated above reactions kinetics and their final (limiting) characteristics.

INTRODUCTION

In spite of the enormous number of papers dealing with the influence of the medium on the rate of chemical reactions (including synthesis of polymers), no strict quantitative theory capable of "universal" application has been put forward up to now. It is now possible to describe the relationship between the reaction rate constants and the equilibrium constants with the nature of the medium in which the reactions take place by means of a single equation.

This important book, for the first time, gives structural and physical grounds of polymers synthesis and curing, and the fractal analysis is used for this purpose. This book presents important aspects on fractal physics of polymer synthesis such as polycondensation, radical polymerization, the branched polymers synthesis, and the curing of cross-linked polymers. The fractal analysis is used for this purpose. The book covers the theoretical fundamentals of macromolecules fractal analysis and then goes on to discuss the fractal physics of polymer synthesis and the fractal analysis and synergetics of catalytic systems. The fractal physics of polymer synthesis presents descriptions of the main reactions of high-molecular substances within the frameworks of fractal analysis and an irreversible aggregation models. Synergetics and percolation theory were also used.

The fractal physics of polymer synthesis is a new topic in the research field of polymer synthesis, which has attracted increasing interest due to its potential applications in the real world, such as modeling of polymeric materials. In this part, basic theory for fractional differential equations and numerical simulations for these equations will be introduced and discussed for polymers of different classes and polymers solutions. In the infinite dimensional dynamics part, we emphasize numerical calculation and theoretical analysis, including constructing various numerical methods and computing the corresponding limit sets, etc. In this book, we show interest in network dynamics and fractal dynamics together with numerical simulations as well as their applications. For each topic the theoretical concepts are carefully explained using examples and applications within the framework of fractal approximations taking into account the hydro-dynamical interactions.

The book covers the theoretical fundamentals of macromolecules fractal analysisand then goes on to discuss the fractal physics of polymer solutions.

CHAPTER 1

POLYCONDENSATION

CONTENTS

1.1   THE SOLVENT NATURE INFLUENCE ON STRUCTURE AND FORMATION MECHANISM OF POLYCONDENSATION POLYMERS

As it is known [1], the following relationship is one from the fractal definitions in reference to a macromolecular coil:

$$R_g \sim N^{1/D_f} \qquad (1)$$

where Rg is macromolecular coil gyration radius, N is polymerization degree, Df is macromolecular coil fractal dimension.

The comparison of the Eq. (1) and the known Flory equation [2]:

$$R_g \sim N^{\nu}, \qquad (2)$$

where ν is Flory exponent, shows that between parameters Df and ν the intercommunication exists Eq. (3):

$$D_f = \frac{1}{\nu} . \qquad (3)$$

Nevertheless, the Eqs. (1) and (2) are valid for different objects. If Flory equation is correct for arbitrary coils, then the fractal Eq. (1) — for only semi-similar ones (by the fractal definition [3]).

The fractal analysis main rules in reference to polymer solutions description can be found in the reviews [4, 5]. The common remark should be made in respect to the Eq. (1). The fractal dimension Df characterizes macromolecular coil structure, defining its elements distribution in space. The increase of Df means Rg decreasing at N = const, i.e., a coil compactness enhancement.

Since the introduction in analysis of macromolecular coil structure, characterized by its fractal dimension Df, is the key moment of polycondensation process fractal physics, then the value Df determination methods are necessary for practical application of polycondensation fractal analysis for solutions. This parameter for macromolecular coil in solution is defined by two groups of interactions: interactions polymer-solvent and interactions of coil elements among them [6]. At present several methods of the first from the indicated groups interactions exist and all of them can be used to a certain extent for Df calculation [5].

The simplest experimental method of Df calculation is the equation [7]:

$$D_f = \frac{3}{1+a_\eta},$$

(4)

where $a\eta$ is the exponent in Mark–Kuhn–Houwink equation, connecting intrinsic viscosity and molecular weight of a polymer.

From the Eq. (4) it follows, that the exponent $a\eta$, earlier assumed purely empirical characteristic, has a clear structural interpretation. One from the calculated methods of Df determination uses the known Huggins equation, which gives the dependence of reduced viscosity ηred on concentration c for diluted polymer solutions [8]:

$$\eta_{red} = [\eta] + k_H [\eta]^2 c + ...,$$

(5)

where $[\eta]$ is intrinsic viscosity, kH is Huggins constant, which characterizes polymer-solvent interactions level.

Besides, the relation between specific viscosity ηsp, c and $[\eta]$ can be obtained, using Shultz–Blashke equation [9]:

$$[\eta] = \frac{\eta_{sp}/c}{1+K_\eta \eta_{sp}},$$

(6)

where $K\eta$ is the constant, accepting in the first approximation equal to 0.28.

The Eqs. (5) and (6) at the condition c=const (the value c is accepted further equal to 0.5 mass. %) allows to obtain the simple expression for kH estimation [10]:

$$k_H = \frac{0.14}{1-0.14[\eta]}$$

(7)

For kH calculation the authors [10] supposed that all the used in work polymers (polyarylates (PAr), poly(vinyl chloride) (PVC), poly(methyl methacrylate) (PMMA), polysulfone (PSF) and polycarbonate (PC)) had the same molecular weight MM = 5 × 105.

Such value MM was chosen because its smaller values give close values kH that increase an estimations error. The values [η], corresponding to the indicated MM magnitude, were calculated according to Mark–Kuhn–Houwink equation [11–13].

In paper [10], the dependence of Df, obtained according to the Eq. (4), on parameter k_H^{-2} is adduced (such form of the dependence was chosen with the purpose of its linearization). A good linear correlation was obtained for 30 different polymer-solvent pairs (correlation coefficient is equal to 0.930 [10]), allowing to predict simply enough the value Df. It is expected that for other MM values the correlation Df (k_H^{-2}) will have a similar form, but another slope. The mentioned dependence Df (k_H^{-2}) allows to make a conclusions number. First, an impression is created that this correlation gives the dependence Df on the polymer-solvent interactions only, characterized by Huggins parameter kH and does not take into account interactions of coil elements among them. However, this correlation linearity itself supposes, that it takes into consideration the second group factors, as well which was mentioned above. For example, it is well known [11], that chain rigidity enhancement results to exponent aη increase in Mark–Kuhn–Houwink equation and, hence, to Df reduction (the Eq. (4)). Simultaneously chain rigidity enhancement results to [η] growth at other equal conditions. Thus, both the chain rigidity increase and the improvement of the solvent theormodynamical quality in respect to polymer give the same effect — [η] increase and, correspondingly, kH growth according to the Eq. (7). This, in its turn, results to k_H^{-2} decrease and Df reduction that is expected. This supposition is confirmed experimentally — in paper [14] the increase of Kuhn statistical segment length A, characterizing the chain thermodynamical rigidity, at the improvement of the solvent thermodynamical quality for two polyarylates, is shown.

Secondly, as it follows from the Eq. (7), the minimum value kH = 0.14 (or maximum value $k_H^{-2} \approx 51$) is reached at [η] = 0. From the plot Df (k_H^{-2}) it follows that Df ≈2.25 corresponds to this value kH. As it is known [6], the screening of the exluded volume interactions results to Df increase and at complete screening (the compensation of the mentioned effects) D_f^c value corresponds to the so-called compensated state. Within the frameworks of Flory's theory, when the compensation is realized by the interactions with other coils, D_f^c=2.5 (for three-dimensional Euclidean space). Another method for the decrease of repulsive interactions among coil elements is the introduction of the attractive interactions.

For this case, corresponding to an isolated coil (a dilute solution) can be written [6] Eq. (8):

$$D_f^c = \frac{4(d+1)}{7},$$

(8)

where d is the dimension of Euclidean space, in which a fractal is considered. It is obvious that in our case d = 3 and D_f^c = 2.286, that practically exactly corresponds to the limiting Df value, obtained from the plot Df (k_H^{-2}).

Another limiting case, corresponding only to the repulsive interactions, can be obtained from the Eq. (7) at [η] = 7.14 dl/g, that corresponds to the condition k_H^{-2} =0. In this case the plot Df (k_H^{-2}) extrapolation to k_H^{-2} = 0 gives Df =1.5, that corresponds to the dimension of the so-called permeable coil [2]. Thus, the limiting Df values, obtained according to the plot Df (k_H^{-2}), completely correspond to the theoretical conclusions [10].

Thirdly, at present the exponent aη in Mark–Kuhn–Houwink decrease at MM growth is well-known [11, 15, 16]. The Eqs. (4), (7) and Mark–Kuhn–Houwink equation combination allow to obtain the following relationship [10]:

$$D_f = \frac{3 \ln MM}{\ln MM + \ln (7.14 k_H - 1) - \ln K_\eta - \ln k_H},$$

(9)

where Kη is constant in Mark–Kuhn–Houwink equation.

The Eq. (9) gives the analytical interrelation between Df (or aη, the Eq. (4)) and MM. The values Df estimations for polyamidoacid at MM = 4 × 10⁴ and 15 × 10⁴, corresponding to two intervals of aη value according to the paper [15] data, allow to obtain the exact correspondence of values Df (or aη), determined experimentally [15] and calculated according to the Eq. (9). Hence, aη decrease (Df increase) at MM growth has law-governed character [10].

One more method of D_f calculation is the usage of Flory–Huggins interaction parameter χ_1, which is determined as follows [8]:

$$\chi_1 = \frac{E_{vap}}{RT} \left(1 - \frac{\delta_p}{\delta_s}\right)^2 + \chi_s,$$

(10)

where Evap is the molar energy of solvent vaporization, R is the universal gas constant, T is temperature, δ_p and δ_s are solubility parameters of polymer and solvent, accordingly, χ_s is an empirical parameter.
Sometimes χ_1 value is defined as follows [8] Eq. (11):

$$\chi_1 = \chi_H + \chi_s \;, \tag{11}$$

where χ_H and χ_s are enthalpic and entropic components of Flory–Huggins interaction parameter, respectively.

The Eq. (10) allows to suppose, that with the aid of the parameter δp interactions of coil elements among themselves can be determined and with the aid of the ratio δ_p/δ_s — interactions polymer-solvent [17]. Besides, the entropic component s takes into account the effect of system interacting elements ordering [8].

In paper [17] the relation Df (χ_1) for 8 polymer-solvent pairs is adduced, which is approximated well enough by a straight line and has expected limits. For $\chi_1 = 0$ (the interactions polymer-solvent absence, $\delta_p = \delta_s$) $D_f = 1.5$, i.e., the permeable coil is formed in solution [2]. It is obvious, that this condition is reached only at χs = 0. For large enough $\chi_1 \geq 1.0$, i.e., for transition solvent — nonsolvent [8], the value $D_f = 2.0$, i.e., it corresponds to θ-condition [2]. The relationship $D_f (\chi_1)$ can be expressed analytically with the aid of the empirical equation [17]:

$$D_f = 1.5 + 0.45\chi_1 . \tag{12}$$

Hence, the proposed above assumption about intercommunication of Df and χ_1 corresponds to the experimental data: Flory–Huggins interaction parameter describes exactly enough interactions system for macromolecular coil in solution, controlling its fractal dimension value. The main problem at the Eq. (12) usage with the purpose of D_f predicting is the empirical parameter χs calculation method absence [17].

Nevertheless, the values δ_p and δ_s knowledge does not allow to define, whelher polymer is diluted in the given solvent or not [18]. Such situation is due to the fact, that the parameters δp and δs are integral characteristic of intermolecular interaction and solubility depends on the availability in molecules of functional groups components, entering or not entering into interaction among themselves. Therefore further development of solubility parameter conception was going on the way of groups influence consideration, which are capable to specific interaction, on substances mutual mixing.

In model [19] two components of the solubility parameter $\vec{\delta}$ (δ_p or δ_s) are identified as components of interaction power field δf and complexing δ_c. The main postulates of the model [19], concerning the considered problem, can be expressed as follows:

The ability of a solvent to dissolve any polymer is determined by two and only two conditions Eq.(13) and (14):

$$\delta_f \geq 0 \ , \tag{13}$$

$$\delta_c \geq 0 \ . \tag{14}$$

The two components δ_f and δ_c for each solvent are related to the value $\vec{\delta}$ as follows:

$$\vec{\delta}^2 = \delta_f^2 + \delta_c^2 \ . \tag{15}$$

The probability of a polymer A being dissolved in a given solvent B increases at decreasing differences of absolute values between their solubility parameter components:

$$\left|\delta_{fA} - \delta_{fB}\right| \text{ and } \left|\delta_{cA} - \delta_{cB}\right| \ . \tag{16}$$

The component δf of solubility parameter includes dispersion interactions energy and dipole bonds interactions energy and the component c — hydrogen bonds interaction energy and interaction energy between an electron-deficient atom of one molecule (acceptor) and electron-rich atom of another molecule (donor), which requires specific orientation of these two molecules. In such treatment there is no need to introduce a separate component for interaction between polar molecules description [19].

On the basis of this model of solubility two-dimensional parameter and analysis of large amount of experimental data the following equation for theoretical Df estimation was proposed in paper [20]:

8

The Fractal Physics of Polymer Synthesis

$$D_f = 1.5 + 0.2(\Delta\delta_f)^{2/(1+\delta_c)},\qquad(17)$$

where $\Delta\delta_f$ is absolute value of components δf difference for polymer and solvent (see the Eq. (16)).

Df value calculation for 16 polymer-solvent pairs according to the Eq. (17) shows a good enough correspondence to calculation according to the Eq. (4) (the average discrepancy makes up approx. 4 %) [20].

The Eq. (17) clearly demonstrates different role of the mentioned above two interaction groups. In Fig. 1 the dependences $D_f(\delta_c)$ for 4 values δ_f (0.5; 0.9; 1.2 and 1.5) are adduced which were calculated according to the Eq. (17). It is easy to see, that $\Delta\delta_f$ increase (or solvent δf increase at constant value δf of polymer) results to D_f considerable growth, i.e., to exluded volume interactions screening. The increase of δ_c results to these distinctions leveling off and at large enough δc the value Df tries to attain asymptotic magnitude of approx. 1.70. Moreover at $\Delta \delta f < 1$ (cal/cm³)1/2, i.e., for good enough solvents, δc increase worsens solubility and for $\Delta\delta f>1$ (cal/cm3)1/2 δc enhancement gives an opposite effect. This example demonstrates clearly polymer-solvent interactions complexity and shows impossibility of strict correlations of Df and integral characteristics δp and δc only obtaining (see the Eq. (15)).

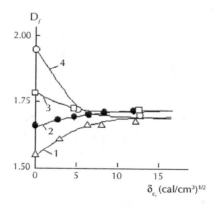

FIGURE 1 The dependences of fractal dimension D_f on solvent solubility parameter component δ_c at $\Delta\delta_f$: 0.5 (1), 0.9 (2), 1.2 (3) and 1.5 (j/m3)1/2 (4).

The experimental express-method of D_f value estimation on the basis of polymeric solutions intrinsic viscosity [η] measurements only was proposed in paper

[21]. In essence, in its basis the same principles are used as at the Eq. (4) derivation. As it is known [22], a macromolecular coil-swelling coefficient α is defined as follows:

$$\alpha = \left(\frac{\langle h^2 \rangle}{\langle h_\theta^2 \rangle} \right)^{1/2},$$ (18)

where $\langle h^2 \rangle$ and $\langle h_\theta^2 \rangle$ are mean-square distances between macromolecule ends in arbitrary and ideal (θ) solvent, respectively.

In its turn, the value α is connected with polymer intrinsic viscosity in arbitrary $[\eta]$ and ideal $[\eta]\theta$ solvents as follows [22]:

$$\alpha^3 = \frac{[\eta]}{[\eta]_\theta}.$$ (19)

The bulk interactions (which are resulted to a coil shape departure from ideal Gaussian one) parameter ε is defined according to the following equation [22]:

$$\varepsilon = \frac{d \ln \alpha^2}{d \ln MM} = \frac{\alpha^2 - 1}{5\alpha^2 - 3}.$$ (20)

In its turn, both ε and D_f depend on the exponent $a\eta$ value in Mark–Kuhn–Houwink equation [23]:

$$\varepsilon = \frac{2a_\eta - 1}{3},$$ (21)

and the dependence $D_f(a\eta)$ is given by the Eq. (4).

The Eqs. (4) and (18)-(21) combination allows to obtain the following relationship for dimension D_f determination by intrinsic viscosities $[\eta]$ and $[\eta]\theta$ values only [21]:

$$D_f = \frac{5([\eta]/[\eta]_\theta)^{2/3} - 3}{3([\eta]/[\eta]_\theta)^{2/3} - 2}.$$ (22)

The value $[\eta]\theta$ can be estimated either directly from an experiment or according to Mark–Kuhn–Houwink equation at the condition $a_\eta=0.5$, which is valid for θ-point, if the constant K_η in this equation is known. Let us note, that such determination D_f method is based on strong dependence of $[\eta]$ on macromolecular coil structure, characterized by the dimension D_f. This dependence is described by Mark–Kuhn–Houwink equation fractal variant [24] Eq. (23):

$$[\eta]=\frac{8.1\left(0.753+0.247/\alpha^3\right)}{m_0^{3/D_f}}, \tag{23}$$

where m0 is monomer link of molecular weight.

The comparison of calculated according to the Eqs. (4) and (22) D_f values showed their good correspondence for biopolymers and polyarylates number. The Eq. (22) also gives D_f(or a_η) dependence on polymer molecular weight [21].

The branching degree (and/or bulk side substituents availability) is a polymer chain important property, to a great extent defining polymers behavior in both solution and condensed state. It has been shown [11, 25], that side substituents molecular weight and branching degree increasing results to systematic decreasing of the exponent a_η in Mark–Kuhn–Houwink equation. In papers [26, 27] the factors, influencing on branched polymers D_f values in solution were studied and variation D_f tendencies at these factors change were defined on the example of two polymers groups: statistically branching polyphenylquinoxalines (PPQX) [25] and bromine-containing aromatic copolyethersulfines (B-PES) [28]. For quantitative characteristic of polymer branching degree the branching factor g was used, which is determined as follows [11]:

$$g^{2-a_\eta}=\frac{[\eta]_b}{[\eta]_l}, \tag{24}$$

where the indices "b" and "l" mean branched and linear macromolecule, respectively.

In Fig. 2 the dependences $D_f(g)$ are adduced for PPQX in N-methylpyrrolydon and chloroform and B-PES in chloroform (points). As one can see, in any case the branching degree, characterized by g decreasing [11, 25], results to D_f growth, i.e., to coil compactness degree enhancement. The thermodynamical quality of solvent in respect to polymer is one more factor, influencing on D_f value. As it follows from Fig. 2 plots, poor solvent for PPQX (N-methylpyrrolydon) results to essentially higher D_f values in comparison with good for both considered polymers solvent (chloroform). Let us note, that the data for linear analogs of these polymers (at g=1) do not correspond to the dependences for the branched polymers, moreover the law for relation of values D_f for branched and linear polymers is absent. So, for the solution of PPQX in N-methylpyrrolydon D_f for the linear analogue is smaller, than D_f for the branched polymer, in chloroform it is larger and D_f for B-PES has the dependence D_f (g), common for both types of polymers [26].

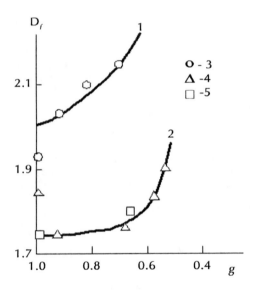

FIGURE 2   The dependences of fractal dimension D_f of macromolecular coil on branching factor g for PPQX (1–4) and B-PES (5). The solvents: N-methylpyrrolydon (1, 3) and chloroform (2, 4, 5). 1, 2—calculation according to the Eq. (25); 3–5—the experimental data.

The plots of Fig. 2 allow to make several conclusions. Firstly, as it was to be expected [6], the value D_f of the branched polymers is controlled by two factors: interactions polymer-solvent and interactions of coil elements among themselves. Secondly, the branching degree g is a prevalent parameter in the second factor definition — we do not observe any correspondence with linear analogs. The parameter ε, determined according to the Eq. (21), can be used for the estimation of character of interaction of macromolecular coil elements among themselves.

In Fig. 3 the dependences $\varepsilon(g)$ for PPQX and B-PES are shown, from which it follows, that polymer chain branching degree increasing defines repulsive interactions among coil elements weakening, i.e., positive values ε reduction. For PPQX coils in N-methylpyrrolydon attractive interactions (negative values ε) are prevalent. Thus, the data of Figs. 2 and 3 suppose, that at the same g values a good solvent (chloroform) screens effectively the attractive interactions between branches and/or side substituents, that results to smaller values D_f of the branched polymers coils in such solvent in comparison with poor solvent [28].

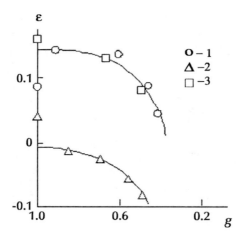

FIGURE 3   The dependences of bulk interactions parameter ε on branching factor g for PPQX (1, 2) and B-PES (3). The solvents: N-methylpyrrolydon (1) and chloroform (2, 3).

To obtain analytical intercommunication between D_f and g the Eqs. (4) and (24) were used. Their combination allows to obtain the following relationship [29]:

$$D_f = \frac{3\ln g}{3\ln g - \left(\ln[\eta]_B - \ln[\eta]_I\right)}. \tag{25}$$

In Fig. 2 the comparison of the dependences $D_f(g)$, obtained according to the Eqs. (4) (points) and (25) (solid curves) is adduced. From this comparison a good correspondence of the dependences $D_f(g)$, obtained by both indicated modes and, as consequence, the possibility of the Eq. (25) application for D_f calculation by the known values g or, vica versa, g values by the known magnitudes D_f follows. The solvent type influence in the Eq. (25) is taken into account by intrinsic viscosities $[\eta]_B$ and $[\eta]_I$ logarithmic difference, adduced in brackets — the better solvent thermodynamical quality in respect to polymer, the larger this difference is and, respectively, the smaller the value D_f is [30]. Let us note an interesting feature of the Eq. (25) — shown in Fig. 2 D_f increasing at polymer branching degree enhancement is not defined by g change, but logarithmic $[\eta]_B$ and $[\eta]_l$ indicated difference value change. As the estimations have been shown, for hypothetical polymer, having the indicated difference constant value, the dependence $D_f(g)$ course would be observed, opposite to the shown one in Fig. 2, i.e., polymer chain branching degree enhancement (g reduction) would result not to growth, but to D_f reduction [28–30].

Let us note in conclusion, that all widespreading in traditional physics-chemistry of polymer solutions parameters are connected in any case with macromolecular coil structure, characterizing its fractal dimension D_f. This circumstance for itself speaks about a very important role of the indicated factor [28].

Having the value D_f estimation methods, the influence of this main structural characteristic of macromolecular coil on polycondensation process basic parameters can be considered. As it is known [31], different organic solvents are used as reactive medium at nonequilibrium polycondensation in solution realization. The solvent type influence on polycondensation main characteristics (conversion degree Q and molecular weight MW) is well-known and is usually explained by different solvent characteristics (dielectric constant, solubility degree, dissolving heat and so on) [31]. However, these effects explanation is not found up to now. Besides, at solvent type influence analysis its correlation with quantitative characteristics of polycondensation process (the same Q and MW) is usually considered, but it not assumed any polymer structure or reaction mechanism changes are not assumed, although secondary reactions occurrence possibility is noted repeatedly [31]. Therefore the authors [32, 33] studied a solvent influence on the enumerated above characteristics on the example of low-temperature polycondensation of chloranhydride of terephthalic acid and phenolphtaleine (polyarylate Ph-2) laws [34]. The fractal analysis methods [3, 35] were used for this purpose reaching. The main postulate of such method is the introduction of the notion of macromolecular coil structure, described by its fractal dimension D_f, which is a strict structural characteristic, and obtaining of the relationships structure-properties in synthesis reactions.

As it has been shown above, polymers macromolecular coils in solution are fractal objects, i.e., self-similar objects, having dimension, which differs from their topological dimension. The coil fractal dimension D_f, characterizing its structure (a coil elements distribution in space), can be determined according to the Eq. (4). The exponent $a\eta$ values for polyarylate Ph-2 solutions in three solvents (tetrachloroethane, tetrahydrofuran and 1,4-dioxane) are adduced in [36]. The values $a\eta$ for the same polyarylate are also given in paper [37]. This allows to use the Eq. (4) for the macromolecular coil of Ph-2 D_f value estimation in the indicated solvents. The estimations showed D_f variation from 1.55 in tetrachloroethane (good solvent for Ph-2) up to 1.78 in chloroform. As it is known [38],

the solvent thermodynamical quality in respect to polymer in the first approximation (similar estimations approximation reasons were considered in work [39])
can be estimated with the aid of their solubility parameters, δ_s and δ_p, difference,
respectively, i.e., $\Delta\delta=|\delta_p-\delta_s|$. Using the literary data for δ_p and δ_s [38], the authors
[33] plotted the dependence $D_f(\Delta\delta)$ for the indicated above four solvents, which
is shown in Fig. 4 and has an expected form — $\Delta\delta$ increasing (solvent thermodynamical quality in respect to polymer change for the worse) results to D_f growth.
The authors [33] used this dependence continuation (shaded line in Fig. 4) as the
calibrating curve, that allows to estimate the macromolecular coil of Ph-2 D_f value
in other solvents, used at its synthesis.

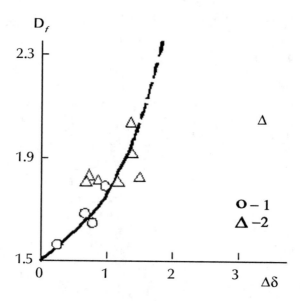

FIGURE 4   The dependence of macromolecular coil fractal dimension D_f on polymer
and solvent solubility parameters difference $\Delta\delta$ for polyarylate Ph-2. D_f calculation: 1 —
according to the Eq. (4); 2 — according to the Eq. (22).

The Eq. (22) gives one more possible mode of D_f value estimation. The
estimations according to the Eqs. (4) and (22) showed, that different solvents
used in low-temperature polycondensation process resulted to various D_f values
(Table 1).

TABLE 1 Comparison of the macromolecular coil fractal dimensions D_f, obtained by different methods, for polyarylate Ph-2, synthesized in different solvents [33].

Solvent	D_f, the Eq. (4)	D_f, the Eq. (22)
N,N-dimethylformamide	1.90	2.03
Nitrobenzene	1.70	1.80
Acetone	1.70	1.81
Dichloroethane	1.63	1.80
Chloroform	1.59	1.90
1,2,4-Trichlorobenzene	1.99	1.80
Benzene	2.28	1.81
Hexane	2.50	2.05

The two obtained results, following from this table data, attract their attention. Firstly, the solvent change defines variation of not only low-temperature polycondensation process quantitative characteristics, i.e., Q and η_{red}, but also variation of macromolecular coil structure, i.e., its fractal dimension D_f. Secondly, although some distinctions by absolute value exist between D_f magnitudes, determined according to the Eqs. (4) and (22), but their change tendencies at solvent variation are identical. In other words, the tendencies similarity supposes, that D_f change, which is due to interaction polymer-solvent change, is fixed up to a certain extent in synthesis process and is transformed in D_f change, which is due to interaction between coil element variation (let us remind, that the values of the reduced viscosity ηred in work [34] were measured in the same solvent — tetrachloroethane). In Fig. 4 the dependence of calculated according to the Eq. (22) (i.e., after synthesis) D_f values on $\Delta\delta$ [38] are adduced. As one can see, this dependence corresponds well to similar dependence on the basis of the Eq. (4), exluding hexane, which does not dissolve Ph-2 and in which synthesis is not practically realized ($Q \approx 2\%$ [34]).

For fractal description of low-temperature polycondensation process the authors [33] made use of the relationship, connecting accessible for reaction active sites number, which is supposed proportional to Q, and D_f value [40]:

$$Q \sim \eta_0 c_0 t^{(3-D_f)/2} , \qquad (26)$$

where η_0 is initial reactive medium viscosity, c_0 is initial reagents concentration, t is reaction running time.

The typical for polycondensation this type value t = 60 min was chosen for all solvents, the value c_0 is also the same for all solvents and therefore is included in the Eq. (26) constant coefficient and the parameter η_0 estimation method should be specially described. In paper [34] the values Q and η_{red} for low-temperature polycondensation realization two conditions are adduced — with stirring (200 rpm) and without reactive mass stirring. In the first case it is supposed, that stirring levels distinctions in η_0 value and then the simplest relationship was used for Q estimation [33]:

$$Q \sim t^{(3-D_f)/2} \quad t = 60 \text{ min.} \tag{27}$$

The Eqs. (26) and (27) are valid for three-dimensional Euclidean space. In Fig. 5 the experimental Q [34] and calculated according to the Eq. (27) Q^T polyarylate Ph-2 conversion degree values comparison is adduced. As one can see, the satisfactory linear correlation, passing through coordinates origin, is observed between Q and Q^T values. This means, that the fractals dimension D_f of macromolecular coil, which, in its turn, is defined by interactions polymer-solvent (Fig. 4), is the only factor, defining value Q at reative mass stirring (for t=const).

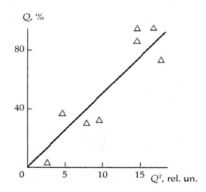

FIGURE 5   The experimental Q [34] and calculated according to the Eq. (27) Q^T conversion degree values comparison for polyarylate Ph-2 synthesis in different solvents (reaction with stirring).

In Fig. 6 the dependence Q^e on D_f value is adduced, from which polymer conversion degree sharp reduction follows at D_f growth. The greatest (and close to 100 %) Q values are reached for D_f values, corresponding to the interval of 1.50–1.67, i.e., permeable coil — coil in a good solvent interval [2]. At transition to the states "coil in θ-solvent" and "coil in poor solvent" ($D_f \geq 2$ [2]) the value Q reduces essentially. Let us note, that reaction occurs in essence with high enough Q values for D_f within the range of approx. 1.50 – 2.20, i.e., within D_f range, corresponding to aggregation mechanism cluster–cluster [41]. At transition to systems universality class particle–cluster ($D_f \approx 2.5$ [41]) reaction is practically stopped.

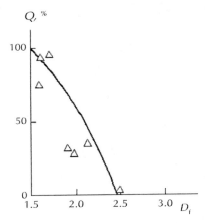

FIGURE 6   The dependence of conversion degree Q [34] on macromolecular coil fractal dimension D_f for polyarylate Ph-2 synthesis in various solvents.

In papers [31, 34] the data about the initial reactive medium viscosity η_0 were not adduced. Therefore the authors [33] estimated this parameter from the condition of the Eq. (26) and relationship [33]:

$$MW \sim \frac{c_0}{\eta_0} \qquad (28)$$

for the four solvents, in which polyarylate Ph-2 synthesis was performed without stirring [34]. As it was expected, for the three from the indicated solvents (nitrobenzene, acetone and dichloroethane) the values η_0 are close enough (2.02–2.66 relative units) and for 1,2,4-trichlorobenzene appreciable reduction η_0 (~ 0.64 relative units) was obtained. In Figs. 7 and 8 the comparison of theoretically calculated Q^T and η_{red}^T and experimentally obtained Q and η_{red} conversion degree and reduced viscosity values, respectively, for Ph-2 synthesis in four solvents without stirring. As was to be expected, the obtained good correspondence of theory and experiment shows parameter η_0 importance in such synthesis kind [33].

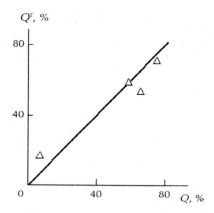

FIGURE 7   Comparison of the experimental Q [34] and calculated according to the Eq. (26) Q^T conversion degree values for polyarylate Ph-2 synthesis in different solvents (reaction without stirring).

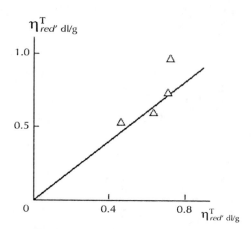

FIGURE 8   Comparison of the experimental η_{red} [34] and calculated according to the Eq. (28) η_{red}^T reduced viscosity values for polyarylate Ph-2 synthesis in different solvents (reaction without stirring).

Hence, the stated above results demonstrated the possibility of quantitative description of solvent nature on polyarylate Ph-2 formation process by low-temperature polycondensation. The proposed within the frameworks of modern physical models treatment showed macromolecular coil structure importance in this process. The indicated structure is defined by interactions polymer-solvent. Besides, the proposed approach allows to take into account such factors influence as stirring availability, initial reactive medium viscosity, reagents initial concentration and reaction duration. Everylling stated above opens polycondensation processes computer simulation perspectives [33].

Gammet's constant $\Sigma\sigma$ is one from physical-chemical parameters, often applied at organic compounds synthesis description [42]. This parameter in quantitative form reflects the character and degree of the given substituent influence on reactive center susceptibility to reagents different types action. The total Gammet's constants can be calculated according to the equation [43]:

$$In\left(\frac{K}{K_0}\right) = \rho\sum\sigma, \qquad (29)$$

where K and K_0 are dissociation constants of substituted and nonsubstituted phenols, respectively, ρ is given reactive series rate constant.

As it follows from the Eq. (29), the value $\Sigma\sigma$ is defined through reaction kinetic characteristics and, as consequence, is the parameter, controlling polycondensation kinetics [44, 45].

Relationships, similar to the Eq. (29), do not take into account polycondensation in solution main element structure — macromolecular coil, although steric and conformational factors influence on polycondensation course is often mentioned. Therefore the authors [46] performed the study, how macromolecular coil structure could influence on the value $\Sigma\sigma$ and, hence, on polycondensation reaction kinetics.

The data for high-temperature polycondensation of different disodium salts of the bisphenols with 4,4'-dichlorodiphenylsulfone (DCDPS) are used [47]. The bisphenols denomination, their conventional signs and also the values of polycondensation rate constants k_p and $\Sigma\sigma$ are adduced in Table 2. Besides, the kinetic curves conversion degree — reaction duration (Q – t) of polycondensation, adduced in the work [47], were used.

The Fractal Physics of Polymer Synthesis

TABLE 2 Designat ions and characteristics of used bisphenols [46].

Bisphenol denomination	Conventional sign	T, K	D_f	k_p, mole/(l·s)	$\Sigma \sigma$ HO–Ar–N
2,2-di-(4-oxyphenyl) propane	DOPP	373	2.05	0.2440	
		398	1.89	1.0550	−0.1107
		423	1.66	3.8700	
di-(4-oxyphenyl) phenylmethane	DOPPM	373	1.68	0.2660	
		398	1.58	1.2610	−0.1292
		423	1.44	3.6200	
3,9-di-(4-oxyohenylethyl)- spirobimethadioxane	DOPES	373	1.63	0.3580	
		398	1.52	1.7800	−0.1277
		423	1.46	3.3700	
2,2-di-(4-oxy-3- methylphenyl) propane	DOMPP	373	1.80	0.3520	
		398	1.60	1.724	−0.1952
		423	1.45	4.3400	
di-(4-oxyphenyl) sulfone	DOPS	373	2.66	0.0064	
		398	2.34	0.0164	0.6589
		423	1.94	0.0853	
1,1-di-(4-oxy-3-chlorophenyl) cyclohexane	DOCPCH	373	2.20	0.0095	
		398	2.05	0.0582	0.2826
		423	1.88	0.2421	

Within the frameworks of fractal analysis the polycondensation kinetics is described by the general Eq. (27). If the indicated relationship describes polycondensation kinetics correctly, then the dependence Q(t) in double logarithmic coordinates should give a straight line, from the slope of which the value D_f, changing within the limits of $1 \leq D_f \leq 3$ [4], is determined. The determined by the indicated mode Df values are also adduced in Table 2.

As the comparative analysis of the values D_f and $\Sigma \sigma$ change character has shown, D_f increase is always accompanied by $\Sigma \sigma$ growth. Therefore in Fig. 9 the dependences $\Sigma \sigma (D_f)$ for three polycondensation temperatures were adduced. As one can see, these dependences are linear and temperature influence is expressed in the curves $\Sigma \sigma (D_f)$ vertical shift. As it is known [24], D_f value is determined by two interactions groups: interactions polymer-solvent and interactions of macromolecular coil elements among themselves.

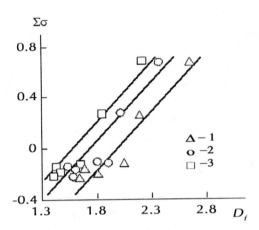

FIGURE 9 The dependences of Gammet's constant $\Sigma\sigma$ on macromolecular coil fractal dimension D_f for polycondensation temperatures T: 373 (1), 398 (2) and 423 K (3).

The appearance of the dependences $\Sigma\sigma(D_f)$, adduced in Fig. 9, assumes that the straight lines vertical shift characterizes the first group of interactions, and the dependences $\Sigma\sigma(D_f)$ for each temperature — the second one. Therefore the common linear dependence (master-straight line) $\Sigma\sigma(D_f)$ can be constructed by the three indicated straight lines superposition and then effective Gammet' constants $\Sigma\sigma_{ef}$ can be determined, proceeding from D_f values. The dependence $\Sigma\sigma(D_f)$ at T=373 K should be chosen as a standard straight line for the following reason. The dependences $\Sigma\sigma(D_f)$ character assumes that $\Sigma\sigma$ increase means indensification attraction interaction among coil elements, resulting to its compactization, i.e., to D_f growth, and on the contrary. Then at $\Sigma\sigma=0$ $D_f=2.0$ it should be expected, i.e., coil dimension in -conditions, where attraction and repulsion are balanced [16, 24]. The dependence $\Sigma\sigma(D_f)$ at T=373 K corresponds to this criterion. Then the generalized dependence analytical form for $\Sigma\sigma_{ef}$ can be written [63]:

$$\Sigma\sigma_{ef} = 0.72D_f - 1.40 .$$

(30)

The Eq. (30) allows to determine the limiting values $\Sigma\sigma_{ef}$ for the studied bisphenols, proceeding from the boundary D_f values. The minimum value $\Sigma\sigma_{ef}^{min}$ =–0.68 is obtained at D_f=1.0 and the greatest value $\Sigma\sigma_{ef}^{min}$=0.76 at D_f=3.0 [63].

Further, using calculated according to the Eq. (30) $\Sigma\sigma_{ef}$ values, the dependence of kinetic parameter K_p on interactions level of the indicated above two groups can be constructed. From the Fig. 10 it follows, that now the dependence lg $K_p(\Sigma\sigma_{ef})$ is approximated well by the sole straight line unlike the works [61, 64], where the individual straight line corresponds to each polycondensation temperature. Thus, Gammet's constant physical sense means, that it describes quantitatively interactions level of macromolecular coil elements among themselves, moreover $\Sigma\sigma$ reduction means intensification of repulsion interaction and $\Sigma\sigma$ increasing — attraction interactions intensification. Both indicated groups combined influence can be taken into account with the aid of the effective Gammet's constant. From the Eq. (29) it follows, that $\Sigma\sigma$ sign and, hence, interactions type is defined by dissociation constants relation: at $K>K_0$ the value $\Sigma\sigma$ is positive and at $K_0>K$ — a negative one. Besides, the Eqs. (29) and (30) combination shows, that D_f increasing results to a rate constant of the given reactive medium ρ reduction at other equal conditions.

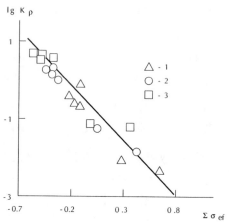

FIGURE 10 The dependence of rate constant of polycondensation K_p on effective Gammet's constant $\Sigma\sigma_{ef}$ in logarithmic coordinates for polycondensation temperatures T: 373 (1), 398 (2) and 423 K (3).

There is one more measure of bulk interactions among randomly approaching coil elements and interactions polymer-solvent. These interactions value (intensity) can be characterized by the parameter ε, defined as follows [25]:

$$\left\langle \overline{h}^2 \right\rangle \sim MM^{1+\varepsilon}, \tag{31}$$

where $\left\langle \overline{h}^2 \right\rangle$ is mean-square distance between macromolecule ends, MM is molecular weight.

In paper [65] the relationship between D_f and ε was obtained, expressed by the equation:

$$D_f = \frac{2}{\varepsilon + 1}. \tag{32}$$

The Eqs. (30) and (32) combination allows to obtain the following approximated relationship between $\Sigma \sigma_{ef}$ and ε [63]:

$$\sum \sigma_{ef} = \frac{1.4\varepsilon}{\varepsilon + 1} \tag{33}$$

Since ε changes from $-1/3$ up to 11 [65], then from the Eq. (33) we obtain: $\Sigma \sigma_{ef}^{min} = -0.69$ and $\Sigma \sigma_{ef}^{max} = 0.70$, that corresponds well to the conducted above estimations of these values.

Hence, the stated above results have shown Gammet's constant treatment possibility as macromolecular coil elements bulk interactions measure. The effective Gammet's constant introduction allows to perform its estimation according to such equivalent parameters as coil fractal dimension D_f and the exponent ε in the Eq. (31). The effective Gammet's constant usage gives the possibility to obtain generalized master — straight line lg $K_p (\Sigma \sigma_{ef})$ for different polycondensation temperatures.

In work [62] it has been shown, that temperature synthesis T increasing at high-temperature polycondensation of polyarylates, obtained by interaction of diphenols of different chemical structure with dianhydride of terephthalic acid, results to reaction rate constant K_p growth (the value K_{p_1} was used for polycondensation first stage up to conversion degree Q≈0.30–0.50). The solvent nature, in solution of which synthesis was performed, influences perceptibly on K_{p_1} value. So, the value K_{p_1} is higher at polycondensation in nitrobenzene environment, than in diphenyloxide environment. The systematic K_{p_1} enhancement was found simultaneously at any solvents and temperatures at the total Gammet's constant $\Sigma\sigma$ reduction. The given effects were explained from the purely chemical point of view.

However, proposed by the authors [62] treatment of results does not allow to obtain the common dependence of K_{p_1} on the indicated above factors, that reduces predicative value of their experimental and theoretical results. Therefore the authors [66, 67] studied the physical sense of factors, controlling the value K_{p_1} at high-temperature polycondensation of polyarylates with fractal analysis methods using [21, 52].

The data of paper [62] for high-temperature polycondensation of polyarylates by interaction of 13 diphenols of different chemical structure with dichloroanhydride of terephthalic acid in high-boiling solvent environment: nitrobenzene and diphenyloxide at 433–473 K. Names, conventional signs and Gammet's constant $\Sigma\sigma$ (–M–C6H4–OH) of diphenols test samples are adduced in Table 3 and values lg K_{p_1} at different synthesis temperatures T and Gammet's constants, taking into account orto-substituents steric hindrances $\Sigma\sigma_{sh}$ — in Table 4. The solvents solubility parameters values δ_s and their evaporation energies Evap are accepted according to the data of papers [36, 56, 68, 69]. The results statistical processing was performed by polynoms approximation by method of least squares [66].

TABLE 3   Diphenols (para- and ortho-derivative of phenol) and their Gammet's constants $\Sigma\sigma$ (–M–C6H4–OH)

No	Name	Conditional sign	$\Sigma\sigma$ (–M–C_6H_4–OH)
	para-derivative of phenol		
1	di-(4-oxiphenyl)sulfone	DOPSn	0.66
2	di-(4-oxiphenyl)sulfide	DOPSd	0.16
3	2,2-di-(4-oxiphenyl)propane	DOPP	−0.11
4	1,1-di-(4-oxiphenyl)cychexane	DOPCH	−0.12
5	3,9-di-(4-oxiphenylethyl)spyrobimetadioxane	Spirole A	−0.13
6	di-(4-oxiphenyl)phenylmethane	DOPPM	−0.13
7	di-(4-oxiphenyl)methane	DOPM	−0.14
	ortho-derivative of phenol		
8	2,2-di-(4-oxi-3,5-dibromidephenyl)propane	TBDPP	0.84
9	1,1-di-(3-chloro-4-oxiphenyl) cyclohexane	DCDPCH	0.28
10	2,2-di-(3-chloro-4-oxiphenyl)propane	DCDOPP	0.27
11	3,3-di-(3-methyl-4-oxiphenyl)phthalide	o–KP	0.09
12	3,3-di-(4-oxi-2-methyl-5-isopropylphenyl) phthalide	TP	0.05
13	2,2-di-(4-oxi-3-methylphenyl)propane	DOMPP	−0.20

In Figs. 11 and 12 the dependences of reaction rate constant K_{ρ_1} in nitrobenzene and diphenyloxide on Gammet's constant $\sum \sigma$ ($-M-C6H4-OH$) at T=473 K are adduced. At the obtained results analysis the systematic K_{ρ_1} reduction at $\sum \sigma$ ($-M-C6H4-OH$) growth is obtained, moreover these dependences can be approximated by two parallel straight lines for para- and ortho-derivatives of diphenols.

TABLE 4 The values of rate constant logarithmic for polyarylates synthesis reactions.

No	Conditional signs of diphenols	$\lg K_{\rho_1}$ Nitrobenzene			Diphenyloxide			$\sum \sigma_*$
		453 K	463 K	473 K	433 K	453 K	473 K	
		para-derivative phenols						
1	DOPSn	−3.3401	−3.0575	−2.7959	–	–	−3.1278	0.66
2	DOPSd	−2.9352	−2.6803	−2.4391	–	–	−2.7498	0.16
3	DOPP	−2.8693	−2.6101	−2.3797	−3.3233	−2.9197	−2.5351	−0.11
4	DOPCH	−2.8086	−2.6068	−2.3889	−3.3098	−2.8739	−2.4479	−0.12
5	Spirole A	−2.7849	−2.5505	−2.3248	−3.2457	−2.8407	−2.4498	−0.13
6	DOPPM	−2.7775	−2.5686	−2.3507	−3.3516	−2.9201	−2.4549	−0.13
7	DOPM	−2.8523	−2.6220	−2.3625	–	–	−2.4299	−0.14
		ortho-derivative phenols						
8	TBDPP	−3.8210	−3.4572	−3.1152	–	–	−3.3862	1.25
9	DCDPCH	−3.3851	−3.1373	−2.9147	−3.9747	−3.4989	−2.9931	0.69
10	DCDOPP	−3.3947	−3.1302	−2.8690	−3.9172	−3.4685	−3.0052	0.68
11	o-KP	−3.1871	−2.9876	−2.7272	−3.7305	−3.2832	−2.8536	0.50
12	TP	−2.9838	−2.7620	−2.6190	−3.6656	−3.2950	−2.8814	0.46
13	DOMPP	−2.9838	−2.7615	−2.5259	–	–	−2.6488	0.21

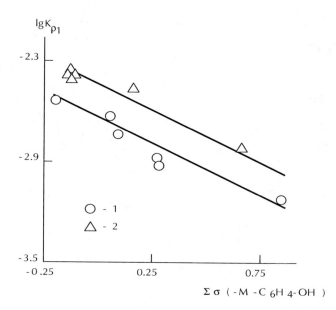

FIGURE 11 The dependences of rate constant K_{ρ_1} of reaction in nitrobenzene on Gammet's constant $\Sigma\sigma$ ($-M-C_6H_4-OH$) in logarithmic coordinates at T=473 K (series 1 — ortho-derivative diphenols, series 2 — para-derivative diphenols).

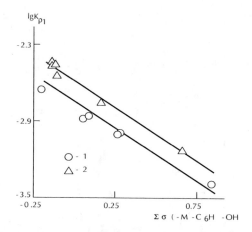

FIGURE 12 The dependences of rate constant K_{ρ_1} of reaction in diphenyloxide on Gammet's constant $\Sigma\sigma$ ($-M-C6H4-OH$) in logarithmic coordinates at T=473 K (series 1 — ortho-derivative diphenols, series 2 — para-derivative diphenols).

These straight lines can be normalized by one straight line displacement up to matching with the second one over any from the two used coordinates axes (for example, $\Sigma\sigma$ axis). The graphical analysis of Figs. 11 and 12 has shown, that for both solvents such matching requires displacement of straight line for ortho-derivative diphenols over $\Sigma\sigma$ (–M–C6H4–OH) axis by value $\Delta\Sigma\sigma{\approx}0.41$ (Fig. 13).

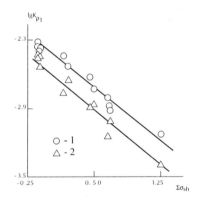

FIGURE 13 The dependences of rate constant κ_{p_1} of reaction in nitrobenzene and diphenyloxide on Gammet's constant $\Sigma\sigma_{sh}$ at T=473 K, obtained by straight line displacement for ortho-derivative diphenols in Figs. 11 and 12 over $\Sigma\sigma$ (–M–C6H4–OH) axis by the value $\Delta\Sigma\sigma{\approx}0.41$ in logarithmic coordinates (series 1 — nitrobenzene, series 2 — diphenyloxide).

Theoretically chemical reaction rate reduction for ortho-derivative diphenol can be easily explained by steric factors. Within the frameworks of fractal analysis it is supposed, that solvents change at polyarylates synthesis should result to interactions polymer-solvent change. In this case the value of macromolecular coil fractal dimension D_f is determined according to the Eq. (12). In its turn, Flory–Huggins interaction parameter can be determined according to the Eq. (10).

At calculation of polyarylates macromolecular coils fractal dimension D_f in nitrobenzene on diphenyloxide the following parameters values were accepted: E_{vap}=40623 and 55072 kj/mole [68, 69], δ_s=10.0 and 13.3 (cal/cm³)1/2 [36, 56], respectively, and δ_p=10.8 (cal/cm³)1/2 [56]. Besides, it is assumed, that the ratio δ_p/δ_s is independent on temperature and χ_s=0, since synthesis was performed in diluted solutions and at high temperatures T. Calculations according to the Eqs. (10) and (12) gave the values D_f=1.530 and 1.838 for nitrobenzene and diphenyloxide, respectively. The Eq. (30) was proposed above for estimation of effective Gammet's constant $\Sigma\sigma_{ef}$. The estimation according to this equation with using of the adduced above values D_f gives the values $\Sigma\sigma_{ef} = -0.298$ and -0.077 for nitrobenzene and diphenyloxide, respectively, and hence, $\Delta\Sigma\sigma_{ef} = 0.222$, that coincides with graphical estimation of this shift (Fig. 14).

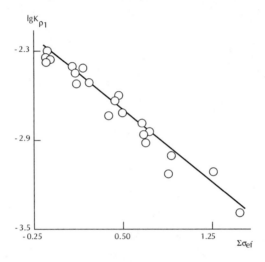

FIGURE 14 The dependence of reaction rate constant K_{p_1} on effective Gammet's constant $\Sigma\sigma_{ef}$ in logarithmic coordinates, obtained by straight line displacement (series 2) in Fig. 13 by $\Delta\Sigma\sigma_{ef} \approx 0.222$.

Let us note, that another δ_p choice for polyarylates gives $\Sigma\sigma_{ef}$ values, differing from the mentioned above, but approximately the same shift $\Delta\ \Sigma\sigma_{ef}$. Thus, the value K_{p_1} change at solvent variation is due completely to the change of macromolecular coil structure, quantitatively described within the frameworks of fractal analysis. In Fig. 14 the straight lines lg K_{p_1} ($\Sigma\sigma_{ef}$) matching with the usage of theoretically estimated shift $\Delta\ \Sigma\sigma_{ef}$ is shown. As one can see, now these dependences for nitrobenzene and diphenyloxide are approximated well by a sole straight line (Fig. 14).

In Figs. 15 and 16 the dependences lg K_{p_1} ($\Sigma\sigma_{sh}$) for polyarylates synthesis in two solvents (nitrobenzene and diphenyloxide) at two different T (453 and 473 K) are adduced. As one can see, these dependences can be approximated again by two straight lines for each from the indicated temperatures and solvents (Fig. 17).

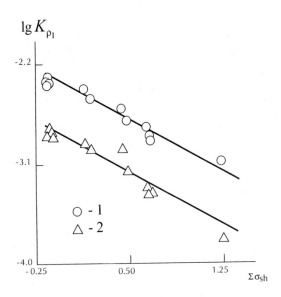

FIGURE 15 The dependences of reaction rate constant K_{p_1} on Gammet's constant $\Sigma\sigma_{sh}$ in logarithmic coordinates at T=473 (series 1) and T=453 K (series 2) in nitrobenzene.

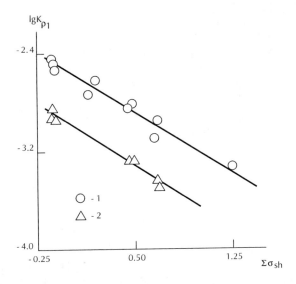

FIGURE 16 The dependences of reaction rate constant K_{p_1} on Gammet's constant $\Sigma\sigma_{sh}$ in logarithmic coordinates at T=473 (series 1) and T=453 K (series 2) in diphenyloxide.

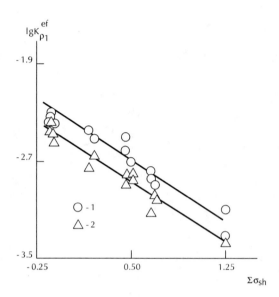

FIGURE 17   The dependence of effective reaction rate constant $K_{\rho 1}^{ef}$ on Gammet's constant $\sum \sigma_{sh}$ in logarithmic coordinates for solutions in nitrobenzene (series 1) and diphenyloxide (series 2), obtained by straight lines (series 2) shift in Figs. 15 and 16 over lg $K_{\rho 1}$ axis by the value Δlg $K_{\rho 1} \approx 0.48$.

Estimations according to the Eqs. (10), (12) and (30) showed, that T change from 453 up to 473 resulted to D_f change by 0.001 only. Therefore it should be assumed, that for matching of straight lines lg $K_{\rho 1}$ ($\sum \sigma_{sh}$) the shift over lg $K_{\rho 1}$ axis is required by the value $\Delta K_{\rho 1} \approx 0.48$, determined in diagram form. This means, that $K_{\rho 1}$ change at T variation is due to polycondensation process thermal activation, described by the general Eq. [62]:

$$\lg k_{\rho 1} = C_1 - \frac{C_2}{T}, \qquad (34)$$

where c_1 and c_2 are constants.

The relationships, similar to the Eq. (34) were adduced in paper [62] for each diphenol and the average values c_1 and c_2 were determined for the calculations according to these data simplification. In this case for polyarylates synthesis the Eq. (34) assumes the following form [67]:

$$\lg K_{\rho 1} = 7.326 + \frac{4700}{T} \quad . \tag{35}$$

Calculation according to the Eq. (35) for T=453 and 473 K gives $\Delta K_{\rho 1} \approx 0.44$, that corresponds well to this shift estimation, conducted in diagram form for the two mentioned synthesis temperatures (Fig. 17).

In Fig. 18, all data $\lg K_{\rho 1}^{ef}$ ($\Sigma \sigma_{ef}$), adduced in Table 4 matching with usage of the two estimated above displacement factors—$\Sigma \sigma_{ef}$ and $\Delta K_{\rho 1}^{ef}$—is shown. All data are approximated well by a sole generalized straight line. The statistical analysis has shown, that this dependence is expressed by the following correlative equation [66]:

$$\lg K_{\rho 1}^{ef} = -2.399 - 0.664 \Sigma \sigma_{ef} \quad , \tag{36}$$

where $\lg K_{\rho 1}^{ef}$ and $\Sigma \sigma_{ef}$ are effective values of these parameters with accounting for the mentioned above shift factors. The correlation 0.977 was obtained for the correlative Eq. (36) by method of least squares [67].

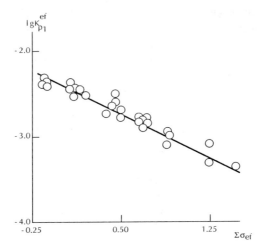

FIGURE 18 The generalized dependence of reaction rate constant $K_{\rho 1}^{ef}$ on effective Gammet's constant $\Sigma \sigma_{ef}$ in logarithmic coordinates, obtained by a straight line (series 2) in Fig. 17 displacement by $\Delta \Sigma \sigma_{ef} \approx 0.222$.

Thus, the stated above results have shown, that polyarylates series high-temperature polycondensation kinetics, characterized by the reaction rate constant K_{p1}, at different temperatures and solvents is controlled by three main factors: process thermal activation (T), diphenol reactiveness ($\Sigma\sigma$) and macromolecular coil structure (D_f). The obtained generalized dependence lg $K_{p1}^{ef}(\Sigma\sigma_{ef})$ allows to perform polycondensation, kinetics prediction, at any rate, for the given reactive series.

In paper [53] it has been shown, that for polyarylates (PAr) series, synthesized by high-temperature (equilibrium) and interfacial polycondensation, the different values of exponent a in Mark–Kuhn–Houwink equation are observed for PAr of the same chemical structure, but synthesized by different modes. Moreover, the value a for PAr, obtained by interfacial polycondensation, exceeds systematically corresponding parameter for PAr, synthesized by high-temperature polycondensation [53]. In paper [53] this distinction was explained by different structure of PAr, obtained by the mentioned polycondensation modes. High-temperature polycondensation hard conditions (high temperature, process of large duration) can cause reaction branched products appearance owing to lacton cycle breaking in phenolphthaleine remainders, that results to branched polymer chains appearance and corresponding a reduction. Such explanation is a special one: similar a distinction is observed for PAr as well on the basis of diane, not having in its structure phenolphthaleine [53]. Hence, it should be assumed, that explanation should be searched for polycondensation modes distinctions, which the considered PAr were synthesized. Therefore the authors [70] proposed general treatment of the values a observed distinctions and, hence, distinctions of macromolecular coil fractal dimension D_f (see the Eq. (4)) for PAr, obtained polycondensation different modes, with fractal analysis methods using.

The values a for seven polyarylates of different chemical structure, obtained by high-temperature (equilibrium) and interfacial polycondensation, determined in three solvents (simm–tetrachloroethane, tetrahydrofuran and 1,4-dioxane) are accepted according to the data of work [53]. The fractal dimension D_f experimental values (D_f^e) in the indicated solvents were determined according to the Eq. (4). The values of solubility parameter s for these solvents are taken from literary sources [25, 36, 56]. The fractal dimension δ_f of solvent molecules structure was determined according to the equation [71]:

$$\delta_f \approx 1.58\left(\delta_s^{1/2} - 2.83\right) , \qquad (37)$$

where δ_s is given in (cal/cm^3)$^{1/2}$.

Polycondensation

Several methods of macromolecular coil fractal dimension determination exist. So, phantom (not taking into account excluded volume effects) fractal dimension D is defined as follows [17] Eq. (38):

$$D = \frac{2d_s}{2-d_s},$$

(38)

where d_s is spectral (fracton) dimension, characterizing coil connectivity degree [72]. The value d_s is equal to 1.0 for linear polymer chains and d_s=1.0–1.33 — for branched chains [72].

The swollen (accounting for excluded volume effects) fractal dimension D_f is determined as follows [17]:

$$D_f = \frac{d_s(d+2)}{d_s+2},$$

(39)

where d is dimension of Euclidean space, in which a fractal is considered (it is obvious, that in our case d = 3).

The Eq. (39) was obtained for the case, when solvent molecules have zero-dimensional (point) structure. If solvent molecules structure is more complex, then D_f value is determined as follows [17] Eq. (41):

$$D_f = \frac{d_s(d+2)}{(1-\alpha_p)d_s+2},$$

(40)

where

$$\alpha_p = \frac{\delta_f}{D}.$$

(41)

From comparison of the Eqs. (39) and (40) it follows, that for δ_f>0 the last equation gives higher D_f values (smaller values a), than the first one. Such relation assumes formal possibility of the description of macromolecular coil structure, obtained by high-temperature polycondensation, with the aid of the Eq. (40) and interfacial one — with the aid of the Eq. (39). The indicated polycondensation modes mechanisms distinction assumes the same treatment: if high-temperature polycondensation is performed in solution, then interfacial one — on phases boundary. This means, that at the first from the indicated polycondensation modes solvent molecules structure influences actively on synthesized macromolecular coil forming structure, screening repulsion interactions among coil elements, whereas in the second from the indicated modes such screening is absent. Therefore the obtained in the presence of solvent with $\delta_f > 0$ coil structure is more compact, than the obtained one without solvent influence. This structural memory effect is maintained at polymers dilution in the same solvent, that defines the exponent a different values for polyarylates of the same chemical structure, but obtained by the indicated above polycondesation modes. For this hypothesis verification in Fig. 19 the comparison of theoretical dependences $D_f(d_s)$, calculated according to the Eqs. (39) and (40), with experimental values D_f^e is performed.

ds values were determined according to the Eq. (40) on the known values D_f^e (the Eq. (4)) and δ_f (the Eq. (37)). As it follows from this comparison, the values D_f^e for PAr, obtained by high-temperature polycondensation, correspond to the Eq. (40) and obtained by interfacial polycondensation to the Eq. (39).

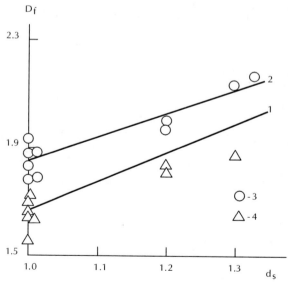

FIGURE 19   The dependences of fractal dimension D_f of PAr macromolecular coil on spectral dimension d_s. 1 — calculation according to the Eq. (39); 2 — calculation according to the Eq. (40) at $\delta_f = 0.60$; 3, 4 — calculation according to the Eq. (4) for PAr, obtained by interfacial (3) and high-temperature (4) polycondensation.

Thus, the data of Fig. 19 confirm the made above assumption about values a (or D_f) distinction causes for PAr, obtained by different polycondensation modes. It is important to note, that this correspondence is correct equally for both linear (d_s=1.0) and branched (d_s =1.20–1.33) PAr, i.e., the supposition about branching chain influence on values a distinction, made in work [53], in the given case is not valid.

In Table 5 the comparison of different parameters for PAr on the basis of phenolphthaleine (Ph-2 according to the designation, accepted in work [53]) for different solvents is adduced, in which Mark–Kuhn–Houwink equation parameters were determined. Firstly, δ_f calculation according to the equation [17]:

$$D_f = \frac{d+2}{2-(\delta_f-2)D} \qquad (42)$$

gave the values δ_f= 0.36–0.84 for Ph-2, obtained by equilibrium polycondensation, and $\delta_f \leq 0$, obtained by interfacial polycondensation. This means, that solvent molecules structure does not influence (does not screen repulsion interactions) on Ph-2 macromolecular coil structure, obtained by interfacial polycondensation, and influences on similar factor for Ph-2, obtained by high-temperature polycondensation. In other words, if solvent influences on coil structure formation in synthesis process, then it will influence on this polymer formed coil structure and at its subsequent dilution and vica versa. Secondly, calculated according to the Eqs. (37) and (42) δ_f values correspond satisfactorily between themselves [73]. Thirdly, if for polyarylate Ph-2, obtained by interfacial polycondensation, D_f values are close to coil dimension in good solvent (~1.67 [16]), then for Ph-2, obtained by high-temperature polycondensation, they increase systematically at δ_f enhancement within the range of D_f^e=1.77–1.94. And at last, D_f^e values for Ph-2, obtained by high-temperature and interfacial polycondensation, correspond to D_f values, calculated according to the Eqs. (40) and (39), respectively, that confirms the made above assumptions.

TABLE 5 The characteristics of macromolecular coil of polyarylate Ph-2 in different solvents.

Solvent	δ_f the Eq. (37)		δ_f the Eq. (42)	D_f^e		D_f the Eq. (40)	D_f the Eq. (39)
	Ph-2t*	Ph-2int**		Ph-2t	Ph-2 int		
Tetrachloroethane	0.36	0	0.60–0.62	1.77	1.55	1.85	1.67
Tetrahydrofuran	0.68	0	0.40–0.55	1.88	1.64	1.83	1.67
1,4-dioxane	0.84	0	0.50–0.72	1.94	1.67	1.89	1.67

* and ** — indices "t" and "int" mean, that Ph-2 was synthesized by high-temperature and interfacial polycondensation, respectively.

Let us note, that D_f value of macromolecular coil in solution defines polymer structure in condensed state [74]. For example, higher values D_f for PAr, of the obtained high-temperature polycondensation, define higher fractal dimension of solid-state PAr [74] and, hence, higher values of fracture strain [75], that is confirmed experimentally [53].

Hence, the stated above results have shown, that the main cause of different macromolecular coil structure of PAr with the same chemical structure, but obtained by different polycondensation modes, is participation (or imparticipation) of solvent molecules with $\delta_f > 0$ in synthesis process. Obtained in this process coil structure maintains at subsequent polymer dilution and solvents influence in this case is similar to their influence in polymer synthesis process. Fractal analysis methods give mathematical apparatus for this problem quantitative study.

1.2   THE LIMITING CHARACTERISTICS OF POLYCONDENSATION PROCESS

The limitingly attainable characteristics of polycondensation process (limiting conversion degree Q_{lim} and molecular weigh MM_{lim} and also molecular weight distribution (MWD)) as a matter of fact define synthesis quality. Therefore to their description much attention is paid, but quantitative relationships for their determination at present are obtained just a little. Proceeding from this, the authors [76] gave quantitative treatment of limiting conversion degree, attainable in polycondensation process, on the example of two series of polymers, synthesized in different solvents: polyarylate Ph-2 (PAr) [51] and copolymers polyurethanearylate (PUAr) [77]. The fractional integration notions were used for this purpose reaching.

D_f values, obtained for PAr and PUAr polycondensation process, showed, that the indicated processes were realized by aggregation cluster–cluster mechanism [49], i.e., by small macromolecular coils joining in larger ones [23]. Thus, polycondensation process is a fractal object with dimension D_f reaction. Such reaction can be presented schematically in a form of "devil's staircase" [80]. Its horizontal parts correspond to temporal intervals, in which reaction is not realized. In this case polycondensation process is described with using fractal time t, which belongs to Kantor's set points [81]. If polycondensation process is considered in Euclidean space, then time belongs to a real number set.

The mathematical calculus of fractional differentiation and integration is used for the description of evolutionary processes with fractal time [81]. As it has been shown in paper [82], in this case the fractional exponent v_{fr} coincides with fractal dimension of Kantor's set and indicates fraction of system states, maintaining during all evolution time t. Let us remind, that Kantor's set is considered in one-dimensional Euclidean space (d = 1) and therefore its fractal dimension d_f<1 in virtue of fractal definition [52]. For fractal objects in Euclidean spaces with higher dimensions (d>1) d_f fractional part should be accepted as v_{fr} or [83]:

$$V_{fr} = d_f - (d - 1), \qquad (43)$$

where d is dimension of Euclidean space, in which fractal is considered.

Let us consider the physical sense and definition of fractional exponent v_{fr} value in the given context. As it is known [84, 85], polycondensation process ceases in gelation point, for which the fractal dimension d_f^g of a forming structure is equal to ~ 2.50. This means, that fractal (macromolecular coil) fraction, not included in evolution (polycondensation) process, at gelation point reaches its maximum value: v_{fr}=1.0. From the said above it follows, that in the considered case at v_{fr}=1.0 d_f=D_f= d_f^g and then the Eq. (43) transforms to the look [76] Eqs. (44) and (45):

$$V_{fr} = D_f - (d_f^g - 1), \qquad (44)$$

or

$$V_{fr} = D_f - 1.5 . \qquad (45)$$

Hence, fraction of macromolecular coil β_e, subjecting to evolution (chemical reaction in polycondensation process) can be defined as follows [76] Eq. (46):

$$\beta_e = 1 - v_{fr}, \tag{46}$$

or,

$$\beta_e = d_f^g - D_f. \tag{47}$$

Further it can be supposed, that limiting achieved in polycondensation process conversion degree Q_{lim} should be equal to β_e, since the remaining coil part, characterized by v_{fr} value, does not participates in macromolecular coil evolution (polycondensation) process. In Fig. 20 the relation between Q_{lim} and $\beta_e=(1-v_{fr})$ for PAr and PUAr is adduced, which demonstrates approximate equality of these parameters (particularly at $Q_{lim}>0.65$), confirming the made assumption. Let us note, that at $Q_{lim}<0.65$ systematical exceeding of β_e values in comparison with Q_{lim} is observed for PAr. This circumstance is due to synthesis the same duration (120 min) of PAr in all eight used solvents. For poor in respect to PAr solvents, where macromolecular coil dimension D_f large values are observed, synthesis process is realized slowly and the indicated polycondensation duration is not sufficient for Q_{lim} reaching [49]. In other words, in this case dynamical scaling conditions for PAr polycondensation are not fulfilled [23].

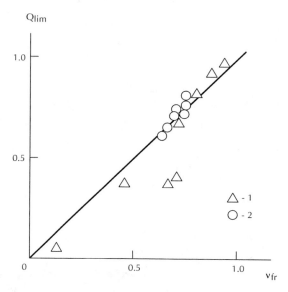

FIGURE 20   The relation between limiting conversion degree Q_{lim} and fractional exponent v_{fr} for polyarylate Ph-2 (1) and polyurethanearylate (2).

From the Eq. (47) it follows, that $\beta_e = Q_{lim} = 1.0$ or the reaction of all initial substances is reached at $D_f = 1.50$. This observation is easily explained. The macromolecular coil with dimension $D_f \leq 1.50$ is called "leaking" [16] or "trasparent" [17], since for such coils their complete interpenetration concerning each other is realized. It is obvious, that in this case any reactive site of coil is accessible for reaction and this allows to obtain maximum possible conversion degree. Hence, the condition $Q_{lim} < 1.0$ for $D_f > 1.5$ is due to "nonleakaging" ("opacity") of macromolecular coils [16, 17].

Hence, the stated above results have shown, that the limiting conversion degree value in polycondensation process is controlled by macromolecular coil structure, characterized by its fractal dimension. The condition $Q_{lim} < 1.0$ at $D_f > 1.50$ is defined by "opacity" of macromolecular coils concerning each other. For "transparent" macromolecular coils ($D_f \leq 1.50$), allowing their complete interpenetration, the value $Q_{lim} = 1.0$ does not depend on macromolecular coil structure within the range of $D_f = 1.0 - 1.50$.

At present the extreme dependence of polymer molecular weight MM on initial reagents concentration is assumed firmly established [48, 51, 86]. However, one should remember, that polymerization processes are kinetic ones, i.e., characterizing them parameters depend on reaction duration. Therefore limitingly reached parameters are realized only in case of polycondensation reaction large of enough duration. The authors [87] demonstrated the necessity of static and kinetic scaling combination for correct dependences molecular weight — initial reagents concentration obtained on the example of polyarylates Ph-2 [51, 86].

At polymerization reactions simulation two scaling types should be discerned: static and kinetic (dynamical) [88]. The first type defines process limiting characteristics irrespective of their achievement way and the second one describes their variation temporal dependence. Both scaling types for polymers molecular weight can be obtained within the frameworks of irreversible aggregation models [58]. The dependence of aggregation process characteristics on aggregating particles concentration c (initial concentration c_0) in some finite space has direct relation to static scaling obtaining. If during cluster (macromolecular coil) growth its density reduces up to environment density (with concentration c at the given moment), then the transition between universality classes of diffusion — limited aggregation (DLA) particle–cluster and Iden's is realized and further macromolecular coil ceases its growth, reaching critical gyration radius R_c [58]. Let us note, that still earlier one more transition between universality classes within the frameworks of DLA models occurs — from aggregation cluster–cluster to aggregation particle–cluster. This reduces strongly reaction rate, but unlike the indicated earlier transition does not cease it [89]. The scale of transition from aggregation particle–cluster to Iden's model depends on particles initial concentration [88]:

$$R_c \sim c_0^{-1/(d - D_f)}, \tag{48}$$

where d is dimension of Euclidean space, in which a fractal is considered (it is obvious, that in our case d = 3).

In its turn, the value R_c is connected with maximum possible polymerization degree N_c according to the Eq. (1). Combination of the Eqs. (1) and (48) with appreciation of intercommunication between N_c and monomer link molecular weight MM_0 ($MM_c = N_c MM_0$) allows to obtain the dependence of limiting molecular weight MM_c (static scaling) in the form [87]:

$$MM_c \sim c_0^{-Df1(d-Df)} . \tag{49}$$

In Figs. 21 and 22 the comparison of values MM^e, obtained experimentally, and MM_c, estimated according to the Eq. (49), for polyarylates Ph-2, synthesized in acetone and dichloroethane is adduced, respectively. The values MM^e were calculated according to Mark–Kuhn–Houwink equation [53]:

$$[\eta] = 0.266 \times 10^{-4} \left(MM^e \right)^{0.938,} \tag{50}$$

for Ph-2 solution in simm-tetrachloroethane.

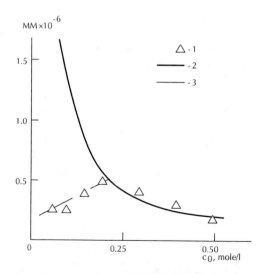

FIGURE 21   The dependences of molecular weight MM^e at t = 90 min (1), MM_c at t = t_c (2) and MM_c at t = 90 min (3) on the initial reagents concentration c_0 for polyarylates Ph-2, synthesized in acetone.

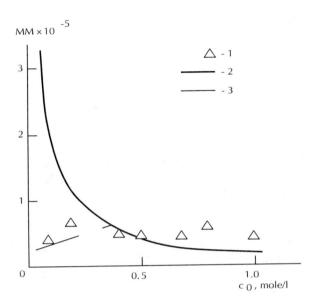

FIGURE 22 The dependences of molecular weight MM^e at t=90 min (1), MM_c at $t = t_c$ (2) and MM_c at t = 90 min (3) on the initial reagents concentration c_0 for polyarylates Ph-2, synthesized in dichloroethane.

As it follows from the data of Figs. 21 and 22, the theoretical curves in both cases are combined well with experimental points at large enough c_0 (>0.25 and >0.35 mole/l for Figs. 21 and 22, respectively), but in small c_0 region theory and experiment strong discrepancy is observed.

For this discrepancy explanation the relationship for kinetic scaling at condition $R_c \ll \xi$ (ξ is transition length scale) should be written [88] Eq. (51):

$$N \sim \left(c_0 t\right)^{D_f /\left(2+D_f-d\right)},$$

(51)

where N is polymerization degree, t is reaction current time.

The condition $R_c \ll \xi$ is chosen by two reasons. Firstly, used at Ph-2 synthesis c_0 values are relatively small and for diluted solutions the value ξ grows as $\bar{n}_0^{-1/(d-D_f)}$ and in the limit of very diluted solutions $\xi \to \infty$ [90]. Secondly, at the obtained for polyarylate Ph-2 values $D_f=1.65$ in dichloroethane and $D_f=1.81$ in acetone [87] macromolecular coils density is large enough and exceeds density of surrounding them solution [88]. These densities become equal at reaction cessation, as it was noted above. Since for the same polymer $N_c \sim MM_c$, then for MM_c reaching reaction duration t_c is required, which is equal [87]:

$$t_c \sim \frac{MM_c^{(2+D_f-d)/2}}{c_0} \qquad (52)$$

From the Eq. (52) it follows, that simultaneous MM_c increase and c_0 reduction resulted to reaction duration t_c sharp growth, which is necessary for MM_c limiting value reaching.

The proportionality constant in the Eq. (52) can be determined by matching of the experimental and theoretical data at the indicated above initial concentrations 0.25 and 0.35 mole/l in supposition, that at these points reaction duration t=90 min is equal to t_c. Then from the Eq. (52) these values MM, which are reached at the given c_0 and t = 90 min, can be obtained. The dependences of the obtained in this way values MM on c_0 at small values c_0 or t < t_c are shown in Figs. 21 and 22 by shaded lines and one can see, that such calculation corresponds completely to the experiment. Hence, the obtained in papers [51, 86] extreme dependences $\eta_{red}(c_0)$ are due completely to the condition t = const = 90 min.

Hence, the adduced above results on the example of the dependences $MM(c_0)$ for polyarylates Ph-2 showed the necessity of polycondensation process both static and kinetic aspects consideration for correct dependences of one or another limiting characteristics. The absence of maximum on curves $MM(c_0)$ means the absence of optimum value c_0, at which maximum value MM is reached. The adduced examples demonstrate clearly correctness and expediency of fractal analysis methods application for polycondensation processes description.

The authors [91] proposed description of organic phase influence on limiting characteristics of polyurethanearylates (PUAr) interfacial polycondensation. As it is known [55], one from the methods of polymer solubility parameter δ_p experimental determination is plotting of the dependence of intrinsic viscosity $[\eta]$, measured in several solvents, on this solvents solubility parameter δ_s value. The smaller difference $|\delta_p - \delta_s|$ or the better solvent thermodynamical quality in respect of polymer is, the larger $[\eta]$ is. The dependences $[\eta](\delta_s)$ have usually bell-like shape and such dependence maximum corresponds to δ_p [55]. In Fig. 23 the dependence of η_{red} on δ_s of solvents, used as organic phase at PUAr interfacial polycondensation is adduced. The dependence $\eta_{red}(\delta_s)$ bell-like shape is obtained again and its maximum corresponds to $\delta_p \approx 10$ (cal/cm^3)$^{1/2}$, that is a reasonable estimation for PUAr [36, 55]. Let us note that all η_{red} values were determined in one solvent, which was not used at synthesis, namely, in mixture phenol-simm-tetrachloroethane. The dependence $\eta_{red}(\delta_s)$, adduced in Fig. 23, allows to make two conclusions. Firstly, the value η_{red}, reached in PUAr interfacial polycondensation process, is controlled by solvent thermodynamical quality and the greatest

η_{red} value can be obtained at $\delta_p = \delta_s$. Secondly, such dependence means macromolecular coil structure fixation depending on the used in synthesis process solvent (structural memory effect [70]).

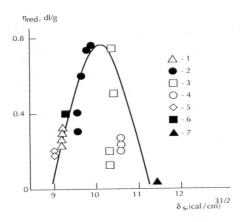

FIGURE 23   The dependence of reduced viscosity η_{red} on solubility parameter δ_s of solvent, using at PUAr synthesis, for n-xylone (1), chlorobenzene (2), simm-tetrachloroethane (3), nitrobenzene (4), toluene (5), chloroform (6) and ethyl ether (7).

The dependence of limiting conversion degree Q_{lim} on δ_s for PUAr, adduced in Fig. 24, has a similar shape. Hence, all made above conclusions are valid for it as well.

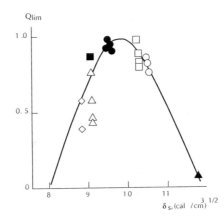

FIGURE 24   The dependence of limiting conversion degree Q_{lim}, obtained at PUAr interfacial polycondensation, on solubility parameter δ_s of used solvents. The designations are the same, as in Fig. 23.

The fractal Eq. (27), where D_f value is determined according to the Eq. (17), can be used for interfacial polycondensation process quantitative description. Calculated according to the Eq. (17) D_f values are adduced in the Table 6, from which large enough interval of their variation follows: D_f=1.592–2.055.

TABLE 6   The characteristics of used at synthesis solvents and macromolecular coil PUAr.

Solvent	δ, (cal/cm^3)$^{1/2}$	δ_p (cal/cm^3)$^{1/2}$	δ_e (cal/cm^3)$^{1/2}$	D_f
Chlorobenzene	9.67	9.39	2.30	1.592
Nitrobenzene	10.62	9.47	4.80	1.606
Simm-tetrachloroethane	10.40	9.25	4.74	1.715
n-xylone	9.16	8.95	1.95	1.733
Toluene	8.93	8.83	1.30	1.747
Chloroform	9.16	8.65	3.01	1.823
Ethyl ether	11.60	9.42	6.78	2.055

Further the value Q_{lim} estimation can be conducted according to the Eq. (27) and compared the obtained thus theoretical values Q_{lim} (Q_{lim}^T) with experimental ones Q_{lim}^e. This comparison is adduced in Fig. 25 (horizontal sections designate the range of the obtained values Q_{lim}^e for PUAr). As it follows from the data of Fig. 25, a good enough correspondence of theory and experiment is obtained, and some data scatter is due to the fact, that at Q_{lim}^T calculation initial reactive medium viscosities distinction is not taken into account (see the Eq. (26)).

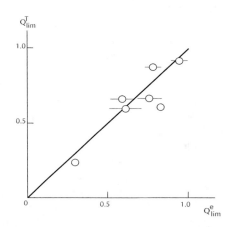

FIGURE 25   The comparison of experimental Q_{lim}^e and calculated according to the Eq. (27) Q_{lim}^T limiting conversion degree for PUAr. Horizontal sections indicate Q_{lim} variation for different PUAr.

The values D_f variation range, adduced in Table 6, assumes, that mechanism of irreversible aggregation cluster–cluster, i.e., the formation of large macromolecular coils from smaller ones, is realized in PUAr interfacial polycondensation process [58]. Therefore, for PUAr limiting molecular weight MM_{lim} estimation the following scaling relationship can be used [45]:

$$MM_{lim} \sim Q_{lim}^{2/(3-D_f)} . \tag{53}$$

In Fig. 26 the comparison of the reduced viscosity experimental values η_{red}^e and calculated according to the Eq. (53) values η_{red}^T is adduced in assumption $\eta_{red}^T \sim MM_{lim}$. As one can see, the good enough correspondence of theory and experiment is obtained and the data scatter is due to the indicated above reason.

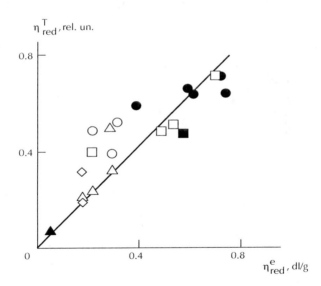

FIGURE 26   The comparison of experimental η_{red}^e and calculated according to the Eq. (53) η_{red}^T adduced viscosity for PUAr. The designations are the same, as in Fig. 23.

Hence, the stated above results assume, that the main limiting characteristics of PUAr interfacial polycondensation at other equal conditions are controlled by thermodynamic quality of organic phase, in which synthesis is realized, with regard to polymer. This factor changes macromolecular coil structure, described within the frameworks of fractal analysis. Fixed in synthesis process coil structure maintains its characteristic features and at polymer subsequent testing, for example, at adduced viscosity measurement in another solvent (structural memory effect).

In a works number [92, 93] the systematic change of molecular weight MM for copolymers aromatic copolyethersulfones (APESF) with their composition was marked, namely, aromatic polyformal blocks contents enhancement results to MM reduction. This observation was explained within the frameworks of special chemical approaches, for example, by cyclic oligomers formation [92].

However, in virtue of high enough community of this tendency of MM reduction it can be assumed, that it has much more key reasons, that it was supposed earlier [92]. Since at present some general description of synthesis processes does not exist, then the authors [94] used the fractal analysis methods for this purpose. They proposed key physical treatment of APESF molecular weight change at their composition variation.

In Fig. 27 the dependences of MM on sulfone blocks contents c_{SF} in APESF were adduced. MM values were calculated according to Mark–Kuhn–Houwink equation for APESF solutions in simm-tetrachloroethane and chloroform. The experimentally determined MM values for two test samples were also used. As it follows from the plot of this Figure, the dependence $MM(c_{SF})$ for APESF is well approximated by linear correlation, that assumes again certain objective regularity, discovering in its basis.

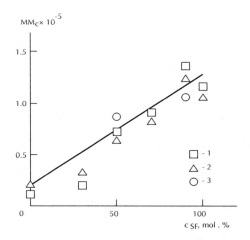

FIGURE 27 The dependence of critical molecular weight MM_c on sylfone blocks contents c_{SF} for APESF. 1, 2 — calculation of MM_c according to Mark–Kuhn–Houwink equation for solutions in simm–tetrachloroethane (1) and chloroform (2). 3 — experimental values of MM_c. The straight line is drawn according to the theoretical data of Fig. 28.

The dependence of aggregation process characteristics on aggregating particles concentration c (initial concentration c_0) in some finite space, considered above, has direct relation to the considering problem. The macromolecular coil critical radius R_c can be estimated according to the Eq. (48). It is obvious, that the value c_0 can be determined through the weight of loaded in reaction course monomers m_r and monomer link weight m_0 [94]:

$$c_0 = \frac{m_r}{m_0}, \tag{54}$$

where the value m_0 is determined as follows [95] Eq. (55):

$$m_0 = \frac{MM_0}{\rho_p N_A}, \tag{55}$$

where MM_0 is polymer elementary link molecular weight, ρ_p is polymer density, N_A is Avogadro number.

In its turn, the value R_c is connected with maximum possible polymerization degree N_c according to the Eq. (1). The combination of the Eqs. (1), (48) and (54) allows to obtain intercommunication of MM_c and m_0 (with appreciation of the equality $MM_c = N_c m_0$) in the following form [94]:

$$MM_c = \frac{m_0^{d/(d-D_f)}}{m_r^{D_f/(d-D_f)}}. \tag{56}$$

Since the value m_r in APESF synthesis process is constant, the value D_f varies not very considerably ($D_f = 1.758–1.800$ for solutions in simm-tetrachloroethane) and total value of exponent for m_r ($D_f/(d-D_f)$) for a good solvent is a little larger than one, then the Eq. (56) assumes, that MM_c variation for APESF (Fig. 27) is due mainly to m_0 change. In Fig. 28 the correlation $MM_c(m_0^{d/(d-D_f)})$ is adduced, which turns out to be linear and passing through coordinates origin. As it was to be expected, MM_c growth at m_0 increasing is observed.

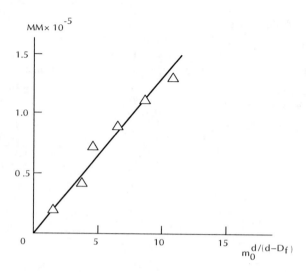

FIGURE 28 The dependence of molecular weight MM on parameter $m_0^{d/(d-D_f)}$ value for APESF.

It is interesting to note, that the same relation between MM_c and m_0 can be obtained, by using a well-known empirical Mark–Kuhn–Houwink equation [25]:

$$[\eta] = K_\eta MM^a ,$$ (57)

where K_η and a are a constant and an exponent of Mark–Kuhn–Houwink equation, respectively.

The value K_η can be expressed through m_0 as follows [53]:

$$K_\eta = \frac{21}{m_0}\left(\frac{1}{2500m_0}\right)^a .$$ (58)

In its turn, the exponent a is connected with fractal dimension D_f according to the Eq. (4). From the combination of the Eqs. (4), (57) and (58) fractal variant of Mark–Kuhn–Houwink equation can be obtained [94]:

$$[\eta] = \frac{21}{m_0}\left(MM\left[\frac{1}{2500m_0}\right]\right)^{(3-D_f)/D_f} .$$

(59)

Expressing MM according to the Eq. (59) as m_0 function, let us obtain (at d = 3) [94] Eq. (60):

$$MM = \frac{2500[\eta]^{D_f/(3-D_f)}}{21^{D_f/(3-D_f)}}m_0^{3/(3-D_f)} ,$$

(60)

that corresponds completely to the dependence $MM(m_0)$, obtained within the frameworks of irreversible aggregation model (the Eq. (56)).

Let us make two remarks in conclusion. As the estimations have shown, the constant coefficient in square brackets of the Eq. (58) does not always give the exact MM estimation and, probably, is adjustable coefficient. Secondly, as Kuchanov pointed out in annotations to paper [96], in polymerization real processes intramolecular reactions resulted to cyclic fragments formation, that makes aggregation models, similar to the considered above, application scarcely probable. However, Kolb [97] demonstrated, that loops (cycles) availability did not influences on the value D_f. Therefore the present model is applicable to polymers, forming cyclic fragments, which aromatic polyformals are [92].

The stated above results on APESF example have shown, that copolymers molecular weight systematic reduction at aromatic polyformal blocks contents increase has key reasons. Molecular coil growth and, hence, molecular weight increase ceases, when coil density reduces up to the environment density. This transition is defined by macromolecular coil fractal dimension D_f and polymer elementary link average weight m_0.

Let us consider in conclusion of the present section molecular weight distribution (MWD) simulation of polyarylates, obtained by polycondensation different modes. For high-molecular compounds the notions of molecule and molecular weight MM have their own features, connected with polymers qualitative distinction from low-molecular compounds [98]. This difference is linked with the fact that polymers are always a mixture of macromolecules with different MM (mixture of polymer — homologs). For such mixture the value MM, found by either mode, is some average value, which depends on the polydispersity degree, the type of MWD function and the method of experimental determination of MM [25]. Therefore a complete description of the molecular weight characteristics of a polymer requires knowledge of its MWD.

Despite its unquestionable importance, the macromolecular coil fractal dimension D_f gives only limited information about an aggregation (polymerization) process. Firstly it is a static value and does not describe the dynamics of an aggregation process. Secondly, the dimension D_f characterizes geometrical properties of only one cluster (macromolecular coil) and cannot be used for the description of a set of clusters [99]. Thus, both classical and fractal approaches indicate the need for polymers MWD study for their characteristic completeness.

The range of D_f values for macromolecular coils in a solution [43] assumes that the polymerization process proceeds according to the mechanism cluster–cluster [58]. The functions of distribution for the indicated mechanism were investigated in a number of studies [99–101]. The authors [102] performed the theoretical description of MWD function within the frameworks of theoretical approach [101] and elucidated the factors, influencing on the shape of this function, on the example of polyarylate (PAr), obtained by three different modes of polycondensation: acceptor-capalytic (PAr-1), high-temperature (PAr-2) and interfacial (PAr-3) ones [102].

In paper [102] the theoretical treatment of a cluster–cluster aggregation accounting for the existence of coalescing of particles or of clusters (monomers or macromolecular coils) to aggregate in real polymerization processes, and their disconnection (destruction) was reported. Macromolecules are in a random environment, influencing the processes of aggregation or destruction in dilute polymeric solution that is described by the stochastic equation [101]:

$$\dot{N} = \lambda N^{a'} - N^b + \xi_t N^{a'} , \quad 0<a<b, \qquad (61)$$

where N is the number of particles in the aggregate, a' and b are exponents, describing the processes of aggregation and destruction, accordingly, $\xi_t = c_t/m$, where c_t is the stochastic contribution to intensity of the aggregation process, due to the influence of the environment, $<c>$ is its average value and m is a coefficient.

The value λ was determined as follows [101] Eq. (62):

$$\lambda = \langle c \rangle / m . \qquad (62)$$

Following the treatment [101], it is believed that the reference times of random influences on fractal aggregation process of macromolecules are much less than the reference times of the aggregation itself and, consequently, it is possible to present ξ_t as white noise of intensity $\langle \xi_t^2 \rangle = \sigma^2$.

The exponent a' in case of cluster–cluster aggregation is connected with fractal dimension D_f of a macromolecular coil according to the following equation [101]:

$$a' = \frac{2D_f - d}{D_f} . \tag{63}$$

The Eq. (61) allows to obtain a density of probability $P_s(N)$, which for a stationary solution looks like [101]:

$$P_s(N) = AN^{-a'} \exp\left\{2\lambda N^{1-a'}\left[1-(1-a')N^{b-a'} / \lambda(b+1-2a')\right] / (1-a'\sigma^2)\right\} \text{ for } a' \neq 1, \tag{64}$$

where A is the normalization constant.

As in the Eq. (64) λ value has no effect on the distribution $P_s(N)$, then the indicated relationship supposes three basic parameters influencing on distribution $P_s(N)$: a', b and σ^2. Each of the indicated parameters characterizes a certain feature of the polycondensation process. The exponent a' is thus essentially defined by the macromolecular coil structure, that directly follows from the Eq. (63). The value b characterizes the type and intensity of the destructive processes. Parameter σ^2 is determined by the stochastic contribution to a polycondensation process and it is possible to assume dependence σ^2 on comonomers initial concentration c_0: the greater c_0, the higher the probability of random collisions. This postulate is particularly important for the mode of interfacial polycondensation, where synthesis proceeds not in all reactive vessel volume, but only in the interfacial layer, for which enhanced in comparison with average value the magnitude c_0 is expected. All experimental MWD curves have the unimodal shape that supposes low mobility for coils in solution or $\sigma^2 < 2/3$ [101].

First of all we shall consider general aspects of the three indicated parameters influence on MWD curves shape. For this purpose we shall construct the theoretical dependences $P_s(N)$ with a serial variation when only one from the indicated parameters varies, whereas the other two are constant. In Fig. 29 the curves $P_s(N)$ are shown for the case when the coil fractal dimension D_f is variable (or a' is variable) and $\sigma^2 = \text{const} = 0.25$, $b = \text{const} = 1.0$. In Figs. 29–31 the normalization constant A was chosen so that the maximum magnitude P_s was equal to about 0.4 for all curves $P_s(N)$. As follows from the data of Fig. 29, D_f increasing results to displacement of maximum of distribution to the higher N side. This effect is most strongly expressed for $D_f = 2.0$, corresponding to θ-conditions [16].

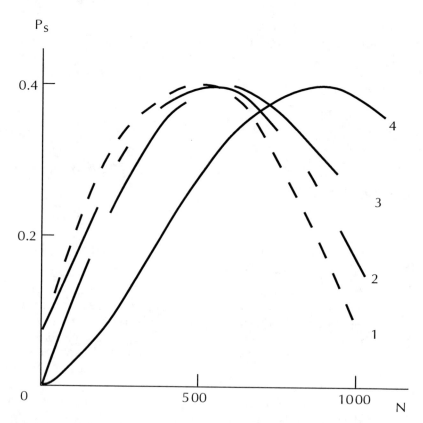

FIGURE 29   Simulation of MWD curves according to the Eq. (64) for $D_f = 1.50$ (1), 1.65 (2), 1.80 (3) and 2.0 (4) b = 1, $\sigma^2 = 0.25$.

In Fig. 30 theoretical $P_s(N)$ curves are shown when σ^2 is variable and $D_f =$ const, b = const. As follows from the data of this figure, the value σ^2 exerts a primary influence on the width of MWD. Strictly monodisperse distribution can be obtained only at $\sigma^2 = 0$. The increase of the stochastic contribution in poly-condensation intensity results to symmetrical broadening of MWD relative to its maximum.

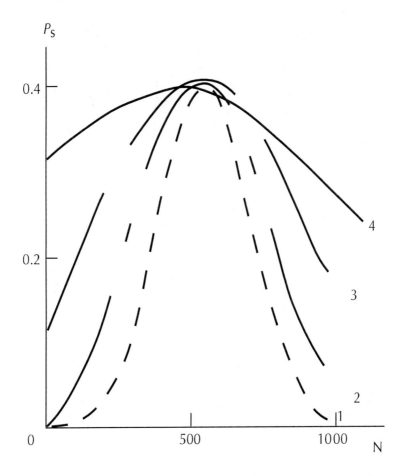

FIGURE 30 Simulation of MWD curves according to the Eq. (64) for $\sigma^2=0.05$ (1), 0.10 (2), 0.25 (3) and 0.60 (4). $b = 1$, $D_f= 1.65$.

At last, in Fig. 31 the theoretical curves $P_s(N)$ are shown, where b is variable and D_f = const, σ^2 = const. The value b = 1 means availability of the channel of disintegration for each bond in a cluster. When the greater part of bonds is fixed (b = 0.2), the maximum in an investigated interval N = 0–1000 is not generally reached. At b = 0.5 (half of the bonds are fixed) $P_s(N)$ curve is fixed at the maximum, but with obvious asymmetry at large N. Further the increasing of b means, as it should be expected, a decrease of the clusters fraction with large N (high-molecular fraction).

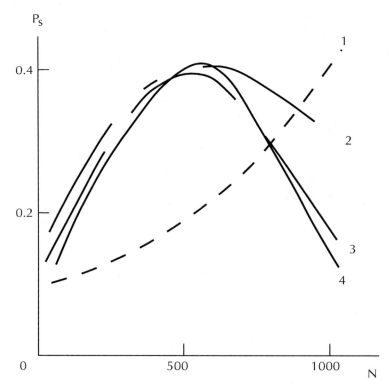

FIGURE 31   Simulation of MWD curves according to the Eq. (64) for b = 0.2 (1), 0.5 (2), 1.0 (3) and 2.0 (4). D_f =1.65, σ^2 = 0.25.

The comparison of theoretical curves $P_s(N)$ (Figs. 29–31) and experimental curves MWD (Fig. 32) allows to make several general conclusions. First, MWD curves width for PAr-3 is much larger, than for PAr-1 and PAr-2. By analogy with Fig. 30 the conclusion can be made, that for PAr-3 the value σ^2 is higher than for the two remaining polyarylates. This effect cause is indicated above. Secondly, for PAr-1 the largest value N, corresponding to MWD curve maximum (Nmax) is observed. The comparison with Fig. 29 assumes, that PAr-1 macromolecular coils have the greatest D_f value from all considered polyarylates. This is the expected enough result, since it is known, that high-temperature (equilibrium) polycondensation forms PAr macromolecular coils with the smallest exponents a in Mark–Kuhn–Houwink equation [53]. The curves MWD (Fig. 32) analysis shows, that the most strong decay of curve in large N region is observed precisely for PAr-1 and the comparison with Fig. 31 demonstrates, that for this polymer the greatest b value is expected. This means, that destruction process is expressed most clearly for PAr-1 and therefore a higher value D_f is expected for it [103].

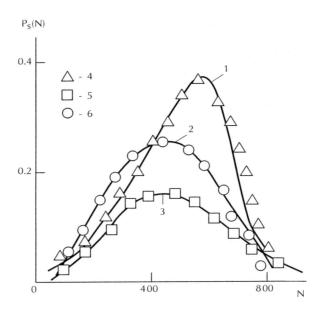

FIGURE 32   MWD curves for PAr-1 (1, 4), PAr-2 (2, 5) and PAr-3 (3, 6). 1–3 — the experimental results, 4–6 — the calculation according to the Eq. (64) after renormalization.

Since polyarylates are formed from two different monomers (diane and di-chloroanhydride of 1,1-dichloro-2,2-di(n-carboxiphenyl) ethylene), then as par-ticle mass for them half of the molecular weight of PAr link was accepted and N calculation was performed according to the indicated weight. Comparison of Figs. 29–31, on the one hand, and Fig. 32, on the other hand, shows, that direct simulation of MWD curves by $P_s(N)$ curves cannot be obtained successfully, since MWD curves have too low Nmax values. Nevertheless, to achieve agreement between the theory and the experiment by use of renormalization is possible for these curves. This is due to the automodality of $P_s(N)$ distribution on the variable

$\tilde{N} = N/N_0$ [101]. Simultaneously it is required to execute the renormalization of white noise of intensity σ^2 as follows [101] Eq. (65):

$$\sigma^2 \rightarrow \sigma_0^2 = \sigma^2 N_0^{2a'-1-b} \ . \tag{65}$$

The indicated renormalization is performed by N_0 definition under conditions at which $\tilde{N}$ is equal to the theoretical value Nmax and N is the experimental value of this parameter. The shown in Fig. 32 excellent agreement between experimental and theoretical MWD curves for PAr confirms the renormalization execution correctness. In its turn, this renormalization reflects one from the main fractals properties — their automodality [52].

As it is known [100], MWD curve for cluster–cluster aggregation depends on the diffusive characteristics of a system in many respects, that is, on clusters mobility expressed by their rate ϑ. The value ϑ can be expressed as follows [100]:

$$\vartheta = m^{\alpha}, \tag{66}$$

where m is mass of the cluster.

In its turn, the value of an exponent α determines the position of a maximum of MWD curve [100]:

$$N_{max} = \frac{-\alpha}{1-\alpha}. \tag{67}$$

The estimation of macromolecular coils in solution mobility according to the Eqs. (66) and (67) has shown, that the value increases in series PAr-1→PAr-2→PAr-3. This increase results to Nmax reduction and enhances random collisions probability, i.e., σ^2 enhancement, that defines MWD curves shape. The increased destruction probability for acceptor-catalytic polycondensation is an additional factor, which somewhat increases the dimension D_f for PAr-1.

Hence, the stated above results have shown, that molecular weight distribution of polyarylates, obtained by polycondensation different modes, can be simulated and predicted within the frameworks of irreversible aggregation cluster–cluster model. The shape and position of MWD curve are controlled by a number of factors, which are common to any synthesis mode, namely, by the macromolecular coil structure, stochastic contribution of coil environment in polycondensation intensity and coil destruction level in synthesis process. Coil mobility in reactive medium influences very strongly on MWD curve shape.

1.3   THE DESCRIPTION OF POLYCONDENSATION KINETICS WITHIN THE FRAMEWORKS OF IRREVERSIBLE AGGREGATION MODELS

The nonequilibrium polycondensation processes kinetics is the problem studied in detail every [48]. However one hasn't got any general description at present yet and this problem is considered on empirical level with using of a large number of chemical sense details. Since such approach does not give perceptible results, then there are reasons to suppose, that the new approach, which does not take into account the indicated above details, is required for the description of the polycondensation kinetics. The scaling (scale invariance), the sense of which contains in abstraction from details of structure (process) and formation of simple features, characteristic for systems of wide class [58], is the obvious pretender to such approach role. Fractal dimensions are often used scaling exponents (indices), which unlike the usual indices of scaling characterize object structure, namely, distribution of its elements in space [58].

The universality hypothesis, the essence of which is formulated as follows: if for different systems formation mechanism the same limiting conditions (system parts interactions) are typical, then these systems get into one universal class of physical phenomena [58]. The models particle–cluster [104] and cluster–cluster [105, 106], distinctions in aggregation mechanisms of which follow from their names, are two broadly known universality classes (aggregate types) in irreversible aggregation processes. These models were developed specially for the description of such practically important processes as flocculation, coagulation, polymerization and so on [107]. Quite enough examples of these models application for a physical processes number description were obtained [12, 108–110]. Therefore the same models using for nonequilibrium polycondensation description represents some interest. Let us note, that application for this problem solution of percolation and some other models did not give the expected result [111, 112].

Accounting for said above appreciation the authors [113] demonstrated principal possibility and expediency of the considered physical principles application for nonequilibrium polycondensation description on the simplest example of polyarylate Ph-2 synthesis in solution. The macromolecular coil fractal dimension D_f value was determined according to the Eq. (4) and spectral dimension d_s — according to the Eq. (39). The estimations have shown, that $D_f=1.55$ and $d_s=1.0$, from which it follows, that in the considered conditions Ph-2 is a typical linear polymer [72].

Proposed at present the scaling models of irreversible aggregation processes [114, 115] are used with different parameters, but in the whole they can be divided into two main groups. The static geometric parameters of forming clusters, the most spreading of which is D_f, included in the first group. The parameters, in either way controlling diffusive processes in cluster formation course, should be included in the second group. The exponent α in the relationship diffusion rate — cluster mass (see the Eq. (66)), diffusivity [114] or reactive medium initial viscosity [108] can be included in their number. The initial particles density, which in the considered case is characterized by the initial reagents concentration c_0, is one more obvious parameter. In the considered case the parameters second group is important, since the value $D_f = 1.55$ allows to make synonymous conclusion, that the given polycondensation process is diffusion-limited aggregation cluster–cluster [58]. Preserving this general set of the parameters, the authors [113] chose for dynamical scaling obtaining (unlike papers [108, 114, 115] the scheme, proposed by Pfeifer a.a. [57], the essence of which is contained in the following. In the indicated work the notion of particles (clusters) accessibility to active centers for realization either reaction was considered within the framework of fractal analysis. In the considered case this notion is linked with the notion of cluster (macromolecular coil) fractality, which is characterized quantitatively by the value D_f. The value D_f shows cluster structure "opening" (accessibility) degree — the smaller D_f, is the more intensive particles (clusters) penetration in cluster internal regions is (the "more it is accessible"). This situation is expressed analytically by the Eq. (27).

The Eq. (27) is the main scaling relation, used by the authors [113] for polycondensation Ph-2 kinetics description. Let us note, that for Euclidean coils (compact globules), for which $D_f = d$ [16], the reaction cannot be realized. It is natural, that in the general case in the Eq. (27) the indicated above diffusive parameters and the value c_0 should be included (see the Eq. (26)), but since in paper [113] only one polymer at a fixed c_0 was considered, then it is supposed, that these parameters are included in a unknown constant of the Eq. (27). Using the indicated relationship and having determined this constant by method of theoretical and experimental data matching, the authors [113] obtained good agreement of these results up to polycondensation reaction duration $t \approx 4$ min (Fig. 33), after that the conversion degree Q the calculated values begins to grow much more rapidly than the experimental one. Let us note, that the same polycondensation reaction duration corresponds to reduced viscosity η_{red} (or molecular weight [48]) growth cessation and the condition $\eta_{red} = $ const achievement. This means, that chemical reaction is ceased up to $t = 8$ min and a sloping part of the curve Q(t) is due to other processes. Let us also note, that this curve correct description up to $t = 4$ min was obtained at using $D_f = 1.55 =$ const. As it is known [115], the dimension of the resulting cluster D_f at aggregation cluster–cluster of two clusters with dimensions D_{f_1} and D_{f_2} can be determined according to the equation:

$$D_f = \frac{d\left(2D_{f1} - D_{f2}\right)}{d + 2\left(D_{f1} - D_{f_2}\right)} \quad . \tag{68}$$

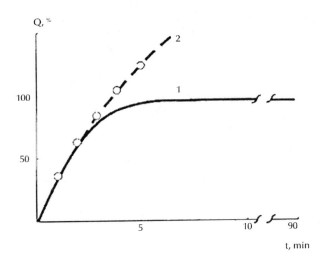

FIGURE 33 The dependence of conversion degree Q on nonequilibrium polycondensation Ph-2 duration t. 1 — experimental data [48], 2 — calculation according to the Eq. (27).

From the Eq. (68) one can see easily, that the condition D_f = const is realized at $D_{f_1} = D_{f_2} = D_f$. In other words, the chemical reaction on the part of t ≤ 4 min is realized by way of small clusters merging in larger ones, the latter — in still larger and so on. Hence, for this reaction the aggregation hierarchical model, considered in paper [116], is valid.

The discrepancy of theoretical and experimental results at t > 4 min and reaction rate reduction is due to physical aspects of process, but not chemical ones, that follows from the mentioned above growth η_{red} cessation. The reaction rate reduction is linked with cluster joining probability p reduction [117]. Let us note, that in this case aggregation process will be realized not by means of chemical reaction, but by interpenetration and physical joining of clusters (macromolecular coils) [113].

Using the Eq. (27) and experimental values Q, the value D_f can be calculated as a function of the polycondensation duration t and obtained thus the dependence $D_f(t)$ is adduced in Fig. 34. As one can see, the monotone D_f growth at t increasing at t > 4 min is observed, that was to be supposed. The theoretical (the greatest) value D_f = 2.11 for chemically limited aggregation [116] is reached at t ≈ 8.8 min. After this system further densification occurs, which is characterized by D_f growth. This process should be continued up to the dimension $D_f ≈ 2.5$, which characterizes polymer gelation [84, 85] and the mechanism of aggregation particle–cluster [104] is realized, i.e., universality class change occurs. The indicated value D_f is reached at $t \approx 48$ min and during time period $t \approx 8.8$–48 min chemical reactions do not occur either, what is confirmed by constant value ηred in this temporal range.

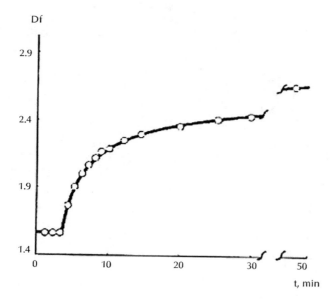

FIGURE 34 The dependence of macromolecular coil fractal dimension D_f on polycondensation process duration t for polyarylate Ph-2.

The possibility of equilibrium process of chemical bonds association-dissociation should be also exluded, since for this process the value D_f does not exceed 2.03 [58]. After $t \approx 48$ min in system neither chemical nor physical changes occur, that is reflected by the conditions $Q = \text{const}$ and $\eta_{red} = \text{const}$. The polycondensation reaction realization rate ϑ_r can be obtained from the Eq. (27), differentiating it by time t. This gives the expression [113]:

$$\vartheta_r \sim t^{\left(1-D_f\right)/2}. \tag{69}$$

Since $D_f \geq 1$ [16], then the Eq. (69) predicts ϑ_r reduction at t growth. This effect is always observed in nonequilibrium polycondensation processes [48].

The constant (and the greatest) value ϑ_r is reached at $D_f = 1$, i.e., for completely straightening chain, when its "accessibility" degree is the greatest.

Let us note in conclusion, that the indicated above and considered factors, the chain destruction appreciation [101], catalyst influence [48] and so on can be required. Nevertheless, the adduced above results demonstrated clearly both possibility and expediency of the considered approach for polymerization processes in general and polycondensation in particular. The conclusion about polycondensation process fractal character should be recognized as particularly important, since it cannot be realized for compact coils (globules) with $D_f = d = 3$ [113].

The authors [118, 119] fulfilled description of the low-temperature polycondensation process of polyarylate Ph-2 in 8 different solvents (N,N-di-methylformamide, nitrobenzene, acetone, 1,2-dichloroethane, chloroform, 1,2,4-trichlorobenzene, benzene and 1,4-dioxane) within the frameworks of irreversible aggregation cluster–cluster model [120].

The critical radius R_c of macromolecular coil was estimated according to the Eq. (48) and current gyration radius of coil R_g can be estimated according to the relationship [108]:

$$ R_g \sim \left(\frac{4c_0 kT}{3\eta_0 m_0} \right)^{1/D_f} t^{1/D_f}, \tag{70} $$

where c_0 is reagents initial concentration, k is Boltzmann constant, T is temperature, η_0 is reactive medium initial viscosity, D_f is macromolecular coil dimension, t is reaction duration.

If in the first approximation all parameters in the right-hand part of the Eq. (70), excluding t, are accepted as constant ones, then at the condition $R_g = R_c$ and $t = t_c$ (t_c is reaction cessation time) let us obtain [121]:

$$ t_c \sim R_c^{D_f}. \tag{71} $$

Since for polyarylate Ph-2 $m_0 =$ constant, then from the Eqs. (1) and (71) at the conditions $R_g = R_c$ and $MM_c = N_c m_0$ let us obtain finally [119] Eq. (72):

$$ t_c \sim MM_c. \tag{72} $$

Further the conversion degree Q (Q_T) can be estimated theoretically according to the Eq. (26). In assumption mp = const and $\eta 0$ = const Q_T can be obtained from simpler Eq. (53). At the comparison of experimental Q_e [51] and calculated according to the Eq. (53) Q_T conversion degree values for polyarylate Ph-2 one can see (Fig. 35), that Q_e and Q_T change tendencies are the same (shaded line), but their quantitative agreement is poor. This is explained by the condition $\eta 0$=const accepting in the Eq. (26). Since the experimental $\eta 0$ values are absent, then the authors [119] conducted rough, but simple this parameter estimation, proceeding from the following considerations. From the Eq. (26) at $Q_T = Q_e$, mr = const = $c0$ and tc = const = 60 min (the typical reaction duration [48]) the values $\eta 0$ were estimated and then the values Q_T were calculated according to the same relationship with variable $\eta 0$. The results of Qe and estimated by the indicated mode QT are shown in Fig. 36, from which much better correspondence of theory and experiment follows in comparison with Fig. 35.

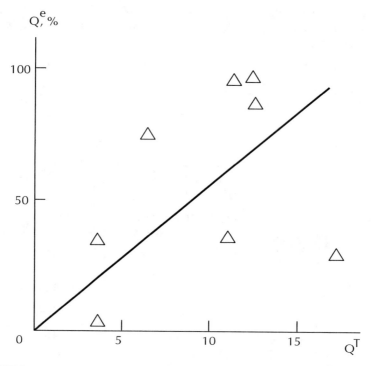

FIGURE 35   The relation between experimental Q_e and calculated according to the Eq. (26) Q_T conversion degree values for polyarylate Ph-2.

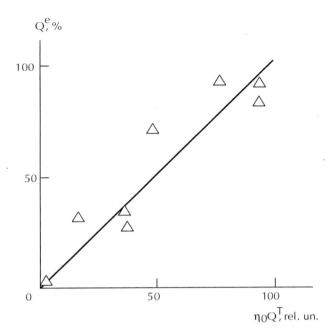

FIGURE 36 The relation between experimental conversion degree values Qe and parameter η0QT for polyarylate Ph-2.

Let us note in conclusion, that polycondensation processes treatment within the frameworks of irreversible aggregation models does not exlude conclusions, made on the purely chemical grounds. Let us explain this on one example. In paper [51] it has been shown, that Ph-2 destruction is observed in N,N-dimethylformamide. Within the frameworks of irreversible aggregation models this means, that two processes are realized in parallel: aggregation and destruction. As it has been shown in work [103], such combination results to the value Df of final aggregates increasing. Actually, D_f value for Ph-2 macromolecular coil in N,N-dimethylformamide is equal to 2.03, that corresponds well to D_f theoretical value for such aggregation type [103].

In the end the conclusion can be made, that reaction cessation in low-temperature polycondensation process is limited by purely physical factor, namely, by macromolecular coil density reaching of reactive medium density (mixture monomer — polymer solution) [121], that is possible for fractal objects only. Such conclusion follows from the simple analysis, adduced below. As it is known [122], the fractal object density ρfr can be calculated according to the Eq. (73):

$$\rho_{fr} = \rho_{Euc}\left(\frac{R_g}{a}\right)^{D_f - d}, \qquad (73)$$

where ρEuc is Euclidean object of the same material, as a fractal object, density, a is a linear scale.

For Euclidean object $D_f = d$ and $\rho fr = const = \rho Euc$. For fractal object R_g increasing (or MM and N) in polycondensation process at a=const results to ρfr reduction in virtue of the condition $D_f < d$ [52]. Therefore there are critical values MMc(Nc) and tc, above which synthesis reaction ceases [118, 119, 121]. This process is simulated within the frameworks of irreversible aggregation models according to the mechanism cluster–cluster.

As it is known [117], in case of different kinds of reaction realization the so-called steric factor p (p ≤ 1.0), showing that not all collisions of reacting molecules occur with these molecules orientation, proper for chemical bonds formation, plays the essential role. In such treatment this factor importance is defined by its reaction rate constant kr proportionality — the smaller p, is the smaller kr is and reaction is realized with the smaller rate [123].

Let us consider the theoretical basis for comparison of different factors influence on the value p. Within the framework of chemical reactions model notions the value kr can be expressed as follows [117]:

$$k_r p\left(\frac{8RT}{3\eta}\right)[A][B], \qquad (74)$$

where R is universal gas constant, T is reaction temperature, η is reactive medium viscosity coefficient, (A) and (B) are two chemical reagents concentration.

Within the frameworks of fractal analysis the notion of particles or cluster "accessibility" to active (reactive) sites of other particles or clusters for either reaction realization was introduced. This notion is linked with cluster fractality notion, which is macromolecular coil in solution. The macromolecular coil fractal dimension Df characterizes its structure "openness" degree — the smaller D_f, the more intensive clusters penetration (or particles penetration) into another then cluster (it is more "accessible"). This postulate is expressed analytically by the Eq. (27). Differentiating the indicated relationship by time t, let us obtain reaction realization rate (the Eq. (69)) and then it can be written [123]:

$$p\left(\frac{8RT}{3\eta}\right)[A][B] = \frac{c_1}{t^{(Df-1)/2}}.$$ (75)

Assuming the multiplier p in the Eq. (75) left-hand part as constant one, the constant c1 in this equation can be estimated from the known boundary conditions of irreversible aggregation model. This model assumes D_f growth at p decreasing [124], that corresponds to tendency, which the Eq. (75) expresses. Assuming, that p = 1 is achieved at very high macromolecular coil "accessibility" degree and estimating this degree by almost minimum value D_f and also estimating polycondensation reaction duration as t = 3 hours or 10800 s, let us get the following simple equation [123]:

$$p = \frac{1.60}{10800^{(Df-1)/2}}.$$ (76)

Another limiting case (chemically limited aggregation model) corresponds to $D_f = 2.11$ [58]. According to the Eq. (76) this value D_f corresponds to $p \approx 0.009$, that is close to p = 0.01 according to the indicated model [125]. Now a different factors effect on the value p (and, hence, kr) can be estimated, using the Eq. (76).

1.3.1 POLYCONDENSATION MODE

As it is known, the same polymer, produced by equilibrium and nonequilibrium, differs by its characteristics, in particular, has different exponents a in Mark–Kuhn–Houwink equation [53]. The values a and D_f are linked between themselves by the Eq. (4). For polyarylate on the basis of phenolphthaleine the values $D_f = 1.96$ (equilibrium polycondensation) and $D_f = 1.80$ (nonequilibrium polycondensation) were obtained, that corresponds to p=0.0185 and 0.039. Thus, polycondensation mode change from equilibrium up to nonequilibrium (interfacial) one results to p increase approximately twice. Approximately the same relation is valid at polycondensation mode change for other polyarylates of different chemical structure.

1.3.2   CHAIN FLEXIBILITY

The comparison of two polyarylates, having conventional signs Ph-1 and Ph-2, has shown that they differ only by a substitution type in the dicarbon acid residual: metha- and para-positions, accordingly. This results to Ph-2 main chain rigidity enhancement in comparison with Ph-1, that is expressed in these polyarylates glass transition temperature values: 603 and 543 K, respectively. The calculation according to the Eq. (4) gives the following results: $D_f = 1.80$ for Ph-1 and $D_f = 1.55$ for Ph-2. The values p for these polyarylates, obtained by the interfacial polycondensation, are equal to 0.039 and 0.124, accordingly. Thus, the chain rigidity increasing at the expence of metha- by para-connecting replacement results to p enhancement in 3.2 times.

For rigid-chain polymer poly-n-phenylenoxadiazole within the range of MM = $(1.0-5.5) \times 103$ the value a = 1.23 was obtained. This gives the values $D_f = 1.35$ and p = 0.315 according to the Eqs. (4) and (76), respectively. Thus, the chain rigidity increasing for poly-n-phenylenoxadiazole in comparison with Ph-1 results to p raising in about 8 times.

1.3.3   SIDE SUBSTITUENTS AVAILABILITY

In the given context two polyarylates on the basis of di-(4-oxiphenyl)-methane and 9,9-di-(4-oxiphenyl)-fluorene (the conditional signs D-20 and D-10, accordingly) were compared. For the first from them molecular weight of substituents at central atom of carbon in bisphenol (side ones in respect to the main chain) makes up MMsub=2, i.e., practically a linear polymer, whereas for D-10 the value MMsub = 152. Calculation according to the Eqs. (4) and (76) gave for D-10 and D-20 the following results: $D_f = 1.83$ and 1.70, |p = 0.062 and 0.339, respectively. Thus, MMsub increasing up to 152 results t p reduction approximately in 5.5 times.

1.3.4   THERMODYNAMICAL QUALITY OF USED AT SYNTHESIS SOLVENT

The value D_f determination for polyarylate Ph-2 synthesis in different solvents war performed according to the Eq. (22). These estimations showed, that the values D_f for Ph-2, synthesized in N,N-dimethylformamide (DMFA) and dichloroethane (DCE) were equal to 2.03 and 1.80, accordingly. Thus, the values for the same conditions of synthesis are equal to 0.0139 and 0.039, respectively. The solubility parameters δ values for Ph-2, DMFA and DCE are equal to 10.7, 12.1 and 9.8 (cal/cm^3)1/2, accordingly [36, 55]. The comparison of δ values of polymer and solvents allows to make conclusion, that in respect to Ph-2 DCE possesses higher thermodynamical affinity than DMFA, i.e., a better solvent. Thus, the improvement of thermodynamical quality of the used at nonequilibrium polycondensation solvent results to p increasing almost in 3 times at the indicated above conditions.

1.3.5   MOLECULAR WEIGHT OF REPEATING LINK

For polyarylates number the molecular weights of repeating link m0 were calculated and the values D_f were estimated according to the Eq. (4). It turns out, that for all polyarylates with metha-substituted residues of dicarbon acid in the main chain the good linear dependence (D_f growth at m0 increasing) is observed and the reasons of deflection from it polyarylates Ph-2 and Ph-7 (para-connections) were considered above. The values p were calculated for two conventional polyarylates with m0 = 300 and 600, which turn out to be equal to 0.071 and 0.020, respectively. Thus, twice rise of m0 results to p reduction approximately in 3.5 times [123].

The reagents relation influence in polycondensation processes of different types is well-known and its description is based on functional groups nonequivalency rule [48]. According to this rule, the surplus of one from the initial substances, if it is not removed in reaction process, results to polymer molecular weight reduction proportionally to this surplus value. In practice this means, that the dependences of molecular weight MM or viscosity η of polymers on reagents relation have extreme character and MM (or η) maximum is reached at equimolar reagents relation [48]. Although this rule was known a long time ago and it was confirmed experimentally many times, nevertheless this effect quantitative description, which is necessary for polycondensation processes prediction, is absent. The authors [126] considered the indicated description possibility on the example of low-temperature polycondensation of chloroanhydride of terephtalic acid (CATA) and phenolphthaleine (PP) in dichloroethane [127].

The adduced in work [127] dependences of conversion degree Q and reduced viscosity η_{red} for polyarylate Ph-2 on relation CATA/PP have extreme character with maximum at equimolar relation CATA/PP = 1:1 and two-fold concentration increase of any from reagents results to Q reduction approximately in 1.5 times and ηred — in about 3 times [127]. For the observed effect quantitative description let us write the relationships, describing Q and η_{red} (or MM) changes in polycondensation process, and consider factors influencing them.

The determined according to the Eq. (4) value D_f for polyarylate Ph-2 is equal to 1.55. Within the frameworks of fractal analysis the conversion degree Q can be described by the Eq. (26) with c0 change on (CATA)(PP), where (CATA) and (PP) is corresponding reagents concentration. Differentiation of the Eq. (26) by t gives the reaction rate ϑ_r [126]:

$$\vartheta_r \sim [CATA][PP]\eta_0 t^{(1-D_f)/2} \quad . \tag{77}$$

On the other hand, reaction rate constant kr is determined according to the Eq. (74).

Let us consider further factors, which can influence on Q and ηred change at relation (CATA)/(PP) variation. First of all, it is the product (CATA)(PP), including in the Eqs. (26), (74) and (77). However, its variation within the limits, considered in paper [127], makes up 0.21-0.25 and is not as large, as to explain the indicated above Q and η_{red} reduction. $\eta0$ change at reagents relation variation is another factor [48]. Let us consider η_0 change within the frameworks of irreversible aggregation models [120]. As it is known [120], such aggregate gyration radius R_g change (the value D_f for Ph-2 indicates, that its formation process is realized by mechanism cluster–cluster [58]) as a function of t is determined according to the Eq. (70) with c0 change on (CATA)(PP). Besides, the value R_g is connected with repeating links number per macromolecular coil (polymerization degree) N by the fractal Eq. (1). In its turn, MM value is written as Nm0. Taking into account the constant values including into the Eq. (70) parameters k, T and t, it can be written [126]:

$$MM \sim \frac{[CATA][PP]}{\eta_0} \quad . \tag{78}$$

From the Eqs. (26) and (78) the values η_0 (in relative units), corresponding simultaneously to the experimental dependences of Q and MM on (CATA)/(PP). Thus estimated values η_0 are adduced in Fig. 37, from which it follows, that they reache minimum at reagents equimolar relation.

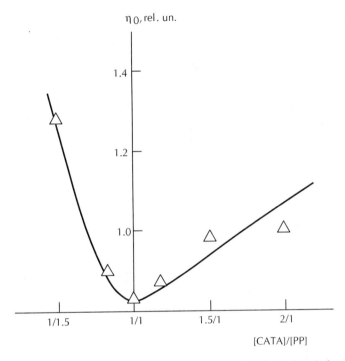

FIGURE 37 The dependence of reagents initial mixture viscosity η_0 (in relative units) on their relation (CATA)/(PP).

And at last, the third factor is defined completely by the macromolecular coil fractal properties. Earlier synthesized macromolecule Ph-2 structure change because of reagents relation variation has not been supposed [48, 127]. Nevertheless, irreversible aggregation models predict such possibility. It is obvious, that at deviation from equimolar relation reagent surplus cannot find "partner" for reaction realization, that will be resulted to p reduction and D_f enhancement [123]. This effect quantitative estimation can be obtained from the Eqs. (74) and (77) combination at the condition, that p reduces proportionally to any reagent surplus. The similar estimations have shown D_f increasing from 1.55 at equimolar relation (CATA)/(PP) up to ~ 1.70 at (CATA)/(PP) = 2:1.

Further, using the obtained changes of the considered above three parameters ((CATA)/(PP), η_0 and D_f) the corresponding to them Q and ηred variations can be estimated with the aid of the Eqs. (26) and (78), accordingly, and also Mark–Kuhn–Houwink equation (the Eq. (50)) and compared the obtained results with the experimental dependences [127]. Such comparison is performed in Fig. 38, from which good enough agreement of theory and experiment follows.

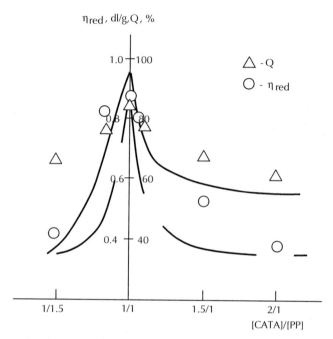

FIGURE 38   The dependences of conversion degree Q (1, 2) and reduced viscosity ηred (3, 4) on reagents relation (CATA)/(PP) for polyarylate Ph-2 (solvent-dichloroethane). 1, 3 — experimental data [127], 2, 4 — calculation according to the Eqs. (26) and (78).

Thus, the stated above results allow to make at any rate two conclusions. The first from them concerns quantitative description complexity of such processes as polycondensation. Actually such obvious enough effect as Q and η_{red} (or MM) change at reagents relation variation requires usaging, as a minimum, of three independent parameters, moreover two from them (η_0 and D_f) are difficult enough to predict. The second conclusion consists of the fact that even the simplest used above model of the considered effect quantitative description has shown its effectiveness and, hence, perspectiveness for polycondensation processes prediction as a whole, including a computer one.

In paper [67] it has been shown, that high-temperature polycondensation kinetics of polyarylates and polysulfones series, characterized by reaction rate constant kr (determined on the first initial part of the kinetic curve Q_{-l}) at different temperatures and reactive mediums is described by three main factors. These factors are: Process thermal activation, bisphenols reactive ability and macromolecular coil structure. The indicated factors can be characterized quantitatively by polycondensation temperature T, bisphenols Gammet's constant $\Sigma\sigma$ and macromolecular coil fractal dimension D_f, respectively. However, the physical sense of "thermal activation" term remains in many respects vaque with the exception of this factor action general tendency: T increasing or process activation energy Eact reduction results to polycondensation rate raising (kr growth). Therefore the authors of work [128] undertook an attempt to elucidate the physical sense of the indicated factor on the example of high-temperature polycondensation of disodium salts of bisphenols number with 4,4'-dichlorodiphenylsulfone [64].

As it was indicated above, at aggregates, consisting of smaller particles (polymeric macromolecules belong to such aggregates), formation the steric factor p (p$\leq$1.0) plays an important role. The value p is linked with polycondensation activation energy Eact as follows [129]:

$$p \sim e - E^{act/RT},\qquad(79)$$

where R is universal gas constant.

From the Eq. (79) it follows, that at small E_{act} (or high T) the value p is close to one and the reaction realizes according to the diffusion-limited regime. And on the contrary, at large E_{act} (or low T) p<<1.0 and the reaction is realized according to the chemically limited regime [129]. Besides, the value p influences essentially on forming macromolecular coil structure, characterized by its fractal dimension D_f [130]. At large p (p $\rightarrow$ 1.0) the coils with open structure and small values D_f are formed and at small p (p $\rightarrow$ 0) compact coils, having large D_f values, are formed [130]. In its turn, the value D_f to a considerable extent defines polycondensation kinetics according to the Eq. (27).

In Fig. 39 the dependence $D_f(p)$ for the indicated polymers at three different temperatures of polycondensation is adduced (the value p was estimated according to the Eq. (79) without proportionality coefficient appreciation). As it was to be expected from the very general considerations, stated above, D_f reduction at p growth is observed. At p = 0 (i.e., the situation, when reaction ceases) the value $D_f \approx 2.5$. This D_f value corresponds to the gelation point and, as consequence, to chemical reaction cessation [84, 85]. For irreversible aggregation models the lower limit $D_f = 1.75$ in case of three-dimensional Euclidean space is accepted [58], however, in the polymers case this limit value should be reduced up to $D_f = 1.0$ [16]. Thus, the range of the obtained according to the Eq. (79) values p = 0–8×10³. Since for real polymers p = 0–1.0 [117], then the Eq. (79) can be written as follows [128]:

$$p \sim e^{-E_{act}/RT} \quad .$$
                                                                                      (80)

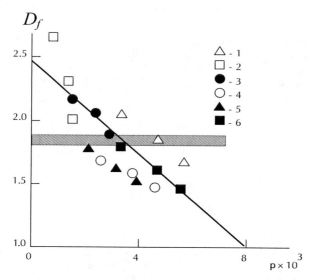

FIGURE 39   The dependence of macromolecular coil fractal dimension D_f on steric factor p, estimated according to the Eq. (79) for DOPP (1), DOPPM (2), DOPES (3), DOMPP (4), DOPS (5) and DOCPCH (6) (see Table 1.2). Df range, corresponding to polycondensation regime change, is indicated by shaded region.

In Fig. 40 the dependence kr(p) (where the value p was calculated according to the Eq. (80)) is adduced, which can be divided into two parts. On the first from them (small p) the value kr grows slowly at p increase and is proportional to the steric factor. On the second part (at $p \geq 0.45$) rapid growth kr begins in a very narrow p range, i.e., the value kr is independent practically on p. Let us note, that in Fig. 40 from the considerations of a more clear notion of the data at small p the greatest values p are not indicated, which were determined theoretically [66]: so, within the range of p = 0.50–0.71 the value kr increases up to ~ 4.3 L/equivmin. This means, that on the first part polycondensation chemically limited regime is realized, for which kr ~ p [117], and on the second one — diffusion limited regime, for which kr is independent on p [117]. In case of bimolecular reaction of the look Eq. (81)

$$A + B \xrightarrow{\; k_r \;} C$$
                                                                                      (81)

and diffusion limited regime the value kr is determined according to the Eq. (74).

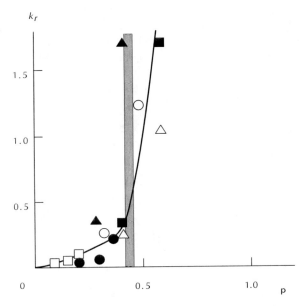

FIGURE 40   The dependence of reaction rate constant kr on steric factor p, calculated according to the Eq. (80). The range p, corresponding to polycondensation regime change, is indicated by shaded region. The conditional signs are the same, as in Fig. 39.

From the Eq. (74) one can see easily, that sharp increase kr in diffusion limited regime (see Fig. 40) is due to two factors: T growth and value η reduction owing to this.

From the data of Fig. 40 the range of p, corresponding to the transition between two indicated polycondensation regimes, can be estimated approximately. This range is narrow enough (p = 0.41–0.44) and it is indicated by shaded region in Fig. 40. The corresponding range of D_f values at transition between these regimes has been also shown in Fig. 39 by shaded region. The indicated range makes up $D_f \approx 1.80$–1.88, that corresponds to the general notions of irreversible aggregation models [131].

From the stated above data it follows, that from the practical point of view polycondensation diffusion limited regime is much more favourable, as for it reaction rate is higher essentially (see Fig. 40). The proposed model allows to construct peculiar "phase" diagram for the considered polymers polycondensation in

E_{act}–T coordinates. If in the Eq. (80) p = 0.41 is accepted as boundary value, then the curve corresponding to this condition, can be calculated from it. Such curve is shown in Fig. 41. Higher and in the left-hand corner of this curve the diffusion limited regime region is settled down (about kr ≥ 1.0 L/equivmole), and lower and in the right-hand corner — chemically limited regime region (about kr≤0.4 L/equiv mole). The adduced in Fig. 41 experimental data have shown a good correspondence with the indicated "phase" diagram.

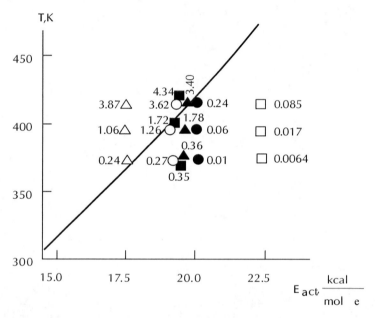

FIGURE 41 The diagram $E_{act}(T)$, showing this space division on two subspaces: corresponding to the diffusion limited (from the left and from above) and chemically limited (from the right and from below) polycondensation regimes. Figures at points show corresponding to them kr values. The conditional signs are the same, as in Fig. 39.

Hence, the stated above results have shown, that the physical sense of "thermal activation" term, expressed in thermal parameters Eact and T changes, consists of steric factor p change, subsequent macromolecular coil structure change and at last polycondensation reaction course change according to the Eq. (27). But the transition from aggregation chemically limited regime up to diffusion limited one is the strongest effect, which results to 15–20 — fold kr increasing.

1.4 THE INFLUENCE OF REACTIVE MASS STIRRING ON MAIN PARAMETERS OF INTERFACIAL POLYCONDENSATION

At present it is well-known [48], that reactive mass stirring in interfacial nonequilibrium polycondensation process tells favourably on polymers synthesis, resulting to increase of both polymer conversion degree Q and its molecular weight MM. A works number exists, confirming this notion experimentally and also this effect several explanations, made on qualitative level [48]. Nevertheless, for quantitative simulation of polymers synthesis processes, including a computer one, such level of the observed effect explanation is unsufficient. Therefore the authors [132] gave the description of the obtained earlier results by stirring influence on the main characteristics of interfacial polycondensation process of polyarylate on the basis of terephthalic acid chloroanhydride and diane (conventional sign D-1) [53].

The Eqs. (26) and (28) were used for synthesis process description within the frameworks of fractal analysis and irreversible aggregation modes [132]. Let us pay attention to the fact, that the fractal dimension D_f value determines to a considerable extent Q value, but molecular weight is independent on this parameter.

It can be supposed, that at any rate two factors exist, influencing on interfacial polycondensation process course and changing at reactive mass stirring — initial viscosity η_0 and macromolecular coil fractal dimension D_f. Let us consider in more detail possible change of η_0 value at stirring. By its physical essence the stirring using is equivalent to shear stress to polymer solution application, that results to reactive mass viscosity η change in comparison with viscosity η_0 of the same solution without stirring. The indicated viscosities ratio can be described by the following relationship [25]:

$$\frac{\eta}{\eta_0} = 1 - \frac{\beta}{a_1\beta + b_1}, \tag{82}$$

where a_1 and b_1 are coefficients, β is parameter, determined according to the equation [25]:

$$\beta = \gamma \frac{\dot{M}M[\eta]\eta_s}{RT}, \tag{83}$$

where $\dot{\gamma}$ is shear rate, $[\eta]$ is intrinsic viscosity of polymer solution, s is solvent viscosity, R is universal gas constant, T is temperature.

Let us pay attention to two important features of the Eqs. (82) and (83). First-ly, the Eq. (82) predicts reactive mass viscosity reduction at application to it shear strain, i.e., at stirring application. Secondly, from the Eq. (83) it follows, that since in synthesis process continuous change MM and $[\eta]$ (and possibly, T) occurs, then the parameter β change will be observed and, hence, reactive mass viscosity ac-cording to the Eq. (82). Therefore it is obvious, that the proposed in paper [132] calculation scheme by synthesis final results is approximate and it can serve only as semiquantitative theoretical grounds. By this reason the authors [132] did not estimate proportionality coefficients (obligatory because of proportionality sign) in the Eqs. (26) and (28) and obtained them by matching of the experimental and theoretical data.

At the estimation of the value η in the Eqs. (82) and (83) the authors [132] assumed, that $\eta 0$ was reactive mass viscosity at stirring with rate 500 rpm (table 152 in paper [48]), the value $\dot{\gamma}$ was accepted proportional to stirring intensity and value $[\eta]$ was estimated according to the known values MM (table 152 in work [48]) according to Mark–Kuhn–Houwink equation for polyarylate D-1 [53] Eq. (84):

$$[\eta] = 2.04 \times 10^{-4} MM^{0.745.} . \qquad (84)$$

Then the values MM were estimated according to the Eq. (28) at the condition c0=const. The comparison of the experimental MMe and calculated according to the Eqs. (28), (82) and (83) MMT molecular weight values for polyarylate D$_{-1}$ is adduced in Fig. 42, from which their good correspondence follows. Thus, the reactive mass viscosity reduction (see the Eq. (28)) is the main factor, defining molecular weight of polyarylate D-1 growth at stirring intensity enhancement. The similar effect was observed for polyamide on the basis of chloroanhydride of terephthalic acid and trans-2,5-dimethylpiperazine (PA). At the stirring intensity increasing of PA from 350 up to 17000 rpm logarithmic viscosity of this polymer in m-crezole is increased from 0.73 up to 2.72 dl/g, i.e., in about 3.7 times. One can see easy, that such stirring intensity enhancement reduces viscosity η about in 2.6 times even without MM and $[\eta]$ increase in synthesis process.

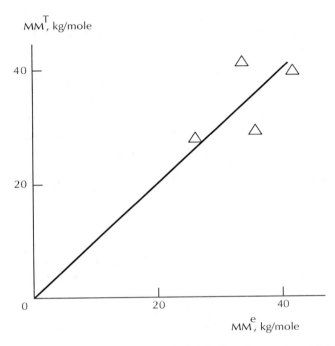

FIGURE 42 Comparison of the theoretical MMT and experimental MMe [48] molecular weight values for polyarylate D_{-1}.

One more important factor, which influences on the value Q according to the Eq. (26), is the fractal dimension D_f change. It is supposed, that at shear stress application macromolecular coil changes its shape from approximately spherical up to very flattened one in shear application direction. This effect is general enough and is realized, for the example, at melt flow [133] or in solid-phase extrusion process [134]. From the coil incompressibility condition its gyration radius R_g increasing in n times in directions, perpendicular to shear direction, results to R_g change in shear direction in n^2 times. Since shear strain is proportional to shear rate and coil size change n is proportional to shear strain [133], then parameter n was determined as stirring intensities ratio. Further the theory, developed for deformed macromolecular coil [135], can be used, according to which:

$$R_g \sim MM^{(d_s-1)/2d_s} ,$$ (85)

where d_s is macromolecular coil spectral dimension.

As the first approximation the authors [132] supposed, that the smallest stirring intensity (500 rpm) did not deform macromolecular coil to some extent significantly and therefore the value D_f for polyarylate D_{-1} could be determined in that case according to the Eq. (4). In its turn, the dimension ds for swollen coil with excluded volume interactions appreciation can be determined according to the Eq. (39).

Thus, by known values n Rg reduction in shear stress direction can be calculated, which is equal to n^2, and then effective value ds can be estimated according to the Eq. (85) and according to the Eq. (39) — the value D_f. As it was assumed above, such calculation showed essential D_f decreasing at stirring intensity growth — from D_f=1.74 for undeformed coil according to the Eq. (4) up to $D_f \approx 1.28$ at stirring intensity 1500–1700 rpm. According to the Eq. (26) such Df reduction should result to corresponding Q growth. The comparison of the calculated theoretically Q_T and obtained experimentally Q_e (table 152 in work [48]) conversion degree values for polyarylate D_{-1} is shown in Fig. 43. The authors [132] performed two variants of theoretical estimation: the first from them assumes, that stirring levels the initial viscosity η0 influence and the second one takes into account the indicated factor. As it follows from the data of Fig. 43, both variants give a good enough correspondence to experiment, but the second from the indicated variants was gave smaller scatter. Nevertheless, to make the final conclusion about reactive mass initial viscosity η_0 effect leveling is impossible, since it prevails at MM determination. The large scatter in this case can be due to the mentioned above given parameter continuous change in synthesis process.

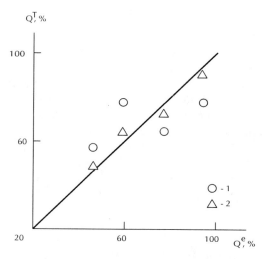

FIGURE 43 Comparison of the theoretical Q_T and experimental Q_e [48] conversion degree values for polyarylate D_{-1}. 1 — QT values were calculated with η_0 appreciation; 2 — without η_0 appreciation.

Hence, the stated above results have shown the principal possibility of fractal analysis and irreversible aggregation models for the description of reactive mass stirring intensity effects in interfacial nonequilibrium polycondensation process. It is obvious, that from the practical point of view the theoretically correct choice of coefficients in the Eqs. (26) and (28) for the substitution in them proportionality sign on equality sign is the most complex question [132].

One more explanation of reactive mass stirring influence on interfacial poly-condensation characteristics is unblended phases contact surface increasing, that accelerates diffusive processes and conduces to polymer formation [48]. Within the frameworks of scaling approach it is also supposed, that stirring reduces re-active medium heterogeneity, which is due to large density fluctuations [136]. Therefore the authors [137, 138] performed semiquantitative analysis of stirring influence on conversion degree Q within the frameworks of the scaling approach on the example of polyarylate D_{-1} interfacial polycondensation [48, 53]. For this polymer stirring rate increasing from 500 up to 1500 rpm enhances the value Q from 0.46 up to 0.93 [48].

Let us consider the reaction in which particles P of a chemical substance diffuse in the medium, containing random located static non-saturated traps (growing macromolecular coils) T. At the contact of particle P with a trap T the particle disappears. It is usually considered that if the concentration of particles and traps is large or the reaction occurs with intensive stirring, the process can be considered as the classical reaction of the first order. In this case it can be assumed that the concentration of particles c(t) decay law will be the following [136]:

$$c(t) \approx \exp(-At),$$ (86)

where A is a constant.

However, it the concentration of random located traps is small, space areas necessarily exist, practically free from traps. Particles entering these areas can reach the traps only after quite a long time and, hence, their number decay in due time will be slower. The formal analysis of this problem shows that the concentration of particles falls down according to the law [136]:

$$c(t) \approx \exp\left(-Bt^{d/(d+2)}\right)$$ (87)

being dependent on the space dimension, where B is a constant.

If the traps can move, their mobility averages the influence of spatial hetero-geneity and in this case [136] Eq. (88):

$$c(t) \sim \exp(-At)\exp\left(-Bt^{d/(d+2)}\right). \tag{88}$$

The authors [137, 138] assumed that the traps moved very slowly, since mass was larger significantly than particles mass and their motion effect could be neglected [97].

The data of work [48] showed that at the stirring rate of 1500 rpm the value $Q \rightarrow 1.0$ and therefore it could be supposed, that the Eq. (86) application was correct at the indicated rate. At the minimum from the used rate of 500 rpm the value Q decreases more than twice and therefore the authors [137, 138] assumed that in that case the Eq. (87) was more applicable. Constants A and B in the indicated equations estimation by the experimental data showed, that their absolute values were close and therefore further their average value would be used, which was equal to 0.15. If to assume, that an exponent (let us designate it as x) in the Eqs. (86) and (87) changes linearly with stirring rate n, then on the basis of the stated above assumptions the calibrating plot, adduced in Fig. 44, can be constructed. The value c(t) can be expressed as (1–Q) and then for Q estimation the generalized equation was used [138]:

$$1-Q \approx \exp\left(-0.15t^x\right), \tag{89}$$

where the value x is determined from the calibrating plot of Fig. 44.

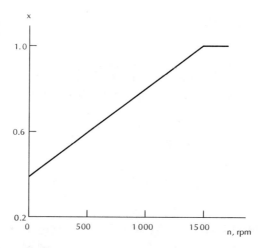

FIGURE 44   The calibrating plot of the dependence of exponent x in the Eq. (89) on stirring rate n. The explanations are in the text.

In Fig. 45 the comparison of calculated according to the Eq. (89) and obtained experimentally [48] the values Q is adduced. As one can see, between the indicated Q values a good correspondence is observed, that confirms correctness of the proposed simple semiquantitative estimation method of reactive mass stirring effect.

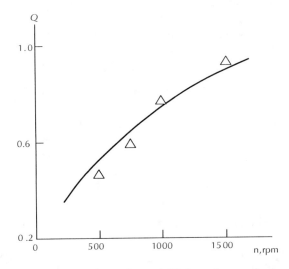

FIGURE 45   Theoretical (1) and experimental (2) dependences of conversion degree Q on stirring rate n for polyarylate D_{-1}.

At stirring rate n = 0 (stirring absence) according to the plot of Fig. 44 the value x ≈ 0.37 is obtained, that is lower appreciably than this exponent value for d = 3. Equaling the exponents in the Eqs. (87) and (89), let us obtain for the case without stirring d ≈ 1.2. This means, that in the indicated case interfacial polycondensation is realized in the space, intermediate between line (d = 1) and plane (d = 2) unlike volume space (d = 3) in case of intensive stirring. It is clear, that in the first case (d≈1.2) contacts particle-trap number will be much less, than in the second case that sharply reduces the reaction rate.

Within the frameworks of fractal analysis the area S of surface with dimension d is determined as follows [139]:

$$S = L^d r^{2-d},$$ (90)

where L is the upper limit of fractal behavior, r is measurement scale.

The estimation according to the Eq. (90) at arbitrary values L and r shows large distinction S at d = 1.2 and d = 3. So, at L = 10 and r = 2 relative units S distinction for the case without stirring and n=500 rpm makes up ~ 20 times, and at L = 100 and r = 2 — even ~ 1150 times. These estimations confirm assumptions about stirring influence made in work [48].

Thus, the entire interval n can be divided into two parts: n=0–500 and n=500–1500 rpm. On the first part Q increasing is reached at the expence of d enhancement and, accordingly, S at n growth and on the second part the limiting value d=3 is achieved and Q growth is due to reactive medium fluctuations scale decreasing (heterogeneity reduction).

In work [48] it has been shown, that at D_{-1} synthesis without stirring the duration of 6, 48 and 168 hours is required for the values Q 0.24, 0.52 and 0.70, accordingly. Calculation according to the Eq. (89) gave for these Q magnitudes the following reaction duration values t: 6, 20.5 and 75 hours, respectively. As one can see, the values t correct order is obtained and their quantitative discrepancy is due to strong (power) dependence of t on the exponent x value in the Eq. (89) and made above approximations.

Thus, the simple scaling approach allows correct not only qualitative, but also quantitative description of the stirring influence on final results of polyarylate D_{-1} interfacial polycondensation. From the physical point of view the interfacial polycondensation rate is determined by spatial fluctuations scale or real dimension of space, in which reaction is realized.

1.5   COPOLYCONDENSATION

The copolymers composition heterogeneity, described by separate macromolecules distribution by their composition, is one from the copolycondensation theory goals [68]. As parameter, charactering this distribution, the microheterogeneity coefficient Kh is usually used, which serves as the quantitative characteristic of links distribution sequence in copolymer chains. The composition heterogeneity study is performed by both statistical and kinetic methods [68, 140, 141]. However, the authors [142] supposed, that this problem understanding can be improved essentially by the usage of fractal analysis methods and irreversible aggregation models, since appreciation of such factors as macromolecular coil structure and its formation mechanism could broaden our notions about the reasons, caused copolymers composition heterogeneity. Therefore the authors [142] considered the indicated factors influence on the example of three copolymers: aromatic copolyethersulfoneformals (APESF), diblock-copolymers of oligoformal 2,2-di-(4-oxiphenyl) propane and oligo-sulfone phenolphthaleine (CP-OFD-10/OSF-10) and diblock-copolymers of oligoformal 2,2-di-(4-oxiphenyl) propane, phenolphthaleine and dichloroan-hydride of isophthalic acid (CP-OFD-10/P-1).

In Fig. 46 the dependences of glass transition temperature Tg, determined by the thermomechanical method, on formal contents cform are shown for the indicated copolymers. As one can see, the dependences Tg(cform) course is different for these copolymers. For APESF the values Tg are situated above additive glass transition temperature T_g^{ad}, for CP-OFD-10/P-1 — lower T_g^{ad} and for diblock-copolymers CP-OFD-10/OSF-10 the dependence Tg(cform) has sigmoid character. Such course of the dependences Tg(cform) for the indicated copolymers supposes different change Kh with copolymers composition. The value KM can be estimated according to the well-known Gordon–Talor–Wood equation [143]:

$$ T_g = K_h \left[T_{g2} - T_{g1} \left(\frac{W}{1-W} \right) \right] + T_{g1} , \qquad (91) $$

where T_{g1} and T_{g2} are homopolymers glass transition temperatures, W is comonomer molar fraction.

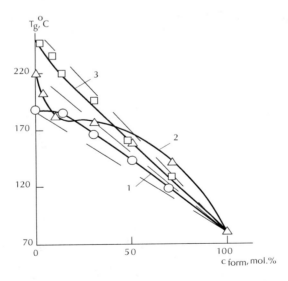

FIGURE 46   The dependence of glass transition temperature T_g on formal contents c_{form} for APESF (1); CP-OFD-10/OSF-10 (2) and CP-OFD-10/P-1.

Let us consider further the causes of value Kh change within the frameworks of irreversible aggregation models. The generalized model of diffusion limited aggregation (DLA) [144] was chosen for copolycondensation process description. The reason of such choice served this circumstance, that experimentally determined values D_f (measured for these solvents, in which synthesis was performed) turned out to be within the range of 1.69–1.89, that corresponded to the universality class of diffusion limited aggregation cluster–cluster [58]. The authors of work [144] proposed the following equation for generalized DLA value D_f determination:

$$D_f = \frac{d^2 + \eta(d_w - 1)}{d + \eta(d_w - 1)},$$

(92)

where d is dimension of Euclidean space, in which a fractal is considered, equal to 3 in our case, η is parameter, the physical sense of which will be considered below, dw is dimension of trajectories of the particles, forming aggregate (macromolecule).

The value d_w was determined according to the Aarony–Stauffer rule [145]:

$$d_w = D_f + 1.$$

(93)

As it has been shown in paper [145], the Eq. (93) is precise enough approximation for aggregation processes, limited by diffusion, that gives the reason for this equation using for dw estimation in case of generalized DLA.

Matsushita [144] in the general case interpreted the parameter η as the ratio n/m (n, m are whole positive numbers), characterizing "chemical" reaction of n random wandering particles with m aggregate perimeter sites. It is obvious, that in the considered case n value characterizes a number of small (monomer and oligomer) molecules (or their intermediate aggregates) and the value m — the number of growing macromolecular coil perimeter sites, accessible for small molecules joining (nonscreened), determined by the parameter n. If to assume, that the value m in copolymerization process remains invariable, then the parameter η change is due to n variation and it should be supposed, that change (more precise, decreasing) n is defined by an intermediate aggregation process. In its turn, the microheterogeneity coefficient Kh characterizes the type of copolymer: for strictly alternating copolymer Kh = 2, for completely statistical — Kh = 1 and for two homopolymers mixture Kh = 0 [140]. By the absolute value Kh deviation from one can be judged quantitatively about links distribution ordering degree in copolymer and by that, to which side this deviation is observed — about monomers (oligomers) tendency either to alternation in chains (Kh>1) or to both comonomers long blocks formation (Kh<1) [68, 140]. In Fig. 47 the relation between Kh and η is shown for the considered copolymers, which turns out to be linear and showing Kh growth at η reduction (n decreasing). As it has been noted above, n decreasing can be linked with intermediate aggregation process. If monomers (oligomers) pair forms intermediate aggregate and already in such a form is included in growing macromolecule, then this will mean n decreasing twice (one molecule is formed instead of two). At Kh = 2.0 the copolymer with strict blocks alternation is formed and this means that all 100% monomer (oligomer) molecules form intermediate aggregates.

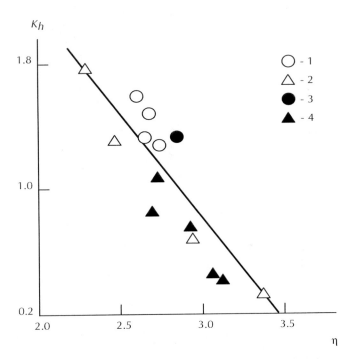

FIGURE 47 The dependence of microheterogeneity coefficient Kh on parameter η for APESF (1), CP-OFD-10/OSF-10 (2), regular copolymer CP-OFD-10/OSF-10 with c_{form} = 30 molar % (3) and CP-OFD-10/P-1 (4).

Let us consider possible reasons of the intermediate aggregates formation. As it is known [117], the steric factor p (p≤1) plays an essential role at chemical reactions different kinds realization, including at that polymers synthesis. It is obvious, that this orientation factor is important for molecules of complex shape and large sizes, which are molecules of the used at copolymers synthesis monomers and the more so oligomers. It has been shown above, that at reaction constant duration t = 60 min the value p can be determined according to the Eq. (76).

In Fig. 48 the dependence $\eta(p)$, calculated according to the Eqs. (76) and (92), is shown. As it follows from the data of this figure, p increasing results to η growth and, hence, to D_f decreasing according to the Eq. (92). Such character of change D_f with p corresponds completely to the theory of aggregates formation by mechanism cluster–cluster [84].

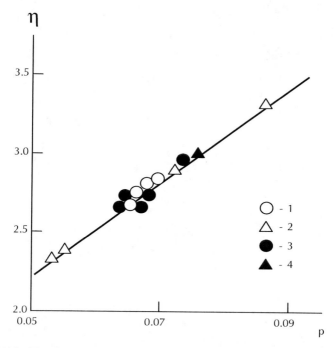

FIGURE 48 The dependence of parameter η on steric factor p for APESF (1), CP-OFD-10/OSF-10 (2), regular copolymer CP-OFD-10/OSF-10 with cform = 30 molar % (3) and CP-OFD-10/P-1 (4).

Hence, the stated above results allow to suppose the following mechanism of copolymers formation within the frameworks of irreversible aggregation models. The steric factor p decreasing means monomers (oligomers) molecules fraction increasing, without forming chemical bond at contact with growing macromolecule and owing to diffusive processes having come back into solution. Such molecules fraction increasing means the probability enhancement of chemical bonds formation among them and, hence, the intermediate aggregates formation probability. Such aggregates formation results to regularly alternating blocks fraction enhancement and, accordingly, to Kh growth. From the plot of Fig. 48 the limiting values η (at p=0 and p=1.0) can be estimated and the limiting values D_f (D_f^{min} and D_f^{max}) can be calculated according to the Eq. (92). At $p = 0$ the value $\eta = 0.7$ and $D_f^{max}=2.35$, that corresponds well to the system fractal dimension in the physical gelation point [85]. At $p = 1.0$ $\eta = 48$ and $D_f^{min} = 1.14$. Such value D_f is typical for rigid-chain polymers. For example, for poly-n-benzamide the value $D_f = 1.11$ according to estimation by the Eq. (4) [146].

As it has been noted above, the value D_f for macromolecular coil in solution is defined by the interactions of two groups: interactions among elements of coil itself and interactions polymer-solvent [24]. Therefore the value D_f can be changed by the solvent variation and, hence, copolymer type, synthesized from the same monomers (oligomers). Increasing D_f means coil compactness enhancement, p reduction and Kh raising.

Hence, the stated above results have shown, that fractal analysis and irreversible aggregation models application allows to obtain the clear physical interpretation of copolycondensation process and estimate its quantitative characteristics. The fractal dimension D_f of macromolecular coil in solution is the main characteristic, controlling this process [142].

The comonomers functional groups activity change is one from the important factors, influencing on the microheterogeneity coefficient Kh value [141]. This change is defined by the ration of reaction rate constant of the first k_{r_1} and the second k_{r_2} functional groups after entry in reaction of the first group. In work [141] Kh increasing at comonomers functional groups activity χ growth has been shown. The authors [147–149] showed the dependence of χ and, hence, Kh on macromolecular coil structure of copolymers in solution, characterized by its fractal dimension D_f, on the example of three mentioned above copolymers: APESF, CP-OFD-10/OSF-10 and CP-OFD-10/P-1.

In Fig. 49 the dependences of D_f on copolymers composition are adduced, about which the following should be said. All values D_f were determined for copolymers solution in chloroform, but copolymers synthesis was realized in other solvents: APESF — in dimethylsulfonoxide and the two remaining copolymers — in methylene chloride. Since the interactions polymer-solvent fix the macromolecular coil structure in synthesis process [46], then it is necessary either D_f determination for the indicated solvents or recalculation of the value D_f in chloroform solution to the values D_f for copolymers solutions in the solvents, used in polycondensation process. The authors [147–149] chose the second from the indicated methods with the Eq. (17) application.

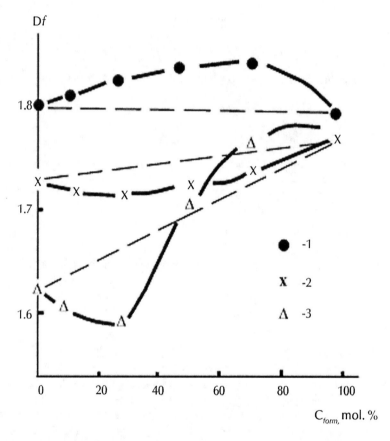

FIGURE 49   The dependences of macromolecular coil fractal dimension D_f, calculated according to the Eq. (17), on formal contents cform for APESF (1), CP-OFD-10/OSF-10 (2) and CP-OFD-10/P-1 (3).

As one can see from the Figs. 46 and 49 plots comparison, the dependences $T_g(cform)$ and $D_f(cform)$ clearly discover expressed similarity in respect to the additive values (shaded lines). In other words, D_f increasing means T_g growth and corresponding Kh enhancement. Within the frameworks of fractal analysis a reaction rate constant kr can be determined as follows [46]:

$$k_r \sim t^{(1-D_f)/2}, \tag{94}$$

where t is the reaction duration.

The definition χ as ratio of k_{r1} and k_{r2} follows from the Eq. (94) [148]:

$$\chi = \frac{k_{r1}}{k_{r2}} \sim t^{(D_{f_1} - D_{f_2})/2}. \tag{95}$$

Let us consider the physical sense of the dimensions D_{f_1} and D_{f_2}. As it is known [141], the value Kh = 1 corresponds to $\chi = 1$ or, proceeding from the Eq. (95), $D_{f_1} = D_{f_2}$. Since Kh=1 means completely statistical copolymer formation, then this supposes the values D_f equality for copolymer and the additive dependence of D_f in respect to homopolymers D_f values (shaded lines in Fig. 49). Proceeding from the said above, D_{f_1} value in the Eq. (95) is the fractal dimension of copolymer macromolecular coil and D_{f_2} is the additive value of this dimension. If $D_{f_1} > D_{f_2}$, then Kh>1 and copolymer have tendency to the same links long sequences formation and at $D_{f_1} = D_{f_2}$, Kh = 1 and purely statistical copolymer is formed as well. The shown in Fig. 50 dependence Kh(ln χ), calculated according to the Eq. (95) at the condition t = 1200 s, confirms the made above estimations. This correlation is similar to the dependence Kh(χ), adduced in work [141], for which the value χ was calculated from the purely kinetic data.

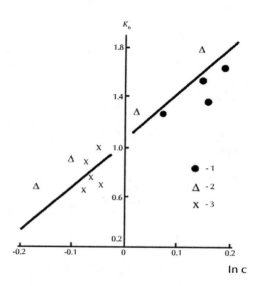

FIGURE 50 The relation between microheterogeneity coefficient Kh and functional groups activity χ in logarithmic coordinates for APESF (1), CP-OFD-10/OSF-10 (2) and CP-OFD-10/P-1.

Hence, the adduced above results showed decisive influence of macromolecular coil structure, characterized by its fractal dimension, on the formed copolymer type. The reason of this is obvious enough — whichever the number of active centres is and whichever their activity, is the access to them is controlled by macromolecular coil structure, namely, the higher D_f is the more compact its structure is and the stronger its surface is screened, that decelerates reaction realization. Let us note one more important aspect — the Eq. (95) introduces the Kh temporal dependence. It is obvious, that the longer reaction duration t is the larger, Kh is at $D_{f_1}>D_{f_2}$ and the smaller — at $D_{f_1}<D_{f_2}$. The reaction duration does not influence on copolymer type only at the condition $D_{f_1}=D_{f_2}$, i.e., at completely statistical copolymer formation.

In work [91] the synthesis was described by interfacial polycondensation method of mixed copolyurethanearylates (PUAr) series on the basis of diphenolurethane (DU), 4,4-dioxiphenyl-2,2-propane (DOPP) and dichloroanhydride of terephthalic acid. Depending on the molar ratio DOPP: DU PUAr were obtained, differing by both final results of synthesis (conversion degree Q and reduced viscosity η_{red}) and properties (for example, glass transition temperature T_g). It was assumed [91], that the values Q and η_{red} are reduced regularly at increasing of less reactive component DU fraction.

The thorough study of PUAr synthesis and test results found out, that for the indicated copolymers the linear correlation was observed (Fig. 51). At the first sight such correlation has occidental character, since the value Q characterizes macromolecular coil in solution behavior and Tg is polymers condensed state property. However, within the frameworks of fractal analysis the genetic inter-communication of macromolecular coil in solution and polymers condensed state structures was substantiated. Within the frameworks of such approach the cor-relation $T_g(Q)$, adduced in Fig. 51, can be explained as follows. As it is known [150], the value Tg is a function of polymer chain flexibility degree: if the chain flexibility is larger, Tg is lower. In this case it can be supposed, that chain flex-ibility enhancement because of PUAr composition change results to Q reduction. The further development of these notions can be obtained within the frameworks of fractal analysis [21, 52]. According to the fractal model [45], the dependence of Q on synthesis duration t is given by the Eq. (27). Comparison of Fig. 51 data and the Eq. (27) allows to suppose that chain rigidity increasing at the expence of DOPP fraction enhancement, resulting to Tg growth, simultaneously results to D_f reduction, that makes macromolecular coil less compact and, hence, more ac-cessible for synthesis reaction. Proceeding from the said above, the authors [151] performed quantitative description of PUAr interfacial polycondensation main characteristics (Q and η_{red}) dependences as a function of polymer chain flexibility, changed at the expence of PUAr components relation change.

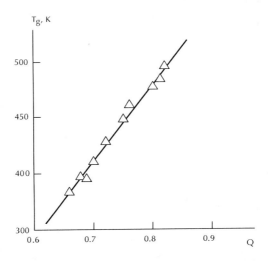

FIGURE 51   The dependence of glass transition temperature T_g on conversion degree Q for PUAr.

The theoretical values Q (Q_T) as a function of the chain statistical flexibility was performed as follows. Within the frameworks of percolation theory it has been shown, that between local order domains (nanoclusters) relative fraction φ_{cl} and T_g the following relationship exists [152]:

$$\varphi_{cl} = 0.03\left(T_g - T\right)^{0.55},\qquad(96)$$

where T is the testing temperature, equal to 293 K.

In its turn, the characteristic ratio C_∞, which is polymer chain statistical flexibility characteristic [25], is determined according to the Eq. (97) [153]:

$$C_\infty = \frac{2}{\varphi_{cl}}.\qquad(97)$$

In Fig. 52 the dependence $Q(C_\infty)$ for the considered PUAr is adduced. As it was supposed, polymer chain statistical flexibility enhancement (C_∞ increasing) results to Q reduction.

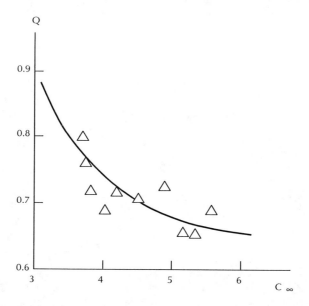

FIGURE 52 The dependence of conversion degree Q on polymer chain statistical flexibility, characterized by characteristic ratio C_∞, for PUAr.

The fractal dimension of polymers condensed state structure d_f is linked with C_∞ by the following equation [153]:

$$C_\infty = \frac{2d_f}{d(d-1)(d-d_f)} + \frac{4}{3} ,$$

(98)

where d is dimension of Euclidean space, in which a fractal is considered (it is obvious, that in our case d = 3).

And at last, further the fractal dimension D_f of macromolecular coil in solution can be determined, using the equation for linear polymers [154]:

$$d_f = 1.5D_f .$$

(99)

In Fig. 53 the dependence of D_f on DOPP contents c_{DOPP} for copolymers PUAr is adduced. As one can see, this dependence is located higher than the additive one. This means, that PUAr macromolecular coils compactness degree, characterized by the value D_f, is also higher than the additive one. The consequence of such dependence of D_f on copolymers composition will be considered below.

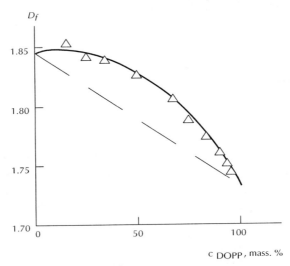

FIGURE 53   The dependence of macromolecular coil fractal dimension D_f on DOPP molar fraction c_{DOPP} for PUAr.

In Fig. 54 the comparison of experimental and calculated according to the Eq. (27) Q dependences on DOPP contents c_{DOPP} (at t=60 min) is adduced. Let us note several conclusions, following from the Fig. 54 plot. Firstly, a good correspondence of theory and experiment is observed. Since the theoretical values Q were calculated according to the known T_g values, i.e., actually by PUAr chain flexibility, then this confirms the proposed above treatment correctness, namely, the dependence of Q on polymer chain statistical flexibility. Secondly, the dependence $Q(c_{DOPP})$ is located lower than the corresponding additive dependence, that is direct consequence of the dependence $D_f(c_{DOPP})$ character, adduced in Fig. 53.

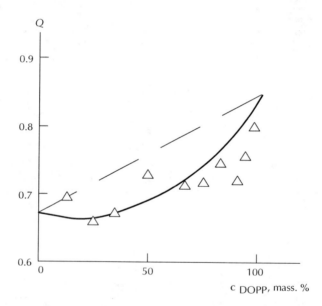

FIGURE 54 The experimental (points) and calculated according to the Eq. (27) (solid line) dependences of conversion degree Q on DOPP molar fraction c_{DOPP} for PUAr.

Since the obtained range of $D_f = 1.747–1.853$ supposes aggregation mechanism cluster–cluster in synthesis process, then for the molecular weight MM estimation the Eq. (53) can be used. Assuming in the first approximation $\eta_{red} \sim MM$, the authors [151] obtained the theoretical dependence $\eta_{red}(c_{DOPP})$ and compared it with the experimental one (Fig. 55). As one can see, a good correspondence of theory and experiment is obtained again.

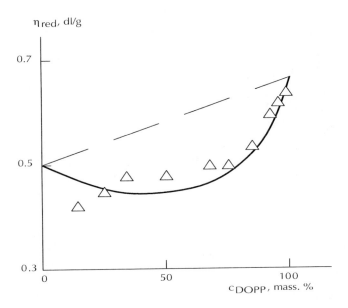

FIGURE 55 The experimental (points) and calculated according to the Eq. (53) (solid line) dependences of reduced viscosity ηred on DOPP contents c_{DOPP} for PUAr.

The microheterogeneity coefficient K_h for PUAr can be determined according to the Eq. (91). The value K_h estimation, performed by the indicated mode, gives $K_h \approx 0.45$. This means, that the copolymers with both comonomers long sequences are formed in PUAr synthesis process [140].

One more important parameter, characterizing copolycondensation process, is functional groups activity χ, which can be calculated according to the Eq. (95). The obtained for PUAr value $K_h < 1$ means the negative value ln χ [141] or the value χ < 1. From the Eq. (95) it follows, that in this case the exponent $(D_{f_1} - D_{f_2})/2$ and $D_{f_2} > D_{f_1}$. The dependence χ(cDOPP) at t=60 min is adduced in Fig. 56, from which it follows, that the extreme (minimum) value χ is reached at equimolar relation DOPP: DU. Such result was to be expected from a very general considerations [143].

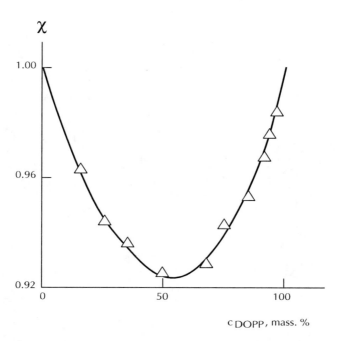

FIGURE 56   The dependence of functional groups activity χ on DOPP molar fraction c_{DOPP} for PUAr.

Hence, the stated above results have shown, that conversion degree and the re-duced viscosity, obtained in PUAr synthesis process, are a function of copolymer chain statistical flexibility: the more rigid chain is, the higher Q and ηred are. The fractal analysis methods allow to make this correlation quantitative treatment. From the chemical point of view the values Q and ηred depend on comonomers functional groups activity χ. The higher χ is, the larger the values Q and ηred are. The value also defines a synthesized copolymer type.

1.6   THE INTERCONNECTION OF MACROMOLECULAR COIL IN SOLUTION AND POLYMER CONDENSED STATE STRUCTURES

At present the fact, is quite doubtless that the structure of the formed in syn-thesis process polymer chain influences essentially on final properties of the polymer in condensed state. For example, the same polyarylate, obtained by equilibrium and nonequilibrium polycondensation, has different structure macromolecular coil in solution and different mechanical properties in condensed state [53]. However, for the similar correlations macromolecular coil structure — solid-phase polymer properties obtaining a hindrances number exists. The first from them consists of structural complexity of polymers condensed state. So, polymers properties change in thermal aging process supposes availability of some intermediate structural organization (let us call it traditionally as supramolecular structure), which can be changed at the fixed initial structure of macromolecular coil [155–157]. The second hindrance is very differing sets of parameters, describing coil in solution and polymer condensed state structures (for example, excluded volume parameter and free volume, accordingly). And at last, that is the absence up to the late time of quantitative conception of polymers condensed state structure.

The development of a new physical approaches and models allows to solve the indicated above problems of late years. So, within the frameworks of fractal analysis the polymer structure on all stages from macromolecular coil up to condensed state can be characterized with the aid of one parameter — fractal dimension — and its change between the indicated states allows to judge about occuring structural changes. The supramolecular structure can be described within the frameworks of the cluster model of polymers amorphous state structure [153], which gives its quantitative description with local order notion drawing and used in it characteristics are linked unequivocally with polymer structure fractal dimension. Therefore the authors [142, 158, 159] proposed the scheme of genetic intercommunication of synthesis products structure and final polymer properties on the example of polyarylates two series [53] with the application of the indicated above approaches.

The data [53] for polyarylates on the basis of di-(4-oxiphenyl)-methane, 2,2-di-(4-oxiphenyl)-propane, 9,9-di-(4-oxiphenyl)-fluorene (the conventional sign of series D) and phenolphthaleine and phenolphthaleine anylide (the conventional sign Ph), produced by nonequilibrium (interfacial) polycondensation, were used.

The fractal dimension Df of macromolecular coil in solution with excluded volume effects appreciation was determined according to the Eq. (39). Since all the considered polyarylates are linear polymers, then for them the spectral dimension is ds = 1.0 [72]. The gelation transition or transition from solution up to the condensed state is characterized by macromolecular coil environment change and now instead of solvent molecules it is in similar coils environment. This results to fractal dimension change and now its value df for the condensed state is determined according to the equation [17]:

$$d_f = \frac{d_s(d+2)}{2} \ . \tag{100}$$

The Eqs. (39) and (100) combination allows to obtain for linear polymers at $d_s = 1.0$ and $d = 3$ the Eq. (99), which demonstrates clearly the genetic intercommunication between reaction products (macromolecular coil in solution) and polymers condensed state structures.

Since the polymer is the substance, consisting of long chain macromolecules, then one from its most important properties is polymer chain flexibility, in many respects defining properties of polymers in solid-phase state [160]. A parameters number, describing the indicated property of polymer chain, exists. So, the rigidity parameter $\tilde{\sigma}$ can be estimated according to the known values of glass transition temperature Tg according to the equation [161]:

$$T_g = 630(\tilde{\sigma} - 1.35) .$$ (101)

One more parameter — the characteristic ratio C∞ — is characteristic of polymer chain statistical flexibility and determined according to the Eq. (98). In Fig. 57 the relation between C_∞ and $\tilde{\sigma}$ for the considered polyarylates is adduced. Since these parameters characterize the same polymer chain property, then between them the correlation is expected [25] that confirms the plot of Fig. 57. Certain scatter of data for two from the considered polyarylates can reflect different values of angles between bonds in chain [25].

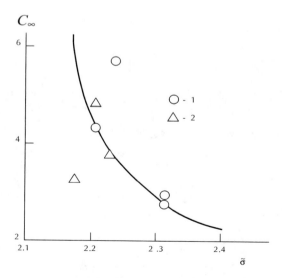

FIGURE 57 The comparison between characteristic ratio C∞ and rigidity parameter $\tilde{\sigma}$ for polyarylates of series Ph (1) and D (2).

The fractal (Hausdorff) dimension (D_f or d_f) characterizes spatial distribution of structure elements for an object [52], but does not give the quantitative description of either these elements, or an entire object. This rule follows from the scaling conception itself [162], where the fractal dimension serves as scaling exponent (index) [58]. The scaling (scale invariability) sence consists in abstraction from structure details and formation of simple universal features, which are specific ones for systems wide class. However, just those structure details, which do not take into account scaling conception, can turn out to be important for separate polymer properties characteristic. This gap can be made up by the cluster model of polymers amorphous state structure [153, 160]. The cluster formation in the indicated model assumes, that it is simultaneously multifunctional junction (with functionality F or statistical segments number per one cluster $n_{cl}= F/2$) of physical entanglements cluster network. The segment length in cluster is accepted equal to chain statistical segment length l_{st} [162] Eq. (102):

$$l_{st} = l_0 C_\infty \ ,$$
(102)

where l_0 is the main chain skeletal bond length, which is equal to 0.149 nm for polyarylates [163].

Within the frameworks of percolation model the clusters relative fraction φ_{cl} is determined according to the Eq. (96). In its turn, between the parameters φ_{cl} and df the following relationship exists [160]:

$$d_f = d - 6 \left(\frac{\varphi_{cl}}{C_\infty S} \right)^{1/2} ,$$
(103)

where S is the cross-sectional area of macromolecule, equal to 0.32 nm² for polyarylates of series D and 0.42 nm² — for series Ph.

And at last, the relation between φ_{cl} and cluster entanglements network density v_{cl} is given as follows [160]:

$$v_{cl} = \frac{\varphi_{cl}}{S l_0 C_\infty} \ .$$
(104)

Using the Eqs. (96) and (104), the value v_{cl} can be calculated. In Fig. 58 the dependence v_{cl} (D_f) is shown, which demonstrates v_{cl} reduction (in the first approximation — local order level) at D_f growth. In other words, Fig. 58 shows the dependence of polymer condensed state structure, characterized by the parameter v_{cl}, on reaction products (macromolecular coil in solution) structure, characterized by the dimension D_f. Let us note, that the dependence v_{cl} (D_f) extrapolation to $v_{cl} = 0$ gives the value $D_f \approx 2.16$, that corresponds to the greatest value D_p, reaching at aggregation by mechanism cluster–cluster [88].

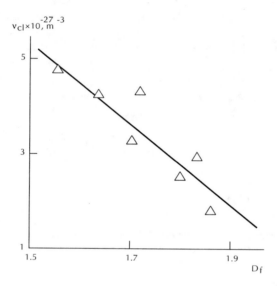

FIGURE 58 The dependence of cluster entanglements network density vcl on macromolecular coil fractal dimension D_f for linear polyarylates.

It has been shown earlier [165], that the characteristic ratio C_∞ is automodelity coefficient of the polymers cluster structure and therefore the distance between clusters R_{cl} can be determined as follows:

$$R_{cl} = l_{st}C_\infty = l_0 C_\infty^2 \ . \tag{105}$$

Besides, the value R_{cl} can be expressed with the aid of the relationship [160]:

$$R_{cl} = 1.80 \left(\frac{v_{cl}}{n_{cl}} \right)^{1/3} , \text{nm.} \tag{106}$$

The Eqs. (105) and (106) combination allows to estimate the segments number by length lst each per one cluster n_{cl}. For the considered polyarylates the value n_{cl} is varied within the very broad limits: $n_{cl} = 2$–40 [158]. Let us consider factors, defining such variation ncl. As it is known [17], macromolecular coil gyration radius R_g is scaled with its molecular weight similarly to the Eq. (1). For the statistical polymer chain, submitted to Gaussian statistics, Flory exponent is equal to 0.5, i.e., $D_f = 2.0$ [24]. However, it was found out experimentally [16], that Flory exponent is varied within the limits of 1/3–1.0 ($D_f = 1.0$–3.0), that is explained by interactions among chain accidentally closer to one another elements and between chain links solvent molecules. These interactions value can be characterized by the intermolecular interactions parameter , which is determined according to the Eq. (31). The parameter ε is also linked with dimension D_f according to the Eq. (32).

From the Eq. (31) it follows, that positive ε values correspond to repulsion forces among macromolecular coil elements and negative ones — to attraction forces. In Fig. 59 the dependence $n_{cl}(\varepsilon)$ for the considered polyarylates is adduced. As it was to be expected, the type and degree of interactions among macromolecular coil elements influence strongly on the value n_{cl}. However, it should not be forgetten, that different macromolecules segments form clusters, therefore one more factor is macromolecular coils intersections number N_{int} [17]. This parameter (for fractals with the same values D_f) is given as follows [17] Eq. (107):

$$N_{int} \sim R_g^{2D_f - d}. \tag{107}$$

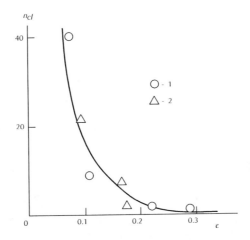

FIGURE 59 The dependence of segments number per one cluster n_{cl} on interaction parameter ε for polyarylates of series Ph (1) and D (2).

Assuming R_g = const, N_{int} (in relative units) can be calculated as a function of D_f. In Fig. 60 the dependence $n_{cl}^{1/2}(N_{int})$ is adduced, which turns out to be linear and passing through coordinates origin. Since N_{int} grows at D_f increasing, then the data of Figs. 58–60 allow to make the following conclusion. The product reaction (macromolecular coil in solution) defines supramolecular (more precisely, supra-segmental) structure of polymer in condensed state, namely, D_f increasing results to local order general degree v_{cl} reduction and formation of clusters more remote from one another, but with a larger number of statistical segments per each cluster.

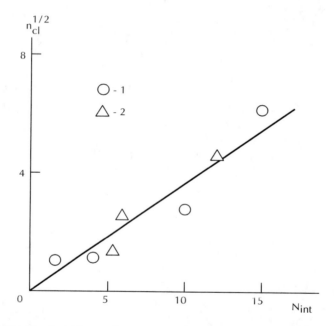

FIGURE 60   The dependence of segments number per one cluster n_{cl} on macromolecular coils intersection number N_{int} for polyarylates of series Ph (1) and D (2).

Let us consider polyarylates properties change with D_f (or df) variation. The Eqs. (99), (96) and (103) combination allows to obtain the analytical relationship between T_g and D_f. In Fig. 61 the comparison of the experimental and obtained by the indicated mode dependences $T_g(D_f)$ is adduced, from which their good correspondence follows. The physical basis of the similar correlation is simple — D_f increasing results to vcl reduction (see Fig. 58) and, hence, to T_g decreasing [166].

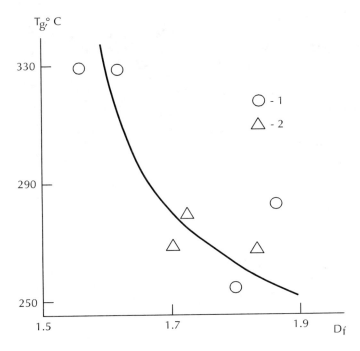

FIGURE 61   The dependence of glass transition temperature T_g on macromolecular coil fractal dimension D_f for polyarylates of series Ph (1) and D (2). The curve is theoretical calculation, the points are experimental data.

Hence, the stated above results assume, that polycondensation in solution products (macromolecular coils) structure defines polymers in condensed state structure and properties. The application of fractal analysis and cluster model ideas allows to both to point out these changes tendencies and to obtain polymers properties quantitative estimation [158].

At present quite enough empirical correlations, linked polymers molecular characteristics and their properties, exist (see, for example, the Eq. (101)). However, as it has been noted above, the absence at present of polymers general structural conception does not allow to obtain the analytical description of genetic connection of structure and their properties on all stages of production from them polymer articles — from synthesis up to condensed solid-phase state. The proposed above methodics give the possibility of both such intercommunication scheme construction and obtaining quantitative analytical relationships for its description. Such possibility realization was performed in works [159, 167] on the example of copolymers polyarylate-polyarylenesulfonoxide (PAASO) [54, 168, 169] and polyurethanearylate (PUAr) [91].

In Fig. 62 the dependence of glass transition temperature T_g on fractal dimension D_f, calculated according to the Eq. (22), is adduced for the five considered copolymers PAASO. As it follows from the adduced plot, T_g linear decay at D_f increasing is observed, that perfectly justifies the proposed by Kargin idea about polymers properties encodion on molecular level [170]. Let us remind, that the value D_f characterizes structure of macromolecular coil in diluted solution, i.e., as a matter of fact, an isolated macromolecule. A solvent variation results to D_f change, that explains polymers properties change at their samples production from solutions in different solvents [171].

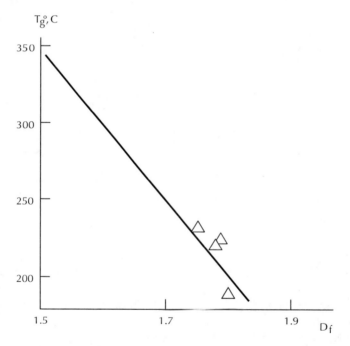

FIGURE 62   The dependence of glass transition temperature T_g on macromolecular coil fractal dimension D_f in tetrachloroethane for PAASO. The points — experimental data, the straight line — theoretical calculation.

However, macroscopic properties, such as T_g, are measured for polymers block (condensed) state, but not for their solutions. The polymers structure fractal dimension d_f for this state can be obtained according to the Eq. (99). In Fig. 63 the dependence of T_g on d_f is adduced which also showed T_g linear decay at d_f growth. Thus, the clear genetic intercommunication between synthesis products (macromolecular coil in solution) structure, characterized by the dimension D_f, condensed state structure, characterized by the dimension d_f, and final polymer properties (in our case T_g) is traced. The synthesis conditions any change, influencing on D_f value, will result to df change and, hence, to the polymer properties change. The properties variation for the same polymer (polyarylate), obtained by different polycondensation modes (equilibrium and nonequilibrium ones), in different solvents and having in virtue of this D_f different values illustrates clearly this postulate [53].

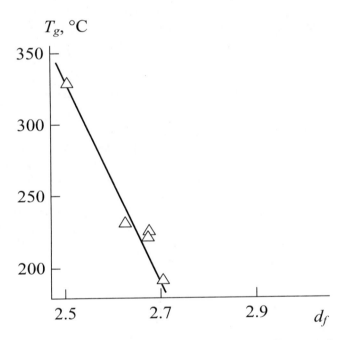

FIGURE 63 The dependence of glass transition temperature Tg on condensed state fractal dimension df for copolymers PAASO. The points — experimental data, the straight line — theoretical calculation.

Nevertheless, the proposed scheme allows to obtain the relationships, similar to the adduced in Figs. 62 and 63, only for one of some polymers type. As it is known [172], the exponent value a in Mark–Kuhn–Houwink equation for polycarbonate (PC) solution in tetrahydrofuran is equal to 0.70 and, hence, let us obtain D_f=1.76 according to the Eq. (4). If to proceed from the plots of Figs. 62 and 63, this gives the value $T_g \approx 498$ K, whereas it is well-known, that for PC the value T_g is smaller essentially and equal to ~ 423 K [161]. This apparent disparity is due to the fact, that the parameter df gives only general polymer structure characteristic and for the last condensed state properties are defined by both molecular characteristics and suprasegmental structure state. This relation can be demonstrated within the frameworks of the cluster model of polymers amorphous state structure [153, 160], where glass transition process is associated with local order domains decay, having thermofluctuational origin, at testing temperature growth [166]. The intercommunication of molecular characteristics C_∞ and S, local order level φ_{cl} and dimension df is given by the Eq. (103). The combination of Eqs. (39), (96), (98)–(100) and (103) allows to estimate theoretically the value T_g. Since for the considered copolymers the value S is unknown, then the authors [159] estimated it as ~ 0.40 nm² by analogy with polymers of similar chemical structure [164]. In Figs. 62 and 63 the theoretical dependences are shown by the solid straight lines, which turned out to be almost linear ones and correspond excellently to the experimental values T_g. The calculation T_g for PC at S≈0.32 nm2 [164] and $C_\infty = 2.4$ [162, 163] gave the value $T_g \approx 426$ K, that is close to the experimental one (423 K [161]).

Hence, the proposed in the present section the scheme of genetica intercommunication of the synthesis products (macromolecular coils) and condensed state structures with polymers properties clearly demonstrates, that the latter basis lays (encoded) already on the synthesis stage and cannot be given by only sole polymer macromolecule chemical structure. The using of the concrete quantitative model of solid-phase polymers structure together with fractal analysis and irreversible aggregation models ideas, gives the possibility of the prediction of not only glass transition temperature T_g, but a number of other important polymers properties [173].

1.7   THE PHYSICAL SENSE OF REACTIVE MEDIUM HETEROGENEITY FOR POLYMER SOLUTIONS

Either fractal objects reactions or reactions in fractal spaces are accepted to call the fractal reactions [174]. The autodeceleration, i.e., reaction rate reduction ϑ_r with it duration t, is a characteristic sign of such reactions. Let us note, that for reactions in Euclidean space the linear kinetics and, accordingly, the condition ϑ_r = const are typical [176]. The fractal reactions in broad sense of this term very often occur in practice (a synthesis reactions, sorption curves, curves stress-strain and so on) [177, 178]. The simplest and clearest relationship for this effect description is the following one [175]:

$$\vartheta_r \sim t^{-h},\qquad(108)$$

where h is the heterogeneity exponent ($0 \leq h \leq 1$), turning into zero in case of classical behavior (reactions in homogeneous or Euclidean mediums) and then ϑ_r=constant, as it was noted above.

The Eq. (108) allows to give the following treatment of physical sense of reactive medium heterogeneity: the higher this heterogeneity degree is, the more rapid ϑ_r decay at reaction realization is. It is obvious, that for this theoretical postulate practical application it is necessary to establish the intercommunication of heterogeneity degree and reactive medium characteristics [178].

Let us consider this question for the most often applied in practice reactive medium type, namely, polymer solutions. The general fractal description of reaction in solution kinetics is given by the Eq. (26). In its turn, the reaction rate ϑ_r can be obtained by the Eq. (26) differentiation by time t, that gives the Eq. (69). The comparison of exponents in the Eqs. (108) and (69) allows to obtain the following equality [178]:

$$h = \frac{D_f - 1}{2} .$$

(109)

Let us estimate the limiting values of macromolecular coil in solution fractal dimension for two cases: $h = 0$ (D_f^0) and $h = 1.0$ (D_f^1). From the Eq. (109) it follows: $D_f^0 = 1.0$ and $D_f^1 = 3.0$. The physical sense of the dimension D_f obtained limiting values in this case is an obvious one. For homogeneous reactive medium ($h = 0$) ϑ_r is the greatest one for all $t > 1$ and $D_f^0 = 1.0$ means, that macromolecule represents completely a stretched polymeric chain, but not macromolecular coil. If the last is a fractal object, for which internal regions are screened by surface and are inaccessible for reaction realization [179], then for the stretched chain macromolecule all reactive centers are accessible and steric hindrances for reaction realization are absent. Therefore the value ϑ_r is constant and the greatest one. The value $D_f^1 = 3.0$ means, that a macromolecule is curdled in compact globule, for which only a small part of the surface is accessible for reaction. Therefore the value ϑ_r is a minimum one and rapidly tries to attain zero [178].

The concrete reaction type is defined by this reaction rate choice. So, in polymers synthesis reactions the high rate is desirable and in thermooxidative destruction process — the low one. As it follows from the Eq. (108), the value ϑ_r is defined by the medium heterogeneity degree. Besides, from the Eqs. (26) and (69) it follows, that the parameters c_0 and η_0 will also influence on a reaction rate.

Hence, the performed brief analysis allows to link macromolecular coil structure and reactive medium heterogeneity degree for process of polycondensation in solution. In its turn, this intercommunication knowledge and the ability to check macromolecular coil structure (for example, by a solvent variation) allow to regulate chemical reaction rate in desirable direction [59-60], [78-79].

REFERENCES

1. Belyi, A. A.; Ovchinnikov, A. A. About nonstationary kinetics of interacting particles diffusion limited reactions. *Reports of Academy of Sciences of SSSR*, 1986, **288**(1), 151–155.
2. Burlatskii, S. F. About influence of reagents concentration fluctuations on biomolecular reactions kinetics in dense polymer systems. *Reports of Academy of Sciences of SSSR*, 1986, **288**(1), 155–159.
3. Aleksandrov, I. V.; Pazhitnov, A. V. About one bistable system with asymmetry chemical intensification. *Chemical Physics*, 1987, **69**, 1243–1247.
4. Burlatskii, S.F.; Ovchinnikov, A. A.; Pronin, K.A. Stochastic aggregation of reacting particles and their destruction kinetics. *Journal of Experimental and Theoretical Physics*, 1987, **92**(2), 625–637.
5. Burlatskii, S. F.; Oshanin, G. S.; Likhachev, V. N. Diffusion limited reactions with polymer chains participation. *Chem. Phys.*, 1988, **7**(7), 970–977.
6. Grassberger, P.; Procaccia, I. The long time properties of diffudion in a medium with static traps. *J. Chem. Phys.*, 1982, **77**(12), 6281–6284.
7. Havlin, S.; Weiss, G. H.; Kiefer, J.E.; Dishon, M. Exact enumeration of random walks with traps. *J. Phys. A*, 1984, **17**(3), L347–L350.
8. Redner, S.; Kang, K. Kinetics of the "scavenger" reaction. *J. Phys. A*, 1984, **17**(3), L451–L455.
9. Kang, K.; Redner, S. Scaling approach for the kinetics of recombination processes. *Phys. Rev. Lett.*, 1984, **52**(12), 955–958.
10. Meakin, P.; Stanley, H. E. Novel dimension-independent behavior for diffusive annihilation on percolating fractals. *J. Phys. A*, 1984, **17**(1), L173–L177.
11. Jullien, R. Fractal aggregates. *Successes of Physical Sciences*, 1989, **157**(2), 339–361.
12. Kaufman, J. H.; Baker, C. K.; Nazzal, A. I.; Melroy, O. R.; Kapitulnik, A. Statics and dynamics of the diffusion-limited polymerization of the conducting polymer polypyrrole. *Phys. Rev. Lett.*, 1986, **56**(18), 1932–1935.
13. Kaufman, J. H.; Melroy, O. R.; Abraham, F. F.; Nazzal, A. I. Growth instability in diffusion controlled polymerization. *Solid State Commun.*, 1986, **60**(9), 757–761.
14. Karmanov, A. P.; Monakov, Yu. B. The fractal structure of lulk- and end-wise-dehydropolymers. *High-Molecular Compounds*, B, 1995, **37**(2), 328–331.
15. Karmanov, A. P.; Matveev, D. V.; Monakov, Yu. B. Dynamics of gvayatsile lignins monomer precursors polymerization. *Reports of Academy of Science*, 2001, **380**(5), 635–638.
16. Baranov, V. G.; Frenkel, S. Ya.; Brestkin, Yu. V. The dimensionality of linear macromolecule different states. *Reports of Academy of Sciences of SSSR*, 1986, **290**(2), 369–372.

17. Vilgis, T. A. Flory theory of polymeric fractals — intersection, saturation and condensation. *Physica A*, 1988, **153**(2), 341–354.

18. Rammal, R.; Toulouse, D. Random walks on fractal structures and percolation clusters. *J. Phys. Lett.* (Paris), 1983, **44**(1), L13–L22.

19. Klymko, P. W.; Kopelman, R. Fractal reaction kinetics: exciton fusion on clusters. *J. Phys. Chem.*, 1983, **87**(23), 4565–4567.

20. Kopelman, R.; Klymko, P. W.; Newhouse, J. S.; Anacker, L. W. Reaction kinetics on fractals: walker simulation and exciton experiment. *Phys. Rev. B*, 1984, **29**(6), 3747–3748.

21. Mandelbrot, B. B. *The Fractal Geometry of Nature*. San-Francisco, Freeman and Company, 1982, 459.

22. Novikov, V. U.; Kozlov, G. V. A macromolecules fractal analysis. *Successes of Chemistry*, 2000, **69**(4), 378–399.

23. Novikov, V. U.; Kozlov, G. V. A polymers structure and properties within the frameworks of fractal approach. *Successes of Chemistry*, 2000, **69**(6), 572–599.

24. Family, F. Fractal dimension and grand universality of critical phenomena. *J. Stat. Phys.*, 1984, **36**(5/6), 881–896.

25. Budtov, V. P. Physical Chemistry of Polymer Solutions. Sankt-Peterburg, *Chemistry*, 1992, 384.

26. Brown, D.; Sherdron, G.; Kern, V. The practical handbook by synthesis and study of the polymers properties. Ed. Zubkov, V. Moscow, *Chemistry*, 1976, 256.

27. Kozlov, G. V.; Temiraev, K. B.; Sozaev, V. A. The macromolecular coil fractal dimension in diluted solution estimation by viscous characteristics. *Journal of Physical Chemistry*, 1999, 73(4), 766–768.

28. Timofeeva29. , G. I.; Dubrovina, L. V.; Korshak, V. V.; Pavlova, S. A. Viscosimetric properties of polyarylates. *High-Molecular Compounds*, 1964, **6**(11), 2008–2013.

29. Samarin, A. F.; Starkman, B. P. Viscosimetric express-method of determination of polymers molecular weight. *High-Molecular Compounds. A*, 1985, **27**(5), 1101–1103.

30. Büller, K. U. Heat- and Thermostable Polymers. Moscow, *Chemistry*, 1984, 1056.

31. Dubrovina, L. V.; Pavlova, S. A.; Bragina, T. P. Properties of a solutions of polyarylates with side alyphatic group. *High-Molecular Compounds. A*, 1988, **30**(5), 995–1000.

32. Vilenchik, L. Z.; Sklizkova, V. P.; Tennikova, T. B.; Bel'nikevich, N. G.; Nesterov, V. V.; Kudryavtsev, V. V.; Belen'kii, B. G.; Frenkel, S. Ya.; Koton, M. M. Chromatographic study of poly-(4,4'-oxydiphenylene) pirromellitimide acid solutions. *High-Molecular Compound, A*, 1985, **27**(5), 927–930.

33. Pavlov, G. M.; Korneeva, E. V.; Mikhailova, N. A.; Anan'eva, E. P. Hydrodynamic and molecular characteristics of fractions of mannane, formed by yeasts phodotorula rubra. *Biophysics*, 1992, **37**(6), 1035–1040.

34. Kozlov, G. V.; Shustov, G. B.; Dolbin, I. V. A fractal dimension and onteractions of a macromolecular coil in a solution. Proceedings of 1-th International Scientific Conf. *Modern Problems of Organic Chemistry, Ecology and Biotechnology*, Luga, 2001, 17–18.

35. Tager, A. A.; Kolmakova, L. K. Solubility parameter, its estimation methods, connection with polymers solubility. *High-Molecular Compounds, A*, 1980, **22**(3), 483–496.

36. Wieche, I. A. Polygon mapping with two-dimensional solubility parameters. *Ind. Engng. Chem. Res.*, 1995, **34**(2), 661–673.
37. Kozlov, G. V.; Dolbin, I. V.; Mashukov, N. I.; Burmistr, M. V.; Korenyako, V. A. Prediction of fractal dimension of a macromolecular coils in diluted solutions on the basis of solubility two-dimensional parameter model. *Problems of Chemistry and Chemical Technology*, 2001, (6), 71–77.
38. Kozlov, G. V.; Dolbin, I. V. Express method of fractal dimension estimation of bio-polymers macromolecular coils in solution. *Biophysics*, 2001, **46**(2), 216–219.
39. Ptitsyn, O. B.; Eizner, Yu. E. Hydrodynamics of polymer solutions. II. Hydrodynami-cal properties of macromolecules in good solvents. *Journal of Technical Physics*, 1959, **29**(9), 1117–1134.
40. Pavlov, G. M.; Korneeva, E. V. Polymaltotrioza. Molecular characteristics and equi-librium chains rigidity. *Biophysics*, 1995, **40**(6), 1227–1233.
41. Kozlov, G. V.; Dolbin, I. V. Fractal variant of Mark-Kuhn-Houwink equation. *High-Molecular Compounds, B*, 2002, **44**(1), 115–118.
42. Korchak, V. V.; Pavlova, S. -S. A.; Timofeeva, G. I.; Kroyan, S. A.; Krongauz, Ye. S.; Travnikova, A. P.; Raubach, Ch.; Schulz, G.; Gnauk, R. Hydrodynamic properties of randomly branched polyphenylquinoxalines with low degrees of branching. *High-Molecular Compounds, A*, 1984, **26**(9), 1868–1876.
43. Burya, A. I.; Kozlov, G. V.; Temiraev, K. B.; Malamatov, A. Kh. The influence of branching on fractal dimension of macromolecular coils in solutions. *Problems of Chemistry and Chemical Technology*, 1999, (3), 26–28.
44. Shustov, G. B.; Temiraev, K. B.; Afaunova, Z. I.; Kozlov, G. V. *A factors, influencing on fractal dimension of branched polymers macromolecular coils in diluted solutions.* Proceedings of KBSC RAS, 2000, (2), 92–94.
45. Kozlov, G. V.; Shustov, G. B. Fractal physics of polycondensation processes. In col-lection: Successes in Polymers Physics-chemistry Field. Ed. Zaikov, G. a.a. Moscow, *Chemistry*, 2004, 341–411.
46. Kozlov, G. V.; Shustov, G. B.; Zaikov, G. E. Fractal physics of polycondensation processes. In collection: *Essential Results in Chemical Physics*. Ed. Goloshchapov, A.; Zaikov, G.; Ivanov, V. New York, Nova Science Publishers, Inc., 2004, 193–241.
47. Kozlov, G. V.; Shustov, G. B.; Zaikov, G. E. The fractal physics of the polycondensa-tion processes. In collection: *Handbook of Polymer Research*. 20. Ed. Pethrich, A.; Zaikov, G. New York, Nova Science Publishers, Inc., 2006, 37–83.
48. Korshak, V. V.; Vinogradova, S. V. Nonequilibrium Polycondensation. Moscow, *Sci-ence*, 1972, 695.
49. Kozlov, G. V.; Temiraev, K. B.; Kaloev, N. I. *Influence of solvent nature on structure and formation mechanism of polyarylate in conditions of low-temperature polycon-densation.* Reports of Academy of Science, 1998, **362**(4), 489–492.
50. Kozlov, G. V.; Burya, A. I.; Shustov, G. B. *Influence of solvent nature on low-temper-ature polycondensation: Fractal analysis.* News of Academy of Sciences (Kazakh-stan), technical sciences, 2007, (5–6), 23–32.
51. Korshak, V. V.; Vinogradova, S. V.; Vasnev, V. A. The study of solvent nature influ-ence on low-temperature polycondensation. *High-Molecular Compounds, A*, 1968, **10**(6), 1329–1335.
52. Feder, F. *Fractals*. New York, Plenum Press, 1990, 246.

53. Askadskii, A. A. Physics-chemistry of Polyarylates. Moscow, *Chemistry*, 1968, 214.

54. Dubrovina, L. V.; Pavlova, S.-S. A.; Ponomareva, M. A. Hydrodynamical and thermodynamical properties of polyblock copolymers solutions. *High-Molecular Compounds, A*, 1985, **27**(4), 780–787.

55. Askadskii, A. A. Structure and Properties of Thermostable Polymers. Moscow, *Chemistry*, 1981, 320.

56. Tager, A. A. Physics-chemistry of Polymers. Moscow, *Scientific World*, 2007, 576.

57. Pfeifer, P.; Avnir, D.; Farin, D. Scaling behavior of surface irregularity in the molecular domain: from adsorption studies to fractal catalysts. *J. Stat. Phys.*, 1984, **36**(5/6), 699–716.

58. Kokorevich, A. G.; Gravitis, Ya. A.; Ozol-Kalnin, V. G. The development of scaling approach at study of lignin supramolecular structure. *Chemistry of Wood*, 1989, (1), 3–24.

59. Breslou, R. Mechanisms of Organic Reactions. Moscow, *World*, 1968, 393.

60. Zhdanov, Yu. A.; Minkin, V. I. *The Correlation Analysis in Organic Chemistry*. Rostov-on-Don, RSU, 1966, 412.

61. Mikitaev, A. K.; Korshak, V. V.; Musaev, Yu. I.; Storozhuk, I. P. Correlation analysis of polycondensation reaction of disodium salts of bisphenols with 4,4'-dichlorodiphenylsulfone. In collection: *Problems of Physics-chemistry of Polymers*. Ed. Mikitaev, A.; Zelenev, Yu. Nal'chik, KBSU, 1972, 29–38.

62. Mikitaev, A. K.; Musaev, Yu. I.; Korshak, V. V. Kinetics and mechanics of high-temperature polycondensation reaction in solution at polyarylates synthesis. In collection: *Polycondensation processes and polymers*. Ed. Mikitaev, A. Nal'chik, KBSU, 1976, 180–211.

63. Kozlov, G. V.; Musaev, Yu. I.; Shustov, G. B.; Burmistr, M. V.; Korenyako, V. A. Physical sense of Gammet's constant for polycondensation reaction of bisphenols with dichlorides. *Problems of Chemistry and Chemical Technology*, 2001, (1), 105–108.

64. Mikitaev, A. K.; Korshak, V. V.; Musaev, Yu. I.; Storozhuk, I. P. Kinetics of polycondensation reaction of disodium salts of bisphenols with 4,4'-DChDPhS. In collection: *Problems of Physics-chemistry of Polymers*. Ed. Mikitaev, A.; Zelenev, Yu. Nal'chik, KBSU, 1972, 4–28.

65. Kozlov, G. V.; Afaunov, V. V.; Temiraev, K. B. *Fractal dimension of biopolymers macromolecular coil in solution as a measure of bulk interactions*. Manuscript is denosited in VINITI of Russian Academy of Sciences, Moscow, 08.01.98, (9-B98).

66. Kozlov, G. V.; Musaev, Yu. I.; Shustov, G. B.; Burmistr, M. V.; Korenyako, V. A. Kinetics of high-temperature polycondensation: influence of synthesis temperature and solvent thermodynamical quality. *Problems of Chemistry and Chemical Technology*, 2001, (2), 106–109.

67. Kozlov, G. V.; Musaev, Yu. I. The fractal analysis of the kinetics of high-temperature polycondensation. In collection: *Fractals and Local Order in Polymeric Materials*. Ed. Kozlov, G.; Zaikov, G. New York, Nova Science Publishers, Inc., 2001, 175–186.

68. *Brief Chemical Encyclopaedia*, 1. Ed. Knunyants, I. Moscow, *Soviet Encyclopaedia*, 1961, 1262.

69. *Brief Chemical Encyclopaedia*, 3. Ed. Knunyants, I. Moscow, *Soviet Encyclopaedia*, 1964, 1112.

70. Kozlov, G. V.; Shustov, G. B. Fractal Analysis of the structural memory of macromo-lecular coil of polyarylates. *Reports of Adygskoi (Cherkesskoi) International Acad-emy of Sciences*, 2007, **9**(2), 138–141.

71. Kozlov, G. V.; Dolbin, I. V. The physical sense and estimation methods of low-molec-ular solvent structural parameters in diluted polymer solutions. Proceedings of Higher Educational Institutions, North-caucasus region, *Natural Sciences*, 2004, (3), 69–71.

72. Alexander, S.; Orbach, R. Density of states on fractals: "fractons". *J. Phys. Lett.* (Paris), 1982, **43**(17), L625–L631.

73. Kozlov, G. V.; Dolbin, I. V.; Zaikov, G. V. *Fractal physical chemistry of polymer solu-tions*. J. Balkan Tribological Association, 2005, **11**(3), 335–373.

74. Kozlov, G. V.; Temiraev, K. B.; Shustov, G. B.; Mashukov, N. I. Modeling of solid state polymer properties at the stage of synthesis: Fractal analysis. *J. Appl. Polymer Sci.*, 2002, **85**(6), 1137–1140.

75. Kozlov, G. V.; Serdyuk, V. D.; Dolbin, I.V. Fractal geometry of chain and deform-ability of amorphous glassy polymers. *Materialovedenie*, 2000, (12), 2–5.

76. Kozlov, G. V.; Shustov, G. B. The simulation of limiting conversion degree at poly-condensation by fractional integral. Proceedings of International interdisciplinary seminar *Fractals and Applied Synergetics*. Moscow, Publishers MSOU, 2001, 155–157.

77. Afaunova, Z. I.; Kozlov, G. V.; Kolodei, V. S. The dependence of molecular weight of polyurethanearylate on acceptor contents in the conditions of acceptor-catalytic polycondensation. Electronic journal *Studied in Russia*, 2001, 150, 1732–1738, http://zhurnal.ape.relarn.ru/articles/2001/150.pdf.

78. Oldham, K.; Spanier, *J. Fractional Calculus*. London, New York, Academic Press, 1973, 412.

79. Samko, S. G.; Kilbas, A. A.; Marichev, O. I. Integrals and Derivatives of Fractional Order and their some Applications. Minsk, *Science and Engineering*, 1987, 688.

80. Halsey, T. C.; Jensen, M. H.; Kadanoff, L. P.; Procaccia, I.; Shraiman, B. I. Fractal measures and their singularities: the characterization of strange sets. *Phys. Rev. A*, 1986, **33**(2), 1141–1151.

81. Bolotov, V. N. The positrons annihilation in fractal mediums. *Letters to Journal of Technical Physics*, 1995, **21**(10), 82–84.

82. Nigmatullin, R. R. Fractional integral and its physical interpretation. *Theoretical and Mathematical Physics*, 1992, **90**(3), 354–367.

83. Kozlov, G. V.; Batyrova, H. M.; Zaikov, G. E. The structural treatment of a number of effective centres of polymer chains in the process of the thermooxidative degradation. *J. Appl. Polymer Sci.*, 2003, **89**(7), 1764–1767.

84. Botet, R.; Jullien, R.; Kolb, M. Gelation in kinetic growth models. *Phys. Rev. A*, 1984, **30**(4), 2150–2152.

85. Kobayashi, M.; Yoshioka, T.; Imai, M.; Iton, Y. Structural ordering on physical gela-tion of syndiotactic polystyrene dispersed in chloroform studied by time-resolved measurements of small angle neutron scattering (SANS) and infrared spectroscopy. *Macromolecules*, 1995, **28**(22), 7376–7385.

86. Vinogradova, S. V.; Vasnev, V. A.; Korshak, V. V. The some laws of low-temperature polycondensation in solution. *High-Molecular Compounds, B*, 1967, **9**(7), 522–525.

87. Kozlov, G. V.; Shustov, G. B. The dependence of limiting molecular weight of polyarylates on initial reagents concentration. Mater. of I-th All-Russian sci.-eng. Conf. *Nanostructures in Polymers and Polymer Nanocomposites*. Nal'chik, KBSU, 2007, 90–94.

88. Hentschel, H. G. E.; Deutch, J. M.; Meakin, P. Dynamical scaling and the growth of diffusion-limited aggregates. *J. Chem. Phys.*, 1984, **81**(5), 2490–2502.

89. Kolb, M. Aggregation phenomena and fractal structures. *Physica A*, 1986, **140**(1–2), 416–420.

90. Muthukumar, M. Dynamics of polymeric fractals. *J. Chem. Phys.*, 1985, **83**(6), 3161–3168.

91. Afaunova, Z. I.; Kozlov, G. V.; Bazheva, R. Ch. The dependence of main characteristics of polyurethanearylates interfacial polycondensation on organic phase nature. Electronic journal *Studied in Russia*, 64, 690–698, 2002. http://zhurnal.ape.relarn.ru/articles/2002/064.pdf.

92. Hay, A. S.; Williams, F. J.; Loucks, G. M.; Pelles, H. M.; Boulette, B. M.; Donahue, P. E.; Johnson, D. S. Synthesis of new aromatic polyformals. *Amer. Chem. Soc. Polymer Prepr.*, 1982, **23**(2), 117–118.

93. Temiraev, K. B.; Shustov, G. B.; Mikitaev, A. K. Synthesis and properties of copolyethersulfonformals. *High-Molecular Compounds, B*, 1988, **30**(6), 412–418.

94. Kozlov, G. V.; Shustov, G. B.; Temiraev, K. B. The dependence of molecular weight of aromatic copolyethersulfonformals on weight of an elementary link: the fractal analysis. In book: *Fractals and Local Order in Polymeric Materials*. Ed. Kozlov, G.; Zaikov, G. New York, Nova Science Publishers, Inc., 2001, 29–35.

95. Kozlov, G. V.; Sanditov, D. S. *Anharmonic Effects and Physical-Mechanical Properties of Polymers*. Novosibirsk, Nauka, 1994, 261.

96. Ernst, M. H. Kinetics of cluster formation at irreversible aggregation. In book: *Fractals in Physics*. Ed Pietronero, L.; Tosatti, E. Amsterdam, Oxford, New York, Tokyo, North-Holland, 1986, 399–429.

97. Kolb, M. Reversible diffusion-limited cluster aggregation. *J. Phys. A*, 1986, **19**(5), L263–L268.

98. Rafikov, S. R.; Pavlova, S. A.; Tverdokhlebova, I. I. *The Methods of High-Molecular Compounds Molecular Weights and Polydispersity Determination*. Moscow, Publishers of Academy of Sciences of SSSR, 1963, 368.

99. Vicsek, T.; Family, F. Dynamic scaling for aggregation of clusters. *Phys. Rev. Lett.*, 1984, **52**(14), 1669–1672.

100. Botet, R.; Jullien, R. Size distribution of clusters in irreversible kinetic aggregation. *J. Phys. A*, 1984, **17**(12), 2517–2530.

10.1 Shiyan, A. A. Stationary distributions of macromolecular fractals masses in diluted polymer solutions. *High-Molecular Compounds, B*, 1995, **37**(9), 1578–1580.

102. Kozlov, G. V.; Batyrova, H. M.; Shustov, G. B.; Mikitaev, A. K. Molecular weight distribution of polyarylates: fractal analysis. Theses of lectures of sci.-pract. conf. *New Polymer Composite Materials*. Moscow, 2000, 44.

103. Botet, R.; Jullien, R. Diffudion-limited aggregation with disaggregation. *Phys. Rev. Lett.*, 1985, **55**(19), 1943–1946.

104. Witten, T. A.; Sander, L. M. Diffudion-limited aggregation as kinetical critical phenomenon. *Phys. Rev. Lett.*, 1981, **47**(19), 1400–1403.

105. Meakin, P. Formation of fractal clusters and networks by irreversible diffusion-limited aggregation. *Phys. Rev. Lett.*, 1983, **51**(13), 1119–1122.
106. Kolb, M.; Botet, R.; Jullien, R. Scaling of kinetically growing clusters. *Phys. Rev. Lett.*, 1983, **51**(13), 1123–1126.
107. Shogenov, V. N.; Kozlov, G. V. *Fractal Clusters in Physics-chemistry of Polymers.* Nal'chik, Polygraphservice and T, 2002, 268.
108. Weitz, D. A.; Huang, J. S.; Lin, M. Y.; Sung, J. Dynamics of diffusion-limited kinetic aggregation. *Phys. Rev. Lett.*, 1984, **53**(17), 1657–1660.
109. Camoin, C.; Blanc, R. Aggregation in a sheared 2D dispersion of spheres with attractive interactions. *J. Phys. Lett.* (Paris), 1985, **46**(2), 167–174.
110. Richetti, P.; Prost, J.; Barois, P. Two-dimensional aggregation and crystallization of colloidal suspension of latex spheres. *J. Phys. Lett.* (Paris), 1984, **45**(23), L1137–L1143.
111. Adam, M.; Delsanti, M.; Durand, D.; Hild, C.; Munch, J. P. Mechanical properties near gelation threshold, comparison with classical and 3d percolation theories. *Pure and Appl. Chem.*, 1981, **53**(6), 1489–1494.
112. Herrmann, H. J.; Landau, D. P.; Stauffer, D. New universality class for kinetic gelation. *Phys. Rev. Lett.*, 1982, **49**(6), 412–415.
113. Kozlov, G. V.; Burya, A. I.; Temiraev, K. B.; Mikitaev, A. K.; Chigvintseva, O. P. The description of low-temperature polycondensation kinetics within the frameworks of irreversible aggregation models and fractal analysis. *Problems of Chemistry and Chemical Technology*, 1998, (3), 26–29.
114. Kolb, M. Unified description of static and dynamic scaling for kinetic cluster formation. *Phys. Rev. Lett.*, 1984, **53**(17), 1653–1656.
115. Kozlov, G. V.; Malkanduev, Yu. A.; Zaikov, G. E. Fractal analysis of polymer molecular weight distribution: dynamic scaling. *J. Appl. Polymer Sci.*, 2003, **89**(9), 2382–2384.
116. Brown, W. D.; Ball, R. C. Computer simulation of chemically limited cluster–cluster aggregation. *J. Phys. A*, 1985, **18**(9), L517–L521.
117. Barns, F. S. The electromagnetic fields influence on chemical reaction rate. *Biophysics*, 1996, **41**(4,) 790–802.
118. Temiraev, K. B.; Shustov, G. B.; Kozlov, G. V.; Mikitaev, A. K. The description of low-temperature polycondensation within the frameworks of irreversible aggregation cluster–cluster model. *Plastics*, 1999, (2), 30.
119. Kozlov, G. V.; Temiraev, K. B.; Ovcharenko, E. N.; Lipatov, Yu. S. The description of low-temperature polycondensation process within the frameworks of irreversible aggregation models. *Reports of National Academy of Sciences of Ukraine*, 1999(12), 136–140.
120. Schaefer, D.W.; Kiefer, K. D. Fractal geometry of silica condensation polymers. *Phys. Rev. Lett.*, 1984, **53**(14), 1383–1386.
121. Kozlov, G. V.; Shustov, G. B.; Zaikov, G. E. The reaction cessation in polycondensation process: fractal analysis. In book: *Progress in Chemistry and Biochemistry. Linetics, Thermodynamics, Synthesis, Properties and Applications*. Ed. Pearce, E.; Zaikov, G. New York, Nova Science Publishers, Inc., 2009, 61–72.
122. Brady, L. M.; Ball, R. C. Fractal growth of copper electrodeposites. *Nature*, 1984, **309**(5965), 225–229.

124. Kozlov, G. V.; Dolbin, I. V.; Shustov, G. B. The polycondensation conditions influence on the probability of chemical bonds formation. Mater. of All-Russian sci.-pract. conf. *Chemistry in Technology and Medicine*. Makhachkala, DSU, 2001, 159–162.

125. Meakin, P. Diffusion-controlled flocculation: The effect of attractive and repulsive interactions. *J. Chem. Phys.*, 1983, **79**(5), 2426–2429.

126. Jullien, R.; Kolb, M. Hierarchical method for chemically limited cluster–cluster aggregation. *J. Phys. A*, 1984, **17**(12), L639–L643.

127. Shogenov, V. N.; Temiraev, K. B.; Kozlov, G. V. The reagents relation influence on low-temperature polycondensation process. In collection: *Physics and Chemistry of Perspective Materials*. Nal'chik, KBSU, 1997, 80–84.

128. Gedgafova, F. V.; Smirnova, O. V.; Storozhuk, I. P. Block-copolymers on the basis of polycarbonate and polyarylates. In collection: *Polycondensation Processes and Polymers*. Ed. Korshak, V. V. Nal'chik, KBSU, 1985, 41–51.

129. Kozlov, G. V.; Musaev, Yu. I.; Mikitaev, A. K. Diffusion and chemically limited regimes of high-temperature polycondensation with bisphenols participation. Mater. of V International sci.-pract. conf. *New Polymeric Composite Materials*. Nal'chik, KBSU, 2009, 112–117.

130. Weitz, D. A.; Huang, J. S.; Lin, M. Y.; Sung, J. Limits of the fractal dimension for irreversible kinetic aggregation of gold colloids. *Phys. Rev. Lett.*, 1985, **54**(13), 1416–1419.

131. Kolb, M.; Jullien, R. Chemically limited versus diffusion limited aggregation. *J. Phys. Lett.* (Paris), 1984, **45**(10), L977–L981.

132. Aubert, C.; Cannell, D. S. Restructuring of colloidal silica aggregates. Phys. Rev. Lett., 1986, 56(7), 738–741.

133. Kozlov, G. V.; Temiraev, K. B.; Afaunov, V. V. The influence of reactive mass stirring on interfacial polycondensation main parameters. *Plastics*, 2000, (2), 23–24.

134. Muller, R.; Pesce, J. J.; Picot, C. Chain conformation in sheared polymer melts as revealed by SANS. *Macromolecules*, 1993, **26**(16), 4356–4362.

135. Aloev, V. Z.; Kozlov, G. V. *Physics of Orientation Phenomena in Polymeric Materials*. Nal'chik, Polygraphservice and T, 2002, 288.

136. Vilgis, T. A.; Haronska, P.; Benhamou, M. Branched polymers in restricted geometry: Flory theory, scaling and blobs. *J. Phys. II* France, 1994, **4**(12), 2187–2196.

137. Djordjevič, Z. B. The observation of scaling in reaction with traps. In book: *Fractals in Physics*. Ed. Pietronero, L.; Tosatti, E. Amsterdam, Oxford, New York, Tokyo, North-Holland, 1986, 581–585.

138. Afaunova, Z. I.; Kozlov, G. V. *The scaling analysis of polyurethanearylate interfacial polycondensation with varied stirring rate*. Bulletin of Kabardino-Balkarian State University, Series Physical Sciences, 2000, (5), 42–44.

139. Kozlov, G. V.; Shustov, G. B. The scaling analysis of interfacial polycondensation with varied stirring rate. Mater. of I All-Russian sci.-engng. Conf. *Nanostructures in Polymers and Polymer Nanocomposites*. Nal'chik, KBSU, 2007, 86–90.

140. Van Damme, H.; Levitz, P.; Bergaya, F.; Alcover, J. F.; Gatineau, L.; Fripiat, J. J. Monolayer adsorption on fractal surfaces: A simple two-dimensional simulation. *J. Chem. Phys.*, 1986, **85**(1), 616–625.

141. Vasnev, V. A.; Kuchanov, S. I. Combined nonequilibrium polycondensation in homogeneous systems. *Successes of Chemistry*, 1973, **42**(12), 2194–2220.

142. Vasnev, V. A.; Vinogradova, S. V.; Markova, G. D.; Voitekunas, V. Yu. Macromolecular design in nonequilibrium polycondensation. *High-Molecular Compounds, A*, 1997, **39**(3), 412–421.

143. Kozlov, G. V.; Shustov, G. B.; Zaikov, G. E. The fractal analysis of copolymerization process. *J. Appl. Polymer Sci.*, 2009, **111**(7), 3026–3030.

144. Aliguliev, R. M.; Oganyan, V. A.; Yurkhanov, V. B.; Ibragimov, Kh. D. Microstructure influence on glass transition temperature of ethylene-propylene copolymers. *High-Molecular Compounds, A*, 1987, **29**(3), 611–615.

145. Matsushita, M.; Honda, K.; Toyoki, H.; Haykawa, Y.; Kondo, H. Generalization and the fractal dimensionality of diffusion-limited aggregation. *J. Phys. Soc.* Japan, 1986, **55**(8), 2618–2626.

145. Sahimi, M.; McKarnin, M.; Nordahl, T.; Tirrell, M. Transport and reaction on diffusion-limited aggregation. *Phys. Rev. A*, 1985, **32**(1), 590–595.

146. Papkov, S. P. Estimation of Kuhn segment value of rigid-chain polymers by viscous properties of diluted solutions. *High-Molecular Compounds, B*, 1982, **24**(11), 869–873.

147. Kozlov, G. V.; Shustov, G. B.; Zaikov, G. E. The macromolecular coil structure influence on functional groups activity at copolycondensation. *Chemical Physics and Mesoscopy*, 2008, **10**(3), 332–335.

148. Kozlov, G. V.; Shustov, G. B.; Zaikov, G. E. The influence of macromolecular coil structure on activity of functional groups at copolycondensation. *Encyclopaedia of Engineer-chemist*, 2011, (6), 9–12.

149. Kozlov, G. V.; Shustov, G. B.; Zaikov, G. E.; Yaryullin, A. F. *The macromolecular coil structure influence on functional groups activity at copolycondensation*. Bulletin of Kazan Technological University, 2012, **15**(9), 101–103.

150. Privalko, V. P.; Lipatov, Yu. S. The influence of macromolecular chain flexibility on glass-transition temperature of linear polymers. *High-Molecular Compounds, A*, 1971, **13**(12), 2733–2737.

151. Afaunova, Z. I.; Kozlov, G. V.; Shustov, G. B. The fractal analysis of acceptor-catalytic polycondensation of polyurethanearylate. Mater. of All-Russian conf. *Catalysis in Biotechnology, Chamistry and Chemical Technologies*. Tver, TSU, 2001, 3–10.

152. Kozlov, G. V.; Gazaev, M. A.; Novikov, V. U.; Mikitaev, A. K. The simulation of amorphous polymers structure as percolation cluster. *Letters to Journal of Technical Physics*, 1996, **22**(16), 31–38.

153. Kozlov, G. V.; Novikov, V. U. The cluster model of polymers amorphous state. *Successes of Physical Sciences*, 2001, **171**(7), 717–764.

154. Kozlov, G. V.; Temiraev, K. B.; Malamatov, A. Kh. The genetic intercommunication of reaction products and polymers condensed state structures and their properties. *Chemical Industry Today*, 1998, (4), 48–50.

155. Kozlov, G. V.; Beloshenko, V. A.; Gazaev, M.A.; Lipatov, Yu. S. Structural changes at heat aging of cross-linked polymers. *High-Molecular Compounds, B*, 1996, **38**(8), 1423–1426.

156. Kozlov, G. V.; Beloshenko, V. A.; Lipatov, Yu. S. The fractal treatment of cross-linked polymers physical aging process. *Ukrainian Chemical Journal*, 1998, **64**(3), 56–59.

157. Kozlov, G. V.; Dolbin, I. V.; Zaikov, G. E. The theoretical description of amorphous polymers physical aging. *Journal of Applied Chemistry*, 2004, **77**(2), 271–274.

158. Kozlov, G. V.; Temiraev, K. B.; Shustov, G. B. The intercommunication of macromolecular coil structure in solution with structure and properties of linear polyarylates condensed state. *Proceeding of Higher Educational Institutions, North-caucasus region, Natural Sciences*, 1999, (3), 77–81.

159. Mashukov, N. I.; Temiraev, K. B.; Shustov, G. B.; Kozlov, G. V. Modelling of solid state polymer properties at the stage of synthesis: fractal analysis. *Papers of the 6-th International Workshop of Polymer Reaction Engng*. Berlin, 5–7 October 1998, 134, 429–438.

160. Kozlov, G. V.; Zaikov, G. E. *Structure of the Polymer Amorphous State*. Leiden, Boston, Brill Academic Publishers, 2004, 465.

161. Privalko, V. P.; Lipatov, Yu. S. About glass transition temperatures of linear polymers. *High-Molecular Compounds, B*, 1970, **12**(2), 102–104.

162. Wu, S. Chain structure and entanglement. *J. Polymer Sci.: Part B: Polymer Phys.*, 1989, **27**(4), 723–741.

163. Aharoni, S. M. On entanglements of flexible and rodlike polymers. *Macromolecules*, 1983, **16**(9), 1722–1728.

164. Aharoni, S. M. Correlations between chain parameters and failure characteristics of polymers below their glass transition temperature. *Macromolecules*, 1985, **18**(12), 2624–2630.

165. Kozlov, G. V.; Novikov, V. U. Synergetics and Fractal Analysis of Cross-Linked Polymers. Moscow, *Classica*, 1998, 112.

166. Beloshenko, V. A.; Kozlov, G. V.; Lipatov, Yu. S. Glass transition mechanism of cross-linked polymers. *Physics of Solid Body*, 1994, **36**(10), 2903–2906.

167. Afaunova, Z. I.; Kozlov, G. V.; Bazheva, R. Ch. The prediction of glass transition temperature of polyurethanearylates obtained by polycondensation different methods. Electronic *Journal Studied in Russia* 73, 809–813, 2001. http://zhurnal.ape.relarn.ru/articles/2001/073.pdf.

168. Dubrovina, L. V.; Ponomareva, M. A.; Shirokova, L. B.; Storozhuk, I. P.; Valetskii, P. M. The study of poly(arylatearylenesulfoxide) block-copolymers synthesis mode influence on their some properties. *High-Molecular Compounds, B*, 1981, **23**(5), 384–388.

169. Dubrovina, L. V.; Pavlova, S.-S. A.; Ponomareva, M. A. Properties of poly(arylatearylenesulfoxide) block-copolymers solutions. *High-Molecular Compounds, A*, 1983, **25**(7), 1536–1543.

170. Kargin, V. A. Selected Works: Structure and Mechanical Properties of Polymers. Moscow, *Chemistry*, 1979, 354.

171. Shogenov, V. N.; Belousov, V. N.; Potapov, V. V.; Kozlov, G. V.; Prut, E. V. The description of stress-strain curves of glassy polyarylatesylfone within the frameworks of high-elasticity conceptions. *High-Molecular Compounds, A*, 1991, **33**(1), 155–160.

172. Shnell, G. Chemistry and Physics of Polycarbonates. Moscow, Chemistry, 1967, 229.

173. Shogenov, V. N.; Kozlov, G. V.; Mikitaev, A. K. The prediction of mechanical behavior, structure and properties of film polymeric samples at quasistatic tension. Collection of Selected Works *Polycondensation reactions and polymers*. Nal'chik, KBSU, 2007, 252–270.

174. Kozlov, G. V.; Bejev, A. A.; Lipatov, Yu. S. The fractal analysis of curing processes of epoxy resins. *J. Appl. Polymer Sci.*, 2004, **92**(4), 2558–2568.

175. Kopelman, R. The exitons dynamics resembling fractal one: geometrical and energetic disorder. In book: *Fractals in Physics*. Ed. Pietronero, L.; Tosatti E. Amsterdam, Oxford, New York, Tokyo, North-Holland, 1986, 524–527.
176. Kozlov, G. V.; Zaikov, G. E. The physical sense of reaction rate constant in Euclidean and fractal spaces at polymers thermooxidative degradation consideration. *Theoretical Principles of Chemical Technology*, 2003, **37**(5), 555–557.
177. Meilanov, R. P.; Sveshnikova, D. A.; Shabanov, O. M. *Sorption kinetics in systems with fractal structure*. Proceeding of Higher Educational Institutions, North-caucasus region, natural sciences, 2001, 1, 63–66.
178. Dolbin, I. V.; Kozlov, G. V. The physical sense of reactive medium heterogeneity for polymer solutions and melts. *Reports of Adygskoi (Cherkesskoi) International Academy of Sciences*, 2004, **7**(1), 134–137.
179. Meakin, P.; Stanley, H. E.; Coniglio, A.; Witten, T. A. Surfaces, interfaces and screening of fractal structures. *Phys. Rev. A*, 1985, **32**(4), 2364–2369.

CHAPTER 2

RADICAL POLYMERIZATION

CONTENTS

2.1   THE PROPERTIES OF CATION POLYELECTROLYTES SOLUTIONS

The cationic water-soluble polyelectrolytes find wide application in different branches of industry, particularly, for the ecological goals solution [1]. It is necessary to determine for these polymers action mechanism and synthesis process understanding their molecular sizes and also water solutions properties [2]. Unlike uncharged and syntactic polymers the cationic polyelectrolytes water solutions are not studied well. Lately it has been shown, that the fractal analysis methods can be used successfully for the description of both synthesis processes [3] and their action mechanism as flocculators [4]. Therefore, the authors [5] studied the cationic polyelectrolytes in solution behavior on the example of copolymers of acrylamide with trimethylammonium methylmethacrylatechloride (PAA–TMACh), using the fractal analysis methods. The data of work [2] for the four copolymers PAA–TMACh with the last contents of 8, 25, 50 and 100 molar % were used for this purpose. The authors [2] obtained for these polymers the equations of Mark–Kuhn–Houwink type (see the Eq. (57) of Chapter 1), which linked the intrinsic viscosity [η] and macromolecular coil gyration radius R_g with polymer average weight molecular weight $\overline{MM}_w$. In the last case the indicated equation like look [2]:

$$\left\langle R_g^2 \right\rangle^{1/2} = K_{R_g} \overline{MM}_w^{\,b}, \tag{1}$$

where K_{R_g} and b are constants for each copolymer.

The macromolecular coil fractal dimension D_f insolution can be determined further according to the Eq. (4) of Chapter 1. In Fig. .1 the dependence of dimension D_f on TMACh contents cTMACh in copolymer PAA–TMACh is adduced, allowing to judged about interactions of macromolecular coil elements among themselves. As it is known [6], the value of macromolecular coil in solution dimension D_f is controlled by two factors groups: interactions polymer-solvent and interactions of coil elements among themselves. Since coils behavior in one solvent (NaCl water solution [2]) is considered, then it should be supposed, that the shown in Fig. 1 D_f variation is due to the factors, related to the second group. Just small contents of component with charged groups results to D_f sharp increasing, i.e., results to essential coil compactization. This means strong attraction interaction between fragments of PAA and TMACh availability. However, at the contents TMACh increasing the reduction Df is observed, moreover within TMACh contents (cTMACh) range of 8–50 molar % this reduction is close to a linear one. Such D_f change supposes strong repulsion interactions between TMACh fragments availability, compensating attraction interactions between PAA and TMACh fragments.

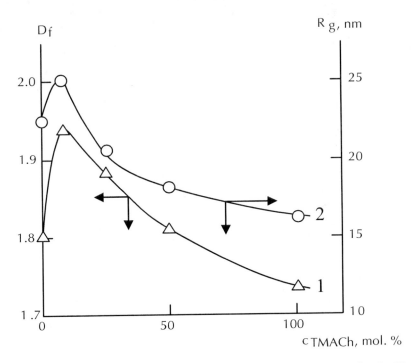

FIGURE 1   The dependences of fractal dimension D_f (1) and gyration radius R_g (2) of macromolecular coil on TMACh contents cTMACh for copolymers PAA–TMACh.

In Fig. 1 the dependence of macromolecular coil gyration radius R_g on cT-MACh is also adduced for the considered copolymers. The attention is paid to the dependences D_f (cTMACh) and R_g(cTMACh) similarity, obvious at their comparison. Let us note, that the values R_g were calculated according to the Eq. (1) in assumption of the same value $\overline{MM}_w=10^5$ for all copolymers. Proceeding from the macromolecular coils compactness degree decreasing owing to D_f reduction at cTMACh increasing, one should expect enhancement R_g, but not its reduction. For this apparent discrepancy explanation the authors [5] used the fractal Eq. (1) of Chapter 1 between R_g and D_f, where the polymerization degree N was determined as $\overline{MM}_w$/MM0, where MM0 is molecular weight of polymer chain "elementary link".

From the Eq. (1) of Chapter 1 and the parameter N determination it follows, that for simultaneous fulfillment of the dependences of D_f and R_g on cTMACh, shown in Fig. 1, N reduction at cTMACh growth is necessary or, that is equivalent, MM0 increasing, which in its turn, means copolymers chain rigidity enhancement. The indicated conditions combination allows to calculate the value MM0 and its dependence on cTMACh for copolymers PAA–TMACh, which is shown in Fig. 2. As it was expected, the essential MM0 increasing (approximately in 4 times) at TMACh contents from 8 up to 100 molar % is observed.

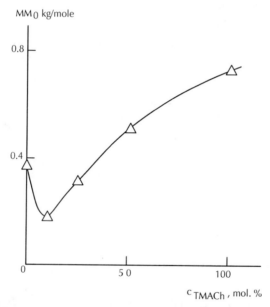

FIGURE 2   The dependence of "elementary link" molecular weight MM0 on TMACh contents cTMACh for copolymers PAA–TMACh.

The same conclusion follows from the coefficient K_η in Mark–Kuhn–Houwink equation change. As the data of work [2] have shown, within the indicated range of cTMACh variation K_η reduction in more than 20 times is observed. The value K_η can be estimated according to the Eq. (58) of Chapter 1 or, equivalently, according to Mark–Kuhn–Houwink fractal variant — to the Eq. (23) of Chapter 1.

In Table 1 the comparison of experimental [2] and calculated according to the Eqs. (58),23 of Chapter 1 K_η values is adduced. As one can see, the experimental values K_η correspond better to the calculation according to the Eq. (23) of Chapter 1, than according to the Eq. (58) of Chapter 1.

TABLE 1   Comparison of the coefficient K_η values in Mark–Kuhn–Houwink equation, obtained by different methods.

TMACh con- tents, molar %	$K_\eta \times 10^4$		
	Experiment [2]	The Eq. (58) of Chapter 1	The Eq. (23) of Chapter 1
0	2.57	2.09	3.20
8	12.80	27.60	20.0
25	6.30	7.10	6.21
50	1.85	1.45	1.86
100	0.60	0.38	0.66

Thus, TMACh contents increasing in copolymers PAA–TMACh has two consequences for macromolecular coil structure — the chain rigidity increasing and the coil fractal dimension D_f reduction are observed. These changes influenced essentially on both synthesis process and low-molecular admixtures flocculation of the considered copolymers.

The authors [5] performed the quantitative estimation of the assumed effects. The reaction rate at polymers synthesis can be estimated according to the Eq. (69) of Chapter 1. The indicated relationship demonstrates, that Df reduction from 1.94 up to 1.74 at TMACh contents enhancement from 0 up to 100 molar % at arbitrarily accepted t value (let us assume 50 min) results to ϑ_r increasing on about 50%.

In work [4] it has been shown, that polymer effectiveness as flocculator is defined by the macromolecular coil in solution density fr — the smaller ρfr is, the higher polymer effectiveness as floccular is. ρfr value can be determined according to the Eq. (73) of Chapter 1, from which it follows, that ρfr value and, hence, polymer effectiveness as flocculator, is dependent on both D_f and R_g. In Fig. 3 two dependences of ρfr on cTMACh are adduced: they are calculated according to the Eq. (73) of Chapter 1, but the first from them uses real values R_g (Fig 1) and the second one the condition $R_g = \text{const} = 25$ nm. As it follows from the plots of Fig. 3, if for the first (real) situation the value ρfr at D_f change from 1.94 up to 1.74 changes by 15% only, then in the second (hypothetical) case the value ρfr decreases about twice. Since polymer effectiveness as flocculator changes in the same proportion [4], then the plots of Fig. 3 demonstrated clearly MM0 or chain rigidity influence on the indicated polymer property.

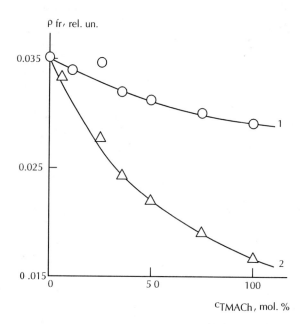

FIGURE 3 The dependences of macromolecular coil density ρfr on TMACh contents cTMACh for copolymers PAA–TMACh. Calculation according to the Eq. (73) of Chapter 1 at real R_g values (1) and at the condition $R_g = const$ (2).

Hence, the stated above results have shown, that the introduction of the changed groups in copolymer changes macromolecular coil in solution structure, reducing its fractal dimension, gyration radius and increasing chain rigidity. Within the frameworks of fractal analysis it was demonstrated, that the indicated factors could influence essentially on both polymers synthesis process and polymer effectiveness as flocculator [5].

2.2   THE FRACTAL KINETICS OF RADICAL POLYMERIZATION

The classical problem in chemical kinetics is the diffusive processes influence on this kinetics. In diffusion-limited reactions their rate is defined by the diffusion duration, which is necessary for, the reagents to reach one another. Similar reactions simulation on Euclidean lattices gave the following results. The reactions were considered [7]:

$$A + A \rightarrow O \, ' \tag{2}$$

$$A + B \rightarrow O \, ' \tag{3}$$

where A and B are reacting particles, O is inert product.

For the indicated reactions the following dependences of reacting particles A density ρA on reaction duration t were obtained [7]:

$$\rho_A \sim t^{-d/2}, \tag{4}$$

$$\rho_A \sim t^{-d/4}, \tag{5}$$

for the Eqs. (2) and (3), accordingly.

As it is known [8], the space type change from Euclidean up to fractal one strongly changes reaction course. In this case the dependences $\rho A(t)$ like look [7]:

$$\rho_A \sim t^{-d_s/2}, \tag{6}$$

$$\rho_A \sim t^{-d_s/4}, \tag{7}$$

for the Eqs. (2) and (3), accordingly, where ds is the spectral dimension of reactive medium (in computer simulation — lattice), defined by its connectivity degree [9].

The authors [10] used the considered above physical model for dimethyldiallylammoniumchloride (DMDAACh) radical polymerization [1] description. As it was shown in work [8], the radical polymerization of DMDAACh was simulated by the diffusion-limited aggregation model according to mechanism cluster–cluster. This means, that the indicated process is realized by small macromolecular coils merging into larger ones. This treatment allows to simulate DMDAACh polymerization as the Eq. (2) with the Eq. (4) or (6) depending on the space type, in which it is realized. In this case the value ρA can be determined as follows [10]:

$$\rho_A = 1 - Q, \tag{8}$$

where Q is the conversion degree.

In Fig. 4, the dependences of $(1 - Q)$ on t in double logarithmic coordinates for DMDAACh polymerization at two values of monomers initial concentration c0: 4.0 and 4.97 moles/l (0.646 and 0.728 by the mass) are adduced. As one can see, in both cases the linear dependences with negative slope, having absolute values 0.26 and 0.38, are obtained. Hence, these dependences correspond to the Eq. (6) and allow to calculate the values ds of reactive medium, which for the used monomer concentrations are varied within the limits of 0.52–2.95.

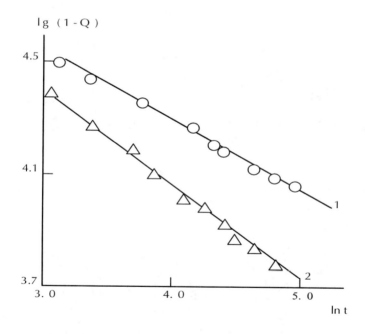

FIGURE 4   The dependences of $(1 - Q)$ on reaction duration t in double logarithmic coordinates, corresponding to the Eq. (6), for DMDAACh polymerization at c0 = 0.646 (1) and 0.728 (2) mass parts.

In Fig. 5 the dependence ds(c0) for DMDAACh polymerization is shown, from which slow ds increasing up to $c_0 = 0.6$ by mass and then spectral dimension rapid growth up to approx. 3, that corresponds to Euclidean space with dimension d = 3 [11], follows. The dependences d_s (c_0) should be extrapolated to

$d_s = 0$ at $c_0 = 0$ according to the following obvious reasons. At monomer absence ($c_0 = 0$) polymerization reaction cannot be realized and according to the Eq. (8) $\rho A = c_{onst} = 1.0$. The dependence $\rho A(t)$ in double logarithmic coordinates, corresponding to the Eqs. (4)–(7), in this case gives $\ln \rho A = 0$ at any t and, hence, zero slope and $d_s = 0$. It is significant that in practice polymerization reaction with $c_0 < 4.0$ moles/l is performed extremely rarely, since its performance takes up too much time. The dependence of ds or reactive medium connectivity degree is very strong and is expressed analytically as follows [10]:

$$d_s \sim c_0^5,$$

(9)

where c_0 is given in mass parts.

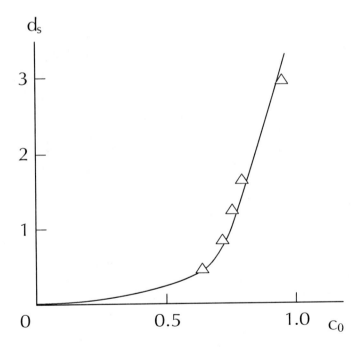

FIGURE 5 The dependences of reactive medium spectral dimension ds on monomers initial concentration c_0 for DMDAACh radical polymerization.

At $c_0 \approx 0.90$ mass parts $d_s = 3$ and value Q reaches approx. 1.0, i.e., the entire monomer complete polymerization occurs.

Let us consider the physical reasons of reactive medium ds change within the indicated above limits. Reactive medium itself presents Euclidean space with dimension d = 3, as and any solution of low-molecular substance in the same solvent [12]. The appearance of dimension d_s with values <3, which are specific for fractal mediums, is due to unevenness of monomer in solution distribution and c_0 decreasing results to law-governed diffusion time increasing, which is necessary for reagent reaching one another. Let us note, that DMDAACh macromolecular coil fractal dimension in water solution is $D_f = 1.65$ [13]. According to Kremer's formula the dimension dm of medium, in which such coil is formed, can be determined [14]:

$$D_f = \frac{2+d_m}{3}, \qquad (10)$$

from which $d_p = d \approx 3$ follows.

The value d_s is linked with exponent θ in the dependence of the diffusion constant D_d on time t as follows [9, 15]:

$$d_s = \frac{2d_m}{2+\theta}. \qquad (11)$$

In its turn, the dependence D_d — distance r, which requires reacting particles to pass up to meeting with one another, has the following appearance [15]:

$$D_d(r) \sim r^{-\theta}. \qquad (12)$$

Assuming $d_m = 3$, from the Eq. (11) according to the known ds values the exponent Q can be calculated, the value of which changes within the limits of 0.034–9.540 at the indicated above c_0 variation. Assuming r = 10 relative units, according to the Eq. (12) the diffusion constant D_d can be calculated, which is varied within the limits of 2.88 × 10–10–0.925, i.e., on 10 orders, at the same c_0. The dependence $\theta(c_0)$ is shown in Fig. 6, from which its sharp decay at c_0 growth follows and at $c_0 \approx 0.90$ mass parts the value θ tries to attain zero, i.e., the transition from anomal diffusion to classical one occurs [16].

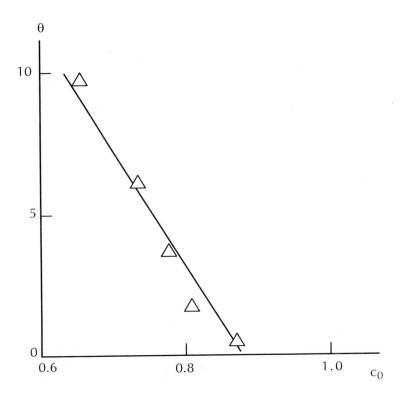

FIGURE 6   The dependence of the diffusion constant exponent θ on monomer initial concentration c_0 for DMDAACh radical polymerization.

As it has been shown in work [1], c_0 increasing within the indicated above range results to the initial reactive medium viscosity η_0 sharp enhancement, moreover the dependence $\eta_0(c_0)$ qualitatively is very resembling the dependence $d_s(c_0)$, shown in Fig. 5. This allows to assume the correlation between d_s and η_0 possibility. The adduced in Fig. 7 plot confirms the supposed correlation $ds(\eta_0)$ availability, which is expressed analytically as follows [10]:

$$d_s \approx 8.17 \times 10^{-2} \eta_0 \,, \tag{13}$$

where η_0 is given in relative units [1].

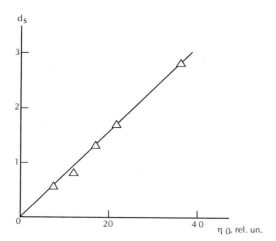

FIGURE 7 The dependence of spectral dimension d_s of reactive medium on initial viscosity η_0 of this medium for DMDAACh radical polymerization.

The reactive medium connectivity degree, controlled by the parameters c_0 or η_0 and characterized by the spectral dimension d_s, defines polymerization reaction course, which follows directly from the Eq. (6). So, in Fig. 8 the dependence of polymerization rate ϑ_r (in relative units according to the data [1]) on the value d_s is shown, which is approximated by a straight line, passing through coordinates origin. The last circumstance points out, that at $d_s = 0$ or $c_0 = 0$ polymerization reaction cannot be realized and its rate ϑ_r is equal to zero.

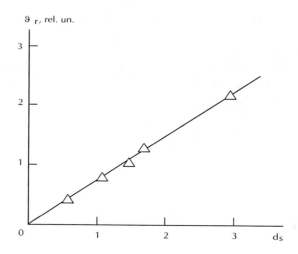

FIGURE 8 The dependence of DMDAACh polymerization rate ϑ_r on reactive medium spectral dimension d_s.

The kinetic curves $Q_{(t)}$ of DMDAACh polymerization at this process duration $t \approx 100$ min reach plateau, after that the value Q changes insignificantly. In addition, the value of conversion degree on plateau Q_{pl} grows at c_0 increasing: at c_0 enhancement from 0.646 up to 0.888 mass parts the value Q_{pl} increases from 0.40 up to about 1.0. In Fig. 9 the dependence $Q_{pl}^2(d_s)$ (such the dependence form was chosen with the goal of its linearization) is shown, which is also approximated by a straight line, passing through coordinates origin and expressed analytically as follows [10]:

$$Q_{pl}^2 = 0.33 d_s.$$

(14)

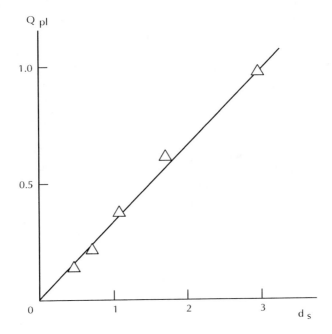

FIGURE 9   The dependence of conversion degree Q_{pl} on kinetic curves plateau on reactive medium spectral dimension ds for DMDAACh radical polymerization.

The obtained result is important from the following point of view. Synthesis reaction realization always gives $Q_{pl} < 1.0$ [17, 18]. The Eq. (14) confirms quantitatively exactly this circumstance and for Euclidean space $d_s = 3$ [11] and $Q_{pl} \approx 1.0$, that is confirmed experimentally [1].

The average viscous molecular weight M_η of DMDAACh estimation can be performed according to the Eq. (28) of Chapter 1. With the Eqs. (8) and (13) the Eq. (28) of Chapter 1 can be written as follows [10]:

$$M_\eta \sim d_s^{-0.8}. \qquad (15)$$

From the Eq. (15) it follows, that M_η reduction at c_0 growth (at $c_0 \geq 0.70$ mass parts) is due to d_s sharp increasing (Fig. 5).

The kinetics of DMDAACh radical polymerization can be described with the aid of the Eq. (26) of Chapter 1. At this relationship inference the assumptions, were used differing from the assumed as the basis in the Eqs. (4)–(7). It was assumed, that a synthesis reaction course is controlled by "accessibility" degree of macromolecular coil for the reaction -the smaller D_f is, the higher the indicated degree is and the polymerization reaction is realized more rapidly. Nevertheless, it should be noted, that in the Eq. (26) of Chapter 1 the parameter η_0 is included, which characterized indirectly reactive medium connectivity degree (the Eq. (13)). Both approaches, taking into account physical properties of both reactive medium (the Eq. (6)) and macromolecular coil (the Eq. (26) of Chapter 1) give adequate description of kinetic curves $Q_{(t)}$ for DMDAACh polymerization.

Hence, the reactive medium at DMDAACh polymerization can be simulated as a certain "virtual" fractal, having the dimension $d_m = d = 3$, but possessing the connectivity degree, characterized by spectral dimension ds, which is typical for fractal objects, i.e., $d_s < 3$ for $c_0 < 0.90$ mass parts or $c_0 < 5.60$ moles/l. Such ds behavior is due to large-scale fluctuations of monomer distribution in solution at $c_0 < 0.90$ mass parts, i.e., by availability of space regions, free from monomer [19]. At big enough $c_0 \geq 0.90$ mass parts these fluctuations disappear and polymerization process behavior is typical for the reactions, realized in Euclidean spaces [8]. The connectivity degree is defined by reactive medium initial parameters (monomer concentration and its viscosity) and, in its turn, checks both polymerization reaction course and its final results (Q_{pl} and MM_η).

The kinetics of DMDAACh radical polymerization in water solutions has the same qualitative features, as radical polymerization kinetics of other polymers [1]. In particular, this is expressed in sigmoid shape of kinetic curve $Q_{(t)}$, which in general case is divided into three sections: initial, of autoacceleration and finish polymerization [20]. For the curve $Q_{(t)}$ the traditional approaches number, considered in detail in works [1, 20], is used. However, the indicated approaches do not take into account the structure of the main element of polymers synthesis in solutions — macromolecular coil and, hence, its possible variations in synthesis process. At present the theory of irreversible aggregation, linked closely to fractal analysis, is developed in details, which was created for the description of exactly such processes as polymerization [21]. The indicated conception usage allows to obtain polymerization processes quantitative description with principally new positions [17, 18]. Proceeding from this, the authors [22–24] developed the model of fractal kinetics of DMDAACh radical polymerization with irreversible aggregation models [8] attraction.

Within the frameworks of fractal analysis the polymerization kinetics is described by the Eq. (27) of Chapter 1 and polymerization formal kinetics within the frameworks of traditional approaches is formulated as follows [20]:

$$\frac{dQ}{dt} = k_r \left(1 - Q\right) , \qquad (16)$$

where k_r is the rate constant of polymerization.

Differentiating the Eq. (27) of Chapter 1 by time t, let us obtain [17]:

$$\frac{dQ}{dt} \sim t^{(1-Df)/2} . \qquad (17)$$

The combination of Eqs. (16) and (17) allows to obtain the equation for D_f estimation according to the kinetic parameters of polymerization process [24]:

$$t^{(D_f-1)/2} = \frac{c_1}{k_r\left(1-Q\right)}, \qquad (18)$$

where c_1 is the constant, estimated from boundary conditions.

As it follows from the Eq. (18), the values kr should be used for the value D_f estimation. However, the authors [24] used the values ϑ_r, proceeding from the following considerations. As it is known [1], at DMDAACh polymerization the following relationship is fulfilled:

$$\vartheta_r = \frac{\vartheta_i k_0}{k_0^{1/2} c_0 \eta_0^{1/2}}, \qquad (19)$$

where k_i is the initiation rate, k_0 is bimolecular break rate constant, c_0 is monomer concentration, η_0 is the relative viscosity of monomer initial solution.

Since the equation is fulfilled [1]:

$$k_0 = \frac{k_0'}{\eta_0^{1/2}},$$

(20)

where k_0' is the elementary break rate constant for chains in standard conditions, when monomer solution viscosity is very small, then for monomer and initiator constant concentrations and also at polymerization constant temperature from the Eqs. (19) and (20) it follows [24]:

$$\vartheta_r \sim k_r.$$

(21)

At the synthesis beginning solution contains only monomer, which within the frameworks of irreversible aggregation models can be considered as particles, uniting later in a cluster (macromolecular coil). As it is known [21], within the frameworks of the indicated models such mechanism is called mechanism particle–cluster and aggregates with fractal dimension $D_f \approx 2.5$ is the result of its action. Besides, the value c_1 was calculated according to the Eq. (18) with the following parameters using: $t = 0.5$ min, $\vartheta r = 4.8$ mol/l·s and $Q = 8.3 \times 10 - 3$.

In Fig. 10 the dependences $D_f(t)$, calculated by the indicated mode, and $Q_{(t)}$, obtained experimentally, are adduced. As it follows from these plots, not only the value D_f strong variation during reaction time $t = 17.5$ min, but also the aggregation mechanism change are observed. The used above boundary condition for c_1 estimation checking can be performed as follows. The value D_f can be estimated according to the Eq. (4) of Chapter 1. In its turn, the exponent a in this equation was obtained from Mark–Kuhn–Houwink equation for DMDAACh [1]:

$$[\eta] = 1.12 \times 10^{-4} MM^{0.82}.$$

(22)

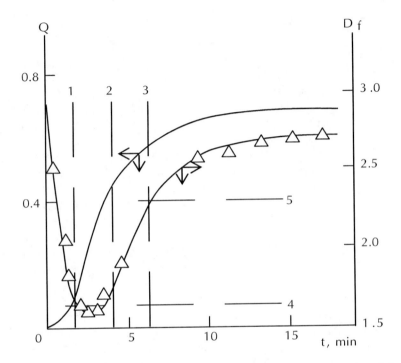

FIGURE 10 The dependences of conversion degree Q and macromolecular coil fractal dimension D_f on reaction duration t for DMDAACh (c_0 = 4.8 mol/l, initiator content 5×10^{-3} mol/l, T = 353 K). The vertical shaded lines 1–3 indicated polymerization stages borders. The horizontal stroke-pointed lines indicate the value D_f of isolated coil in diluted solution and critical value D_f^c.

From the Eqs. 4 of Chapter 1 and (22) the value D_f = 1.65 for DMDAACh in diluted solution can be estimated. In Fig. 10 this value is pointed out by a stroke-pointed line 4. As one can see, this D_f magnitude corresponds well to minimum value D_p, estimated according to the Eq. (18) for more concentrated solution. This correspondence confirms boundary conditions choice correctness, since the value D_f for macromolecular coil in diluted solution is a minimum one [25].

Now, using the data of Fig. 10 the general description of DMDAACh radical polymerization fractal kinetics can be given. In the initial synthesis point (t=0) monomer is contained in the solution, which, as it was noted above, within the frameworks of irreversible aggregation models should be considered as particles, forming aggregate. The unification of these particles in clusters on the first stage (in Fig. 10 the stages are indicated by vertical shaded lines, moreover as the line 0 ordinates axis is accepted), i.e., on the stage 0–1, occurs according to the mechanism particle–cluster, that results to dense macromolecular coils formation with the dimension $D_f \approx 2.5$, corresponding to the indicated mechanism.

The reaction rate on this initial section of polymerization is precisely small in virtue of large enough D_f value, that follows directly from the Eq. (17). Then these clusters begin to consolidate, forming some larger clusters, i.e., the mechanism particle–cluster is changed by mechanism cluster–cluster. Let us remind, that to the autoacceleration section beginning the polymerization degree N reaches considerable value (N = 200–315) [1]. In the section 1–2 the value $D_f \approx 1.65$, that corresponds to diffusion-limited mechanism cluster–cluster [21]. This stage corresponds to the autoacceleration section, i.e., to the section, where the reaction rate is the greatest one. Thus, within the frameworks of the proposed treatment the transition from the initial section to the autoacceleration section is due to aggregation mechanism change and, as consequence, to sharp D_f reduction. In other words, the line 1 in Fig. 10 corresponds to cessation of the reaction by the monomer — coil joining mode and on the section 1–2 reaction is realized completely at the expense of macromolecular coils consolidation only. The large reaction rate on the section 1–2 (of autoacceleration) defines macromolecular coils great number appearance and now each from them is in similar coils environment, that results to these coils compactization and, hence, to D_f growth (solution becomes "dense" according to the definition [25]). Such process continues until the coil does not reach the critical dimension D_f^c, defined according to the Eq. (8) of Chapter 1. The estimation according to this equation at d = 3 gives $D_f^c = 2.29$ (the horizontal stroke-pointed line 5 in Fig. 10). This point corresponds to the stage 2–3 (finish polymerization) completion. The stage 2–3 correspondence to "dense" solution transition correctness can be checked with the aid of the model [25], where the relation between D_f and macromolecular coil fractal dimension $\overline{D}_f$ in "dense" solution is given as follows:

$$D_f = \frac{(d+2)\overline{D}_f}{2(1+\overline{D}_f)}. \tag{23}$$

If to accept the value D_f, corresponding to the stage 2–3 beginning ($D_f \approx 1.76$, Fig. 10), then according to the Eq. (23) let us obtain $\overline{D}_f = 2.38$, that corresponds well to the made above estimation.

The last stage of curve $Q_{(t)}$ is gelation, which represents a joining system cluster (spreading from one end of reactive bath up to the other) formation. Since DMDAACh is linear polymer, then the physical gelation is implied, i.e., included in joining cluster macromolecular coils are consolidated by macromolecular entanglements network. As theoretical [26] and experimental [27] studies have shown, the dimension $d_f \approx 2.5$ corresponds to the gelation transition, that corresponds completely to the data of Fig. 10. On this stage the synthesis reaction is practically near completion. As it has been shown in work [28], the fractal dimensions of macromolecular coil in solution and condensed state (gel) df for linear polymers are linked by the Eq. (99) of Chapter 1. If, as earlier, to accept the value $D_f \approx 1.76$, corresponding to the stage 1–2 ending, then according to the Eq. (99) of Chapter 1 let us obtain $d_f \approx 2.64$, that again corresponds well to the data of Fig. 10.

Let us note one more important aspect. As the experimental studies [1] have shown, the curves $MM_{(t)}$ reaching of the saturation section occurs earlier, than the kinetic curves $Q_{(t)}$. This observation is also explained within the frameworks of irreversible aggregation models. As it is known [24], the critical (the greatest) macromolecular coil gyration radius R_c is determined according to the Eq. (48) of Chapter 1. At this level reaching the macromolecular coil ceases its growth and, hence, the molecular weight increasing is ceased. Further insignificant MM and Q enhancement can be realized only at the expence of growth of coils, having gyration radius Rg smaller than the critical one, i.e., at $R_g < R_c$.

In Fig. 11 the dependence $\vartheta_r(t)$ for DMDAACh, plotted according to the data of work [1], is adduced. The plots of Figs. 10 and 11 comparison shows, that the dependence $D_f(t)$ and $\vartheta_r(t)$ are antibate ones, although the maximum on the dependence $\vartheta_r(t)$ is expressed more clearly, than the corresponding to it minimum on the curve $D_f(t)$. This is explained by the fact, that the value ϑ_r is defined not only by D_f but also by its two variables (t and Q) according to the Eq. (18). This important factor is routinely missed in the conventional approaches of description of radical polymerization processes [1, 20].

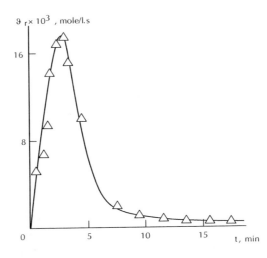

FIGURE 11   The dependence of reaction rate ϑ_r on reaction duration t for DMDAACh.

Further we shall consider the method of theoretical description of the curve $Q_{(t)}$ within the frameworks of fractal kinetics based on application of the Eq. (28) of Chapter 1. Since with respect to the described process of DMDAACh synthesis the value D_f is a variable, then the increment Q (ΔQ_i) after termination of the arbitrarily chosen temporary interval $t_i = t_i + 1 - t_i$ was estimated. For the indicated interval it is possible to write [24]:

$$\Delta Q = \Delta t_i^{(3-D_i)/2}.$$ (24)

Then the value Q_i, corresponding to time t_i, was determined by summation [24]:

$$Q_i = Q_{i-1} + \Delta Q_i.$$ (25)

In Fig. 12 the comparison of experimental and calculated within the frameworks of fractal kinetics (according to the Eq. (24) and (25)) kinetic curves $Q_{(t)}$ for DMDAACh is shown. As one can see, excellent correspondence between the theory and experiment is obtained: the discrepancy does not exceed 6%, i.e., it is not higher than Q definition error [1].

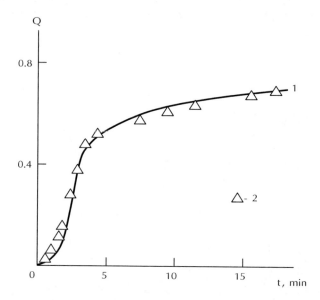

FIGURE 12   Comparison of experimental (1) and obtained within the frameworks of fractal kinetics (2) kinetic curves conversion degree-reaction duration Q_{-t} for DMDAACh.

Hence, the proposed technique allows to estimate variation of macromolecular coil structure of DMDAACh, characterized by its fractal dimension D_f, during the entire polymerization reaction. It has been shown, that exactly this factor, is not taken into account in conventional theories, and defines the most important characteristics of polymerization process: rate, conversion degree, molecular weight. Besides, the fractal kinetics methods allow the quantitative description of polymers synthesis, particularly, they give the correct shape of the kinetic curve $Q_{(t)}$.

Let us consider the kinetics of DMDAACh initial polymerization in more details. The autoacceleration effect in radical polymerization (the so-called gel-effect) plays an important role in polymers synthesis processes and during years number was a topic of intensive studies [20, 29-32]. At present it is assumed, that the autoacceleration realization is due to a structural changes in polymer solution [31, 32]. The detailed interpretation of this point of view is adduced in the indicated above works. However, it should be noted, that polymerization auto-acceleration beginning is accompanied by a number of other important effects, which get much less attention. Let us indicate the some from them. The attention is paid to very high molecular weights MM of polymer, which are realized on polymerization initial section at low reaction rates, small conversion degrees (up to 5%) and small reaction durations (of order of 5 min) [32]. Secondly, as it has been shown in work [33], the reaction rate on initial section is proportional to $\eta_0^{1/2}$ (where η_0 is monomer initial mixture viscosity) and on autoacceleration section — to η_0. Thirdly, the question arises, if macromolecular entanglement network formation is the cause of autoacceleration [32] or its consequence. And at last, definite doubts arise about correctness of the term "gel-effect" application in connection with autoacceleration beginning. The authors [34] considered these problems with the positions of irreversible aggregation models on the example of DMDAACh [1].

In Fig. 13, a kinetic curve conversion degree — reaction duration Q_{-t} is shown schematically, on which the technique of estimation of the characteristic times t_1 and t_2, corresponding to termination of initial section and autoacceleration section of curve Q_{-t} are indicated. The calculated according to the Eq. (4) of Chapter 1 D_f value, which is equal to 1.65, defines unequivocally polymer formation mechanism as diffusion-limited aggregation cluster–cluster. However, as it was indicated above, in polymerization process beginning at $t = 0$ monomers (particles) solution was the initial reactive mixture, where macromolecular coils were absent. Let us remind that for such coil formation a macromolecule should consist of, as a minimum, 20 monomer links [35]. Therefore, it is obvious, that on the first stage of polymerization mechanism particle–cluster should be realized and this mechanism will act until in solution macromolecular coils (clusters) sufficient number is not formed for mechanism cluster–cluster realization [36]. Hence it follows that the main assumption — the transition from the initial section to autoacceleration section of kinetic curve Q_{-t} is due to aggregation mechanism change, namely, the transition from mechanism particle–cluster to mechanism cluster–cluster or to system universality class change. Within the frameworks of this supposition all further analytical conclusions will be obtained.

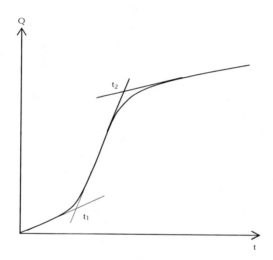

FIGURE 13 The scheme of determination of termination times of initial section t_1 and autoacceleration section t_2 on kinetic curve conversion degree — reaction duration Q_{-t}.

Let us consider the correctness of the term "gel-effect" application. In modern physics under the term "gelation" a cluster formation, spreading from one end of reactive bath up to the other (or from one end of lattice up to the other in computes simulation) is understood [21]. This formulation supposes sharp change of the main system parameters, both static and dynamic, at gelation point. The fractal dimension value changes from 1.65–2.12 up to approx. 2.50, that is confirmed both theoretically [26] and experimentally [27]. If the process kinetics is described by the Smoluhovski equation [26], then the scaling exponent of this equation kernel $\omega = 1/2$ is gelation point characteristics. At $\omega < 1/2$ the general scaling is valid for entire clusters distribution and average cluster size grows regularly as a function of time, whereas for $>1/2$ such simple scaling violates, the greatest (joining) cluster grows more rapidly than others [26]. The diffusion rate of clusters with mass 9 can be expressed by the Eq. (66) of Chapter 1. Up to the gelation point the exponent α in this equation is smaller than zero and this means, that smaller clusters diffuse more rapidly and behind gelation point $\alpha > 0$. The temporal dependence of gyration radius R_g of clusters (macromolecular coils) likes look [37]:

$$R_g \sim t^{1/z} , \tag{26}$$

where the exponent z is determined according to the equation [37]:

$$z = D_f (1-\alpha) - (d-2) , \tag{27}$$

where d is the dimension of Euclidean space, in which a fractal is considered (it is obvious, that in our case d = 3).

The value R_g is linked with polymerization degree N and for the same polymer with molecular weight MM according to the Eq. (1) of Chapter 1. In Fig. 14, the dependences $MM_{(t)}$, plotted according to the data of work [32], for DMDAACh with c_0 = 4.5 and 5.5 mol/l are adduced. Using these data together with the Eqs. (1), (26) and (27) of Chapter 1 the values z, α can be calculated and then the values ω according to the equation [26]:

$$2\omega = \alpha + \frac{d-2}{D_f} \tag{28}$$

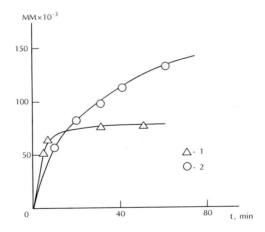

FIGURE 14 The dependences of molecular weight MM on reaction duration t for DMDAACh polymerization with initial monomers concentration c_0: 4.5 (1) and 5.5 (2) mol/l.

For DMDAACh the following values of the indicated parameters were obtained: $z = 3.10$, $\alpha = -1.48$ and $\omega = -0.435$. Thus, a polymer on autoacceleration stage is far enough from gelation point and even from transition sol-gel (see Fig. 1 of reference [26]). For DMDAACh with $c0 = 5.5$ mol/l the following values were obtained: $z = 8.63$, $\alpha = -4.84$ and $\omega = -2.12$. Thus c_0 increasing from 4.5 up to 5.5 mol/l does not bring closer the autoacceleration section to the gelation point, but on the contrary removes this section from gel state.

Hence, the obtained above data totality indicates unequivocally on incorrectness of the term "gel" application to DMDAACh autoacceleration section. Very often the gelation definition as transition of liquid (easy mobile and viscoelastic) systems in solid-like state of gel or jelly is used [38]. In this case the gelation is due to the origin of spatial phase or molecular net (network) in bulk liquid system, which deprives a system of fluidity and gives it some properties of a solid body (elasticity and so on). It is obvious, that the adduced above definition of gelation point [26] is more precise than the given one in Ref. [38]. Therefore, the appearance in autoacceleration section beginning of some solution elasticity [32] is impossible to consider as gelation proof. In virtue of the indicated reasons and ambiguity appearance the authors [34] assume, that the term "gel-effect" should be avoided and the term "autoacceleration" is more precise.

Let us consider now the temporal frameworks of the transitions initial section-autoacceleration and autoacceleration-finish polymerization, i.e., the factors, influencing on the values t_1 and t_2 (Fig. 13). With this purpose the authors [34] used both static and dynamic scaling of polymerization reaction. The first scaling type defines a process limiting characteristics irrespective of their reaching way and the second gives the temporal dependence of these characteristics variation [39]. The dependence of aggregation processes on aggregating particles concentration c (initial concentration c_0) in some limited space has a direct relation to the considered problem. The static scaling assumes, that if during cluster (macromolecular coil) growth its density reduces up to the environment density (with concentration c at the given moment), then the transition occurs between universality classes of diffusion-limited aggregation (DLA) and macromolecular coil reaches its limiting gyration radius R_c. Such transition scale is dependent on initial particles (monomers) concentration c_0 and is defined by the Eq. (48) of Chapter 1. In its turn, dynamical scaling defines the dependence of R_g on aggregation (reaction) duration t in the form [39]:

$$R_g \sim c_0^{1/(d-D_f)} t .\qquad(29)$$

Supposing, that the value R_c is reached at duration t_c, let us obtain [34]:

$$t_c \sim \frac{1}{C_0^{2/(d-D_f)}}. \tag{30}$$

Thus, using for the initial section $D_f \approx 2.45$ [27] and for the autoacceleration section $D_f = 1.65$ (the Eq. (4) of Chapter 1), theoretical values t_1 and t_2 (t_1^T and t_2^T, accordingly) can be determined and compared with these parameters experimental values t_1^e and t_2^e. Such comparison is performed in Fig. 15, from which the theory and experiment good correspondence follows, confirming the made above main assumption of paper [34].

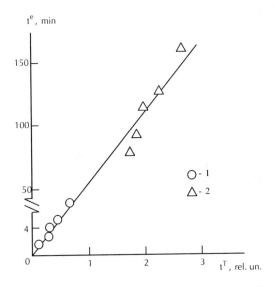

FIGURE 15 The relation between experimental t_1^e, t_2^e and theoretical t_1^T, t_2^T values t_1 (1) and t_2 (2) for DMDAACh polymerization.

Let consider further DMDAACh molecular weight rapid growth reasons on the initial section of the kinetic curve $Q_{(t)}$. The following relationship of dynamical scaling can be written for polymerization degree N or MM [39]:

$$N \sim MM \sim c_0^{1+d/(d-D_f)} t^d. \tag{31}$$

Let us perform the estimation of MM change on the initial section ($D_f = 2.45$ and $t_1 = 4.5$ min) and the section of autoacceleration ($D_f = 1.65$ and $t_2 = 30$ min) for DMDAACh with $c_0 = 4.5$ mol/l. The estimation according to the Eq. (31) gives for t_1 MM $\approx 1.49 \times 10^6$ relative units and for t_2 MM $\approx 3.43 \times 10^6$ relative units. Thus, on the initial section approx. 30% polymer common MM is formed. The experimental data (Fig. 14) suppose, that on the initial section approx. 38% polymer common MM is formed, that is close to the theoretical estimation. The similar results were obtained for other values c_0 [34].

Let us consider the reasons of reaction rate ϑ_r different dependences on reactive medium initial viscosity η_0 on the initial section and autoacceleration. As it was noted above, on the first from the indicated sections $\vartheta_r - \eta_0^{1/2}$ and on the second one $\vartheta_r - \eta_0$. Strictly speaking, the dependence of ϑ_r on η_0 is linked with the dependence of ϑ_r on translational friction coefficient f_0 of macromolecular coils in solution [40]. As it is known [40]:

$$f_0 \sim \eta_0 . \qquad (32)$$

For clusters (macromolecular coils) the value f_0 can be written as follows [40]:

$$f_0 \sim K_f N^{1/D_f} , \qquad (33)$$

where the coefficient K_f is determined according to the equation [40]:

$$K_f = \frac{1}{a\rho^{1/D_f}} . \qquad (34)$$

In its turn, in the Eq. (34) a is an aggregating particles (monomers) size, ρ is dimensional constant.

Since for the same polymer (in the given case — DMDAACh) a and ρ are constants, then from the Eqs. (32)–(34) it follows [34]:

$$f_0 \sim \eta_0 \sim N^{1/D_f} . \qquad (35)$$

Hence, for the initial section $\eta - N0.408$ and for the autoacceleration section $\eta - N0.606$, that defines the dependences of ϑ_r on η_0 for the indicated sections (at the same N).

Polymerization kinetics can be described by the fractal Eq. (26) of Chapter 1. Replacing for the initial section the value η_0 by N0.408 at N = 300 and for the autoacceleration section η_0 by N0.606 also at N = 300 and changing accordingly the constant coefficient in the Eq. (26) of Chapter 1, the transition from the first section to the second one can be described within the frameworks of this equation, but with different value D_f for the indicated sections. Such description example is adduced for DMDAACh with c_0 = 4.97 mol/l. As one can see, the polymerization process kinetics is predicted by the fractal Eq. (26) of Chapter 1 at the known process constants c_0, f_0 and D_f availability.

The reaction rates ϑ_r can be calculated according to the plot $Q_{(t)}$ data in transition point from the initial section to the autoacceleration section at t = 2.5 min for c_0 = 4.97 mol/l (Fig. 16) as this plot slopes from the right and from the left from the indicated point. These ϑ_r values are equal to 0.65 and 3.1%/min, i.e., their ratio is equal to approx. 0.21. The value ϑ_r^T can be determined theoretically as Q derivative by time and then from the Eq. (26) of Chapter 1 it can be obtained [34]:

$$\vartheta_r^T \sim \eta_0 c_0 t^{\left(1-D_f\right)/2} . \qquad (36)$$

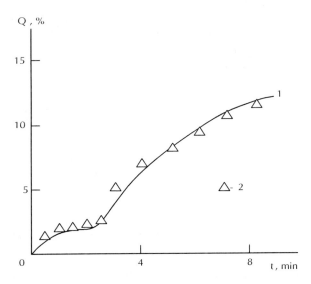

FIGURE 16 The simulation of kinetic curve conversion degree-reaction duration Q_{-t} for DMDAACh polymerization with c_0 = 4.97 mol/l. 1-the experimental curve, 2-calculation according to the Eq. (26) of Chapter 1.

The comparison of theoretical values ϑ_r^T on the initial section and on the auto-accleration section gives the values 4.43 and 21.1 relative units, accordingly, that corresponds completely to the experimental values ϑ_r ratio. Thus, the reaction rate increasing on the autoacceleration section is due to D_f decreasing from approx. 2.50 up to 1.65, that induces corresponding f_0 enhancement according to the Eq. (35) as well.

Let us consider in conclusion the reason of macromolecular entanglements network formation at the transition to the autoacceleration section. It is obvious, that such network formation requires contact of macromolecular coils in solution and their interpenetration. It should be supposed, that the higher coils interpenetration distance l is the denser, the network of macromolecular entanglements is. Within the frameworks of fractal analysis the value l is determined as follows [41]:

$$ l \sim b \left(\frac{R_g}{b} \right)^{2(d-D_f)/d}, \tag{37} $$

where b is a lower linear scale of fractal behavior.

Assuming b = 1, the value l at D_f = 2.45 (initial section) and D_f = 1.65 (auto-acceleration section) can be calculated. In this case the values l, equal to 5.3 and 90.9 relative units, accordingly, were obtained. Hence, D_f change at the transition to the autoacceleration section results to l growth in about 17 times, that should tell accordingly on macromolecular entanglements density increasing.

Hence, the stated above results demonstrated, that both autoacceleration effects and accompanied it effects within the frameworks of irreversible aggregation models can be explained with the aid of one obvious enough assumption, namely, by the aggregation (polymerization) mechanism change from mechanism particle–cluster (monomer-macromolecular coil) to mechanism cluster–cluster (coil–coil). This transition is due to reaching by coil gyration radius of the critical value that makes impossible this mechanism further action. All effects, accompanied autoacceleration beginning, are also defined by this aggregation mechanism change and, accordingly, macromolecular coils structure, characterized by its fractal dimension.

The authors [42–44] continued to study autoacceleration effect at DMDAACh radical polymerization within the frameworks of fractal and scaling approaches. In paper [39], two limiting asymptotic regimes for the description of dynamical scaling of the aggregates growth in bath with particles initial concentration c_0 were considered. For these regimes definition the authors [39] introduced the length characteristic scale ξ, defining the transition from aggregate fractal behavior to nonfractal one. The value $=R_c$ is determined according to the Eq. (48) of Chapter 1. The two regime $R_g \ll \xi = R_c$ and $R_g \gg \xi = R_c$ were considered in dynamical scaling, where R_g is the macromolecular coil gyration radius. The first regime is characterized by coil density, which is much greater reactive medium density, and aggregate size (polymerization degree for macromolecular coil) N is determined as follows [39]:

$$N(c_0, t) \sim (c_0 t)^{D_f/(2+D_f-d)} . \tag{38}$$

The second regime is characterized by approximately the same aggregate and reactive medium densities and for it scaling is described by the Eq. (31). The relation between N and Q within the frameworks of irreversible aggregation models is described as follows. The temporal evolution of Rg in the model DLA cluster–cluster can be described by the Eq. (70) of Chapter 1. For the considered case all parameters excepting c_0, t and η_0 which in the Eq. (70) of Chapter 1 right-hand part can be accepted as constant ones and then [44]:

$$t \sim \left(\frac{\eta_0}{c_0} R_g \right)^{D_f} . \tag{39}$$

Since R_g and N are linked by the Eq. (1) of Chapter 1, then the Eq. (39) can be rewritten in the form [43]:

$$t \sim \left(\frac{\eta_0}{c_0} \right)^{D_f} N . \tag{40}$$

The substitution of the Eq. (40) in Eq. (26) of Chapter 1 gives the intercommunitation between N and Q [42]:

$$Q \sim c_0^{(D_f-1)/2} \eta_0^{(5-D_f)/2} N^{(3-D_f)/2} . \tag{41}$$

Now, using the Eq. (41), the dependence $Q_{(t)}$ in the two considered above regimes: $R_g \ll \xi$ and $R_g \gg \xi$, designated further as regime 1 and regime 2, accordingly, can be estimated. For each from the indicated regimes the dependences $Q_{(t)}$ were calculated for three values c0: 2.0, 4.0 and 5.5 mol/l. The values of all parameters at these calculations were accepted in relative units [42].

In Fig. 17 the dependences $Q_{(t)}$ in both regimes for $c_0 = 4.0$ and 5.5 mol/l, calculated by the indicated mode, are adduced. Let us note first of all, that the non-linearity is typical for all calculated dependences, i.e., the reaction rate increasing at t growth or autoacceleration effect, which is expressed more clearly in case of regime 2 and at higher values c_0. It is obvious, that the plots $Q_{(t)}$ nonlinearity is due to the power dependences in the Eqs. (31), (38) and (41), and therefore there are no reasons to consider autoacceleration effect as a critical phenomenon. The dependences $N_{(t)}$, adduced in Fig. 18, find out similar tendencies.

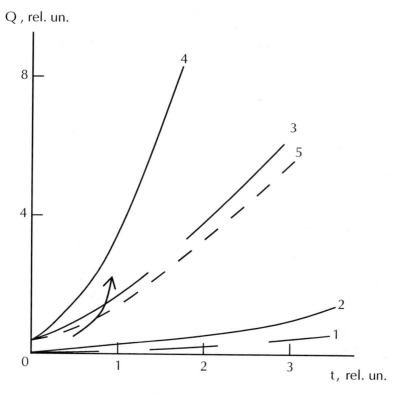

FIGURE 17   The dependences of conversion degree Q on reaction duration t for initial monomers concentration c_0: 4.0 (1, 2) and 5.5 (3, 4) mol/l. The calculation for regime 1 (1, 3) and regime 2 (2, 4). 5 — the calculation for regime 2 and $c_0 = 5.5$ mol/l at $D_f = 2.0$.

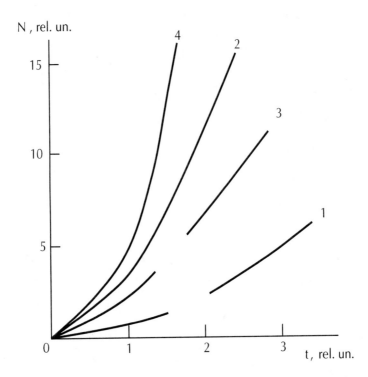

FIGURE 18 The dependences of polymerization degree N on reaction duration t for initial monomers concentration c_0: 4.0 (1, 2) and 5.5 (3, 4) mol/l. The calculation for regime 1 (1, 3) and regime 2 (2, 4).

Let us consider several features of autoacceleration effect. Its availability is independent on the initial monomer concentration. In Fig. 19, the dependences $Q_{(t)}$ for $c_0 = 2.0$ mol/l are adduced and one can see, that for this low enough monomer the reaction rate increasing at t growth is observed. The sole distinction of the dependences $Q_{(t)}$ for $c_0 = 2.0$ and 5.5 mol/l is these dependences approaching for the regimes 1 and 2 in case of smaller monomer initial concentration c_0. The value c_0 reduction up to 0.5 mol/l eliminates the distinction between regimes 1 and 2 (Fig. 20). The impossibility of regime 2 in diluted solutions realization is the cause of this effect. As it is known [25], for such solutions $\xi \to \infty$ and therefore the condition of regime 2 realization $R_g \gg \xi$ becomes impossible [43].

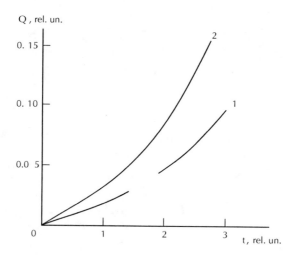

FIGURE 19   The dependences of conversion degree Q on reaction duration t for initial monomer concentration c_0 = 2.0 mol/l. The calculation for regime 1 (1) and regime 2 (2).

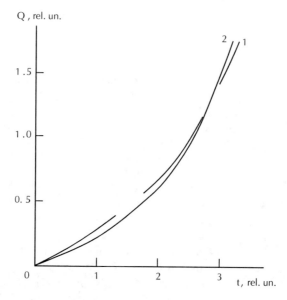

FIGURE 20   The dependences of conversion degree Q on reaction duration t for initial monomer concentration c_0 = 0.5 mol/l. The calculation for regime 1 (1) and regime 2 (2).

One from the autoacceleration effect causes the interpenetration of compressed up to θ-size macromolecular coils with stationary entanglements network formation is assumed [32]. Let us consider this effect influence within the frameworks of the proposed treatment. As it is known [25], in θ-conditions the excluded volume effects are screened completely and the coil fractal dimension $\bar{D}_f$ in this case is determined according to the Eq. (23), which for $D_f = 1.65$ and $d = 3$ gives $\bar{D}_f = 2.0$, i.e., the coil dimension in θ-conditions [45]. The dependence Q(t) at the value $\bar{D}_f$ instead of D_f using in the Eqs. (31) and (41) for regime 2 and $c_0 = 5.5$ mol/l is adduced in Fig. 17 (a dotted line). As one can see, the effective fractal dimension increasing from $D_f = 1.65$ up to $\bar{D}_f = 2.0$ results to aggregation regime change from regime 2 to regime 1. The reasons of this are obvious — the indicated dimension increasing is defined by coil compactization and enhancement of its density ρfr, expressed by the Eq. (73) of Chapter 1. The simple estimation at arbitrary $R_g = 10$ relative units shows that fractal dimension increasing from 1.65 up to 2.0 results to approximately two-fold ρfr growth, i.e., to ρfr enhancement in respect to reactive medium density, that is typical for regime 1. Let us note in this connection, that the opposite effect, namely, autoacceleration sharp intensification, can be due to the transition from regime 1 to regime 2, as it is shown by the arrow in Fig. 17.

Besides, the proposed treatment allows to taken into account monomer type influence on autoacceleration effect by the monomer length l_m using. In this case ξ definition looks like [39]:

$$\xi \sim l_m \left(c_0 l_m^d \right)^{-1/(d-D_f)}. \tag{42}$$

The simple estimation according to the Eq. (42) shows, that lm increasing from 5 up to 10 relative units results to ξ reduction in about 2.3 times, that means that the regime of aggregation approaches to asymptotic regime 2 (at $R_g = $ const).

In Fig. 21 the experimental dependences $Q_{(t)}$ for DMDAACh with $c_0 = 4.0$ and 5.5 mol/l (the solid lines) are adduced. The comparison of Figs. 17 and 21 demonstrates complete qualitative analogy of the experimental and theoretical dependences $Q_{(t)}$. What is more, quantitative relation of these curves can be obtained easily by Fig. 17 temporal scale normalization (by the multiplication by 5). The normalized theoretical data are shown in Fig. 21 by points and one can see that they correspond well to the experimental results.

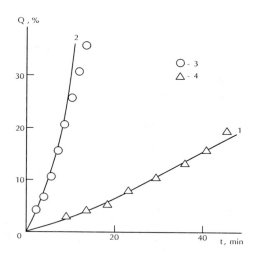

FIGURE 21   The comparison of the experimental (1, 2) and theoretical (3, 4) dependences $Q_{(t)}$ for initial monomer concentration c_0: 4.0 (1, 3) and 5.5 (2, 4) mol/l.

At present the autoacceleration effect absence is supposed at certain synthesis conditions, namely, at c_0 and 0 small values [1]. The last parameter is not included in the Eqs. (31) and (38) at all, although the curves $N_{(t)}$ discover the autoacceleration effect (Fig. 18). As to monomers concentration c_0, even for small c_0 the autoacceleration effect is expressed clearly enough (Fig. 19), although at these conditions it can be realized at large t. It is necessary to indicate, that the proposed in work [39] and used here technique does not take into account monomer exhausting in reaction course, that is displayed inevitably in practice and results to reaction rate essential reduction. In other words, the effects of autoacceleration and monomer exhausting act in opposite directions and at small c_0 the possibility of the first effect by the second one suppression.

Hence, the stated above results have shown that the autoacceleration effect is not presented as the critical phenomenon and is due to the power dependence $Q_{(t)}$ specific character, which can be obtained within the frameworks of irreversible aggregation models. Its manifestation extent is defined by the parameters c_0 and D_f (and also η_0), i.e., by the parameters, which control diffusive processes in solutions. In the general case structure formation conceptions cannot be used for the autoacceleration effect explanation.

In work [46], the temporal dependence of concentration of 1,1'-dimethyl–4,4'-bipyridine monocations, diffusing freely in solution and are low-molecular substances, was studied. Since macromolecular coils in solution are fractal objects, then the question arises, for how much the obtained in work [46] temporal dependences of reagent concentration decay are applicable to high-molecular substances and how macromolecular coil fractality can change the obtained in work [46] relations concentration-reaction duration. The authors [47] elucidated these questions on the example of DMDAACh radical polymerization in water solutions.

As the radical polymerization model the authors [47] considered the reaction, in which polymer macromolecular coils P were diffused in solution, consisting of statistically located nonsaturating traps T, which were also DMDAACh coils. At coils P and T contact the coil P disappears, forming the larger coil T. Such reactions can be described with the aid of the Eqs. (86) and (88) of Chapter 1.

In Fig. 22, the dependence of $c_{(t)} = (1-Q)$ on reaction duration t at three values of the initial monomers concentration c0 is adduced. As one can see, c_0 increasing results to much stronger decay of $(1-Q)$ and, besides, to much stronger curvature of these plots. To considered the compliace of the experimental plots with the Eq. (88) of Chapter 1 in Fig. 23 the dependence of $(1-Q)$ on t is adduced in logarithmic coordinates for DMDAACh at the same values c_0. As one can see, at $c_0 = 0.646$ kg/l this dependence is practically linear, at c_0 enhancement up to 0.826 kg/l the weak curvature appears and at $c_0 = 0.888$ kg/l the plot curvature is expressed clearly. Let us consider now the possibility of the adduced in Fig. 23 dependences description within the frameworks of the Eq. (88) of Chapter 1. The exponent in the Eqs. (87) and 88 of Chapter 1 is defined by the dimension of Euclidean space d only and at d = 3 it is equal to 0.6. In case of fractal objects the necessity of introduction in the exponent of their fractal dimension, in the given case — a macromolecular coil fractal dimension D_f, is obvious. This follows from the key rule [11]: fractal objects behavior description is correct with the aid of three dimensions only, in a number of which the fractal (Hausdorff) dimension D_f is included. In work [48] it has been shown, that in case of aggregation cluster–cluster the exponent for reaction rate is equal to $(2D_f-d)/D_f$ and this means, that for the conversion degree it will be equal to $(3D_f-d)/D_f$. Since $(1-Q)$ change is identical to Q change, then for it the exponent $(3D_f-d)/D_f$ should also be accepted, that at d=3 and $D_f=1.65$ gives 0.847. Let us note, that for Euclidean objects (low-molecular substances) their behavior description requires d knowledge only that the Eqs. (87) and (88) of Chapter 1 express. For fractal objects, besides the used d and D_f, a the third dimension (spectral (fracton) dimension ds [11]) application is required formally. However, for macromolecular coils the values D_f, ds and d are linked by the Eq. (39) of Chapter 1. In virtue of this circumstance the two from them only are independent parameters, that defines d and D_f application correctness on the considered above exponent.

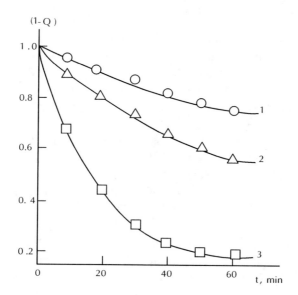

FIGURE 22 The dependence of monomers concentration decay $(1 - Q)$ on reaction duration t for DMDAACh at initial monomers concentration c_0: 0.646 (1), 0.826 (2) and 0.888 (3) kg/l.

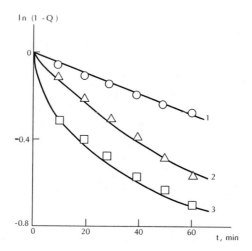

FIGURE 23 The dependences of monomers concentration decay $(1 - Q)$ on reaction duration t in logarithmic coordinates, corresponding to the Eq. (88) of Chapter 1, for DMDAACh at initial monomers concentration c_0: 0.646 (1), 0.826 (2) and 0.888 (3) kg/l. The curves were calculated according to the Eq. (88) of Chapter 1, the points are experimental data.

In Fig. 23, the comparison of the calculated with exponent 0.847 instead of approx. 0.6 [46] application dependences ln $(1-Q)$ on t according to the Eq. (88) of Chapter 1 with experimental data is also adduced. As one can see, at the coefficients A and B proper choice the good correspondence of theory (the Eq. (88) of Chapter 1) and experiment is obtained. Let us note, that exponent increasing in the Eq. (88) of Chapter 1 from 0.6 for Euclidean objects up to 0.847 for fractal ones means more rapid decay of monomers contents with time and, hence, more rapid polymerization reaction realization at other equal conditions.

In work [46], it has been shown, that the coefficient A value in the Eqs. (86) and (88) of Chapter 1 should be proportional to diffusivity D_d. The value D_d of macromolecular coil in solution can be estimated as follows [49]:

$$D_d = \frac{kT}{6\pi\eta_0 R_g \alpha} \text{ ,}$$ (43)

where k is Bolzmann constant, T is temperature, η_0 is reactive medium initial viscosity, R_g is macromolecular coil gyration radius, d is coefficient, defined by boundary conditions on coil surface (accepted usually equal to one [49]).

Since in the Eq. (1) of Chapter 1 proportionality sign is used, then the value D_d is determined in relative units at the condition $T = \text{const}$ [47]:

$$D_d \sim \frac{1}{\eta_0 R_g} \text{ .}$$ (44)

In Fig. 24, the dependence $A(D_d^{-1})$ is adduced, which turns out to be linear. As it is known [17, 18], η_0 increasing and, hence, D_d reduction (the Eq. (43)) results to Q growth and, accordingly, to $(1-Q)$ reduction. Therefore, D_d decreasing should result to A absolute value growth, that is confirmed by Fig. 24 plot.

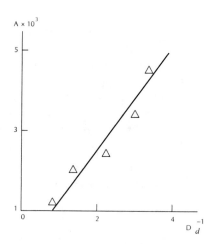

FIGURE 24   The dependence of coefficient A in the Eq. (88) of Chapter 1 on diffusivity D_d for DMDAACh.

In Fig. 25 the dependence of B absolute value on η_0^2 and the linear correlation is obtained here again. B value characterizes the system spatial fluctuations (heterogeneity) degree. η_0 increasing reduces macromolecular coils mobility (the Eq. (43)) and fixes the indicated fluctuations, that explains the adduced in Fig. 25 correlation $B(\eta_0^2)$.

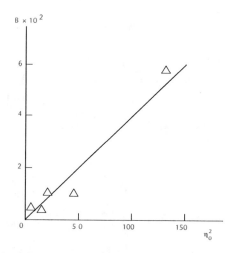

FIGURE 25   The dependence of coefficient B in the Eq. (88) of Chapter 1 on monomers solution initial viscosity η_0 for DMDAACh.

Let us note, that for two lower monomer concentrations, adduced in Figs. 22 and 23 (c_0 = 0.646 and 0.826 kg/l) the contributions of singular and nonsingular parts of function ln (1–Q)(t) are practically equal, whereas for c_0 = 0.888 kg/l the contribution of the second one predominates and makes up approx. 94 %. As it follows from the Figs. 22 and 23 data, this results to significantly more rapid (1–Q) decay and, hence, to polymerization reaction rate enhancement.

Hence, the stated above results confirmed the possibility of radical polymerization description with the aid of the general Eq. (88) of Chapter 1. The initial monomer concentration increasing results to system spatial fluctuation intensification owing to diffusivity reduction and function (1–Q) (t) singular part contribution increasing. Reacting objects (macromolecular coils) factuality results to an exponent in the Eq. (88) of Chapter 1 singular part change and, hence, to radical polymerization rate and completeness degree change.

Many important processes, realized in disordered systems, have nonexponential behavior on large times [50]. This postulate is related completely to the reactions, realized in the indicated systems. Such behavior to a considerable extent is defined by both spatial and temporal disorder availability, which can be described within the frameworks of fractal models [50]. In case of polymerization reactions these processes are linked with large-scale fluctuations (heterogeneity) of macromolecular coils distribution in solution. In its turn, this heterogeneity should be controlled by such generally accepted in practice characteristics as the initial monomer concentration c_0 and its solution initial viscosity η_0. In the long run just these parameters define polymerization process characteristics: polymer conversion degree Q and molecular weight MM. The authors [51] performed the research of large-scale spatial fluctuations appearance possibility in DMDAACh radical polymerization process and their influence on the final synthesis process characteristics within the frameworks of fractal models [46, 50].

For DMDAACh radical polymerization process simulation the authors [51] chose bimolecular reaction [50]:

$$A + B \rightarrow 0 , \qquad (45)$$

at the condition $D_A = D_B$, where D_A and D_B are reagents A and B diffusivities, accordingly.

Further the authors [51] supposed, that DMDAACh aggregation mechanism corresponds to the hierarchical model [52], in which two similar clusters are consolidated, forming a larger one and so on. Within the frameworks of such model the conditions, which are necessary for the Eq. (45) correctness, are fulfilled automatically. Further it is supposed, that two clusters A and B consolidation results to their annihilation, since they on this stage cease to exist. Each new stage of hierarchical aggregation can be considered separately and successively.

Using this model, the dependence of particles (monomers) concentration decay, equal to (1–Q), on temporal scale of polymerization t can be constructed. Such dependence in double logarithmic coordinates for four values c_0 is presented in Fig. 26. As one can see, the adduced dependences at large enough t can be described by the following relationship [51]:

$$1-Q \sim t^{d_s/4} \ , \tag{46}$$

where d_s is spectral dimension, characterizing reactive medium connectivity degree [15].

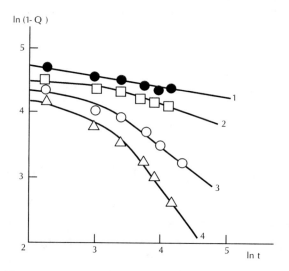

FIGURE 26   The dependences of monomers concentration decay (1–Q) on reaction durationt in double logarithmic coordinates for DMDAACh with initial monomers concentration c_0: 0.646 (1), 0.727 (2), 0.808 (3) and 0.888 (4) kg/l.

c_0 increasing within the indicated in Fig. 26 range results to growth of the values ds, calculated from the slope of linear plots, from approx. 1.0 up to 5.3. This means, that at some c_0 values the critical value $d_s = 4$ is reached [50]. Then the exponent in the Eq. (46) is written as $\gamma d_s/4$, where γ is characteristic of temporal (energetic) disorder and, since $d_s = 4$, then the Eq. (46) can be written as follows [51]:

$$1 - Q \sim t^{-\gamma} . \tag{47}$$

The transition from the Eq. (46) to the law (2.47) at DMDAACh polymerization reaction description is realized within the range of $c_0 \approx 0.727\text{–}0.808$ kg/l. As it is known [50], at the critical value ds=4 large-scale fluctuations appear in the system. Their influence on DMDAACh synthesis process will be considered lower.

The sharp increasing ds at c_0 growth should be linked with the same enhancement of the initial solution viscosity η_0 [1]. The diffusivity D_d is determined according to the Eq. (43), from which it follows, that η_0 increasing in about 4 times at c_0 change from 0.646 up to 0.888 kg/l [1] will sharply decrease D_d and, hence, will form spatial fluctuations [46], that should result to d_s enhancement. In Fig. 27 the dependence $d_s^{1/2}(\eta_0)$ is adduced, confirming this supposition, which is expressed analytically as follows [51]:

$$d_s = 10^{-2} \eta_0^2 . \tag{48}$$

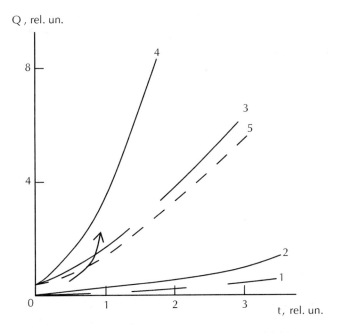

FIGURE 27 The dependence of spectral dimension d_s on initial monomers solution viscosity η_0 for DMDAACh.

Let us consider now, how the found above transition to large-scale spatial fluctuations, linked with increasing c_0 (and, hence, η_0) and having a place within the range of $c_0 = 0.727–0.808$ kg/l, can influence on synthesis characteristics Q and MM. In Fig. 28, the dependences $Q_{(c0)}$ and $MM_{(c0)}$ are shown and the shaded region shows the indicated above transition range.

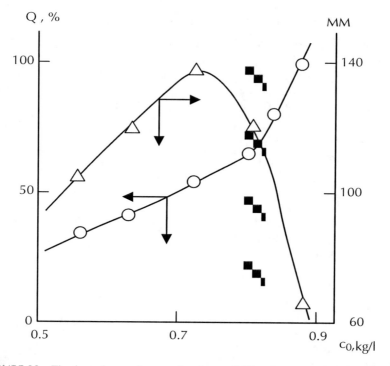

FIGURE 28   The dependence of conversion degree Q (1) and average-viscous molecular weight MM (2) on initial monomers concentration c_0 for DMDAACh. The shaded region indicates the transition range.

As it follows from the adduced in Fig. 28 data, the combined consideration of the curves $Q_{(c0)}$ and $MM_{(c0)}$ allows to reveal two essentially differing regions of their change. At large-scale spatial fluctuations absence, i.e., at $c_0 < 0.727$ kg/l, the similar and slow enough Q and MM increasing at c_0 growth is observed. At the indicated fluctuations availability, i.e., $c_0 > 0.808$ kg/l, more rapid change of both dependences is observed and, what is more, now they are changed antibately. Let us also note, that in the transition region (shaded region) the dependence $Q_{(c0)}$ fold is observed. Thus, the transition to large-scale spatial fluctuations, which is due to c_0 growth (and, as consequence, by η_0 increasing and D_d reduction) results to principal changes of the dependences $Q_{(t)}$ and $MM_{(t)}$: in the first case rapid growth Q at t increasing is observed and conversion degree reaches the limiting value, close to 1.0, and MM begins to reduce at c_0 growth monomer sharply enough at that.

The monomer surplus formation in one region of reactive medium and its deficiency in the other, that is spatial fluctuation, is such effect cause. Large-scale fluctuations mean the indicated regions large enough sizes. In other words, in regions of the first type a sharp local growth c_0 occurs. Such effect should result to Q growth and macromolecular coil gyration radius reduction [53] or polymer molecular weight decrease that is observed experimentally (Fig. 28).

Hence, the stated above results demonstrated the possibility of appearance in DMDAACh radical polymerization process of temporal (energetic) disorder. The transition to density large-scale fluctuations is defined by the initial solution viscosity sharp increasing at monomer concentration growth that so sharply moderates diffusive processes, fixing spatial and temporal heterogeneity. In its turn, these key changes influence strongly on polymerization process main parameters: conversion degree Q and molecular weight MM.

The dependences of self-diffusivity D_{sd} of chain Gaussian macromolecules study on different factors has large significance for both the development of notions about macromolecules mobility in solution and at polymers physics and chemistry diverse tasks solution [54]. However, up to now the main attention has been paid to the dependence of D_{sd} on solution concentration c and macromolecule molecular weight MM. Nevertheless, there is one more very important factor — macromolecular coil structure, characterized by its fractal dimension D_f. Therefore, the authors [55] obtained the relation between D_{sd} and D_f and also elucidated the influence of D_f change in DMDAACh radical polymerization process on this reaction rate variation. These studies were performed for two DMDAACh syntesises, differing by the initiator (ammonium persulphate (PSA)) contents cPSA. The value cPSA makes up 5×10^{-4} mol/l for DMDAACh-1 and 5×10^{-3} mol/l for DMDAACh-2 [1].

In Fig. 29, the dependences of D_f on reaction duration t for DMDAACh polymerization at both indicated above conditions are adduced. As one can see, the two dependences $D_f(t)$ are similar qualitatively and their distinction consists of only in more slow reaction realization for DMDAACh-1 (the kinetic parameters for DMDAACh synthesis adduced conditions differ on about two orders [1]). Let us note, that for DMDAACh-2 the aggregation mechanism change is observed twice: within the range of $t < 2$ min mechanism particle–cluster is changed by diffusion-limited mechanism cluster–cluster and within the range of $t = 5$–10 min opposite change occurs [21]. It is important to note, that the last mechanisms change occurs at constant initial monomers concentration c_0 and practically constant value MM [1].

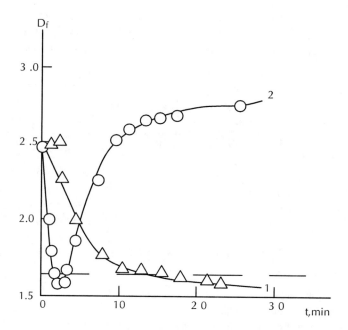

FIGURE 29   The dependences of macromolecular coil fractal dimension D_f on reaction duration t for DMDAACh-1 (1) and DMDAACh-2 (2).

The value of self-diffusivity D_{sd} for macromolecular coil can be estimated according to the following simple relationship [56]:

$$D_{sd} \sim A_1 \left(\frac{\varphi^*}{\varphi} \right)^{\beta} D_R, \tag{49}$$

where A_1 is the numerical coefficient, φ^* is the threshold value of concentration φ, D_R is macromolecule friction coefficient, for which intramolecular hydrodynamical interaction among chain elements is not taken into account, β is the exponent, determined as follows [56]:

$$\beta = \frac{2-\nu}{3\nu-1}, \tag{50}$$

where v is Flory's exponent in the relationship gyration radius — chain po-lymerization degree (the Eq. (2) of Chapter 1). The value φ^* can be deter-mined according to the equation [56]:

$$\varphi^* = 0.35c^* [\eta] = 0.35\left[0.64 \ln^2 \left(2\sqrt{N}\right)\right], \qquad (51)$$

where c^* is the monomers critical concentration, $[\eta]$ is intrinsic viscosity, N is polymerization degree. Estimating the value $[\eta]=0.80$ dl/g, N=315 [1], the value φ^* can be determined. From these estimations the ratio value $\varphi^*/\varphi = 1.32$ was obtained [55].

The Eqs. (1), (2) and (50) of Chapter 1 combination allows to express the exponent β as D_f function [55]:

$$\beta = \frac{2D_f - 1}{3 - D_f} \qquad (52)$$

Now, assuming A_1 = const and D_R = const, the value D_{sd} in relative units can be expressed as a function of D_f, using the Eqs. (49) and (52). In Fig. 30, the dependence of the polymerization rate $9r$ on D_{sd} is adduced for DMDAACh-2. As one can see, the correlation $9_r(D_{sd})$ is linear and passes through coordinates origin. Such dependence was expected, since the existing polymers synthesis theories provide the direct pro-portional relationship between the reaction rate constant k_r (or 9_r) and D_{sd}. So, in work [57] two equations, obtained in the chaotic phases approximation:

$$k_r = 4\pi D_{sd} R_g \left(\frac{9}{5} + \frac{R_g}{aN}\right)^{-1} \qquad (53)$$

and in the average density approximation:

$$k_r = 4\pi D_{sd} R_g \left[1 - \left(\frac{R_g}{4\pi aN}\right)^{1/3}\right] a \qquad (54)$$

are adduced, where R_g is a coil gyration radius, a is a reacting particles reac-tive radius.

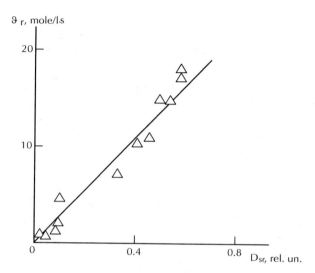

FIGURE 30 The dependence of polymerization rate ϑ_r on macromolecular coil self-diffusion coefficient D_{sd} for DMDAACh-2.

For DMDAACh radical polymerization it can be written [57]:

$$a = 2R_g . \tag{55}$$

In this case the members in brackets of the Eqs. (53) and (54) give very weak correction to the directly proportional dependence k_r on D_{sd}, particularly at large enough N [55].

As it is known [58], in case of a chemical reactions number, including polymers synthesis, at that the so-called steric factor p (p $\leq$ 1) plays an essential role. Within the frameworks of model notions the value k_r with the steric factor p appreciation can be described according to the Eq. (74) of Chapter 1. The combination of the Eqs. (74) and (17) of Chapter 1 allows to obtain the following relationship [55]:

$$p\left(\frac{8RT}{3\eta_0}\right)[A][B] = \frac{c_2}{t^{(D_f-1)/2}}, \tag{56}$$

where c_2 is the constant.

Assuming that the multiplier at p in the Eq. (56) left-hand is constant, the constant c_2 in this equation can be estimated from the known boundary conditions of the irreversible aggregation model [52]. This model supposes D_f growth at p decreasing, that corresponds to the tendency, which the Eq. (56) expresses. Assuming also that $D_f \approx 1.60$ corresponds to $p \approx 0.7$ [52], let us obtain the following simple relationship [55]:

$$p \approx \frac{3.06}{t^{(D_f-1)/2}} \,,$$

(57)

where reaction duration t is given in s.

In Fig. 31, the dependence $p_{(t)}$, calculated according to the Eq. (57), is adduced for DMDAACh. As it should be expected from the theoretical considerations within the frameworks of irreversible aggregation models [21], the value p is changed antibately to D_f (Fig. 29). At $D_f \approx 1.60$ $p \approx 0.69$, i.e., diffusion-limited aggregation (the diffusive regime) is observed and $D_f \approx 2.11$ $p \approx 0.05$, i.e., chemically limited aggregation (the kinetic regime) is observed.

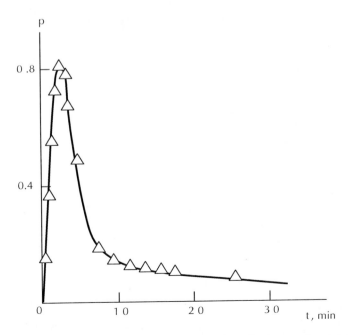

FIGURE 31 The dependence of steric factor p on polymerization reaction duration t for DMDAACh-2.

In Fig. 32, the dependence $\vartheta_r(p)$ is adduced for DMDAACh-2. ϑ_r (or k_r) linear growth p increasing is observed. This results have been expected too, as it has been shown above by the Eq. (74) of Chapter 1.

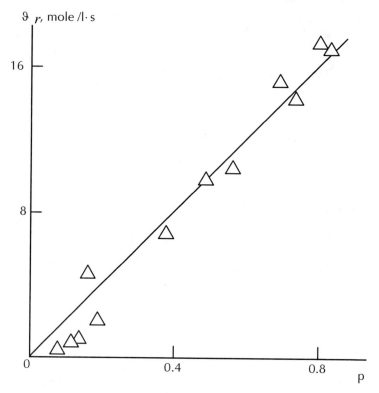

FIGURE 32   The dependence of polymerization rate ϑ_r on steric factor p for DMDAACh-2.

Hence, the adduced above results have shown the essential dependence of DMDAACh self-diffusion coefficient D_{sd} in solution on macromolecular coil structure, characterized by fractal dimension D_f. It is important to note, that D_{sd} change, which is due to coil structure variations, can occur at the constant solution viscosity and polymer molecular weight. The aggregation mechanism change results to fractal dimension D_f and steric factor p essential changes. The indicated changes induce strong variation of polymerization rate.

As it is known [1], the kinetic curves Q_{-t} in the radical polymerization case have sigmoid shape and break down into three sections, usually called an initial stage, autoacceleration and finish polymerization sections (see Fig. 13). The conversion degree values Q at transition from one stage to another can be denoted as Q_1 and Q_2 (accordingly to times t_1 and t_2 of the indicated stages termination, Fig. 13) and the final conversion degree — as Q_f. It has been shown earlier [59] that at self-organization from previous point of structure unstability to subsequent on the adaptation universal algorithm is realized:

$$A_m = \frac{z_i}{z_{i+1}} = \Delta_i^{1/m},$$

(58)

where z_i and $z_i + 1$ are critical values of governing parameter, controlling structure formation; their ratio determines the system adaptation measure Am to structural reshaping; m is reshaping's number, Δ_i is structure stability measure, maintaining constant at m change from m = 1 up to m = m_{max}.

Assuming, that in radical polymerization process, including DMDAACh polymerization, the conversion degree is governing parameter, the value Am can be determined for the two last from the indicated stages as the ratio of Q boundary values, i.e., $A_m^{aut} = Q_1/Q_2$ (autoacceleration stage) and $A_m^{fin} = Q_2/Q_f$ (finish polymerization stage). The calculations showed these parameters large distinction: $A_m^{aut} = 0.232$ and $A_m^{fin} = 0.915$. The self-similarity constants of evaluating systems, determinated by the gold proportion law, can be used for values Δi and m estimation. The tabulated data allow to obtain $\Delta i = 0.232$ and m = 1 for autoacceleration stage and i = 0.232 and m = 16 for finish polymerization stage. Thus, the system stability is the same on both stages, but its adaptivity measure changes sharply at the expense of possible reshaping's number variation [60].

One can suppose, that DMDAACh macromolecular coil adaptivity to polymerization reaction realization will be defined by its structure, characterized by fractal dimension D_f: the smaller D_f is, the easier and more intensively the indicated reaction realizes and the smaller adaptivity measure A_m is. In Fig. 33 the dependence $A_m(D_f)$ is adduced for the initial reagent three concentrations c_0 in DMDAACh case.

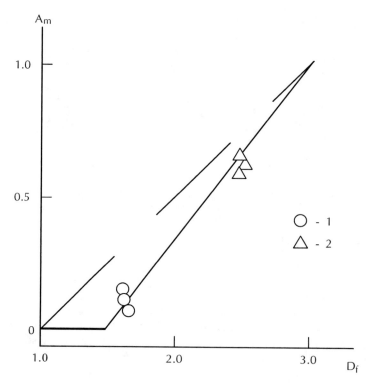

FIGURE 33 The dependence of adaptivity measure Am on macromolecular coil fractal dimension D_f for DMDAACh. The stages of autoacceleration (1) and finish polymerization (2).

As one can see, the linear correlation, extrapolating to $A_m = 0$ at $D_f = 1.5$ (transparent coil [45]) and to $A_m = 1.0$ at $D_f = 3.0$ (compact globule [45]) is obtained. In other words, the factor, defining macromolecular coil adaptivity measure to reactive medium, is a purely physical parameter: its fractal dimension D_f [60]. The indicated limiting values, obtained from the extrapolation data, were expected from the most general considerations: at $D_f = 1.5$ leaking coils are transparent for one another, that gives polymerization the greatest efficiency and for compact globule chemical reactions are impossible. Let us note that reagents chemical parameters variation this picture does not change: this picture the limiting values Q_1, Q_2 and Q_f can only be changed, that is observed at c_0 variation. The relation between Am and D_f can be written analytically as follows [60]:

$$A_m = 0.667\left(D_f - D_f^e\right).$$

$$(59)$$

As it is known [17], within the frameworks of fractal analysis polymerization kinetics is described by the general Eq. (27) of Chapter 1. The combination of the Eqs. (27) and (59) of Chapter 1 allows to obtain the following expression [60]:

$$Q \sim t^{0.75(1-A_m)} . \tag{60}$$

As it was expected, at $A_m = 0$ reaction was realized most intensively (Q_{approx}. t0.75) and at $A_m = 1.0$ reaction does not realize at all ($Q \sim$ const). Let us note, that the known condition for reaction in polymeric solutions $D_f \leq 2.28$ [6] applies restrictions on the value m: $m \leq 2$, i.e., restricts structure adaptivity.

In Fig. 33 the hypothetical dependence $A_m (D_f)$ within the frameworks of the Eq. (27) of Chapter 1 is shown by a shaded line, which assumes $A_m \neq 0$ at $D_f = 1.5$ ($A_m = 0$ at $D_f = 1.0$ and $A_m = 1.0$ at $D_f = 3.0$). The discrepancy of the dependences $A_m (D_f)$, obtained within the frameworks of synergetics and fractal analysis, supposes, that self-organizing structures are macromolecular coils (or their totalities), but not monomer molecules, the reaction of which is realized within the range of $D_f = 1.0–1.5$ [60].

2.3 THE LIMITING CHARACTERISTICS OF RADICAL POLYMERIZATION

General for physical processes notions as scaling and universality classes are used widely in irreversible aggregation models [21]. The two main scaling types, considered in irreversible aggregation models (static and dynamical ones) exist [37]. The first from them gives process final characteristics irrespective of its kinetics and the second one describes process parameters change in aggregation course, as a rule, as a time function. The authors [61] applied the dynamical scaling for the description of DMDAACh radical polymerization in water solutions at different initial monomer concentrations kinetics.

The main scaling relationship, used for DMDAACh polymerization kinetics description, is the Eq. (27) of Chapter 1. Let us consider the necessary parameters, which should be included in the indicated relationship proportionality coefficient. First of all, the initial monomers concentration c_0, expressed in mass parts, is such a parameter — it is obvious, that c_0 increasing causes Q enhancement. The second parameter should characterize monomer solution properties, which control diffusion processes in reaction course. The monomers solution relative viscosity η_0, determined in work [33], was chosen as such a parameter. Then the Eq. (27)

of Chapter 1 becomes suitable for quantitative description of DMDAACh kinetic curve Q_{-t} in autoacceleration and finish polymerization sections [61]:

$$Q = 0.32 c_0 \eta_0 t^{(3-D_f)/2}, \qquad (61)$$

where Q is given in per cent, η_0 is dimensionless value, c_0 is given in mass parts, t — in minutes and the numerical coefficient 0.32 was obtained by the theoretical and experimental data matching method and served for the agreement of parameters with different dimensions, including in the Eq. (61).

In Fig. 34, the dependence of Q on $c_0 \eta_0 t$ for all used values c_0 is adduced for the reaction autoacceleration section in double logarithmic coordinates for dynamical scaling, described by the Eq. (61), correctness checking. As one can see, the data set well enough on one common straight line with the slope of approx. 0.7, that confirms the Eq. (61) application correctness. From this equation it follows, that the straight line slope is equal to $(3-D_f)/2$ and this allows to determine the value D_f with the adduced above slope appreciation as approx. 1.60. This result corresponds well to the value $D_f = 1.65$, determined according to the Eq. (4) of Chapter 1, that confirms in addition the proposed scaling correctness.

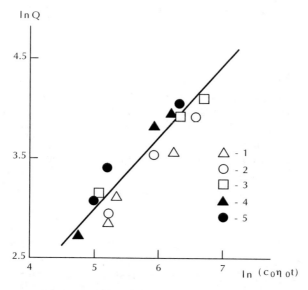

FIGURE 34   The dependences of conversion degree Q on parameter $c_0 \eta_0 t$ value for DMDAACh in double logarithmic coordinates at initial monomer concentration c_0: 0.646 (1), 0.729 (2), 0.808 (3), 0.829 (4) and 0.888 (5) kg/l (autoacceleration section).

In Fig. 35, the dependence, similar to the shown one in Fig. 34 is adduced, but for the final polymerization stage. In comparison with the plot of Fig. 34, the adduced in Fig. 35 dependence has two principal distinctions. Firstly, this figure data are not described by a sole straight line, but break down into a parallel straight lines family. This situation is explained by the fact, that final polymerization section begins with not zero value Q, but with some finite value Q_1, corresponding to the conversion degree on autoacceleration stage end. Since for different c_0 the values Q_1 have different magnitudes [32], then this results to parallel displacement of straight lines for different c_0. Besides, this means, that scaling for autoacceleration section should be written as follows [61]:

$$Q = Q_0 + 0.32c_0\eta_0 t^{(3-D_f)/2},\qquad(62)$$

where Q_0 is the conversion degree, corresponding to the initial section termination and for the final polymerization section as follows [61]:

$$Q = Q_0 + Q_1 + 0.32c_0\eta_0 t^{(3-D_f)/2}.\qquad(63)$$

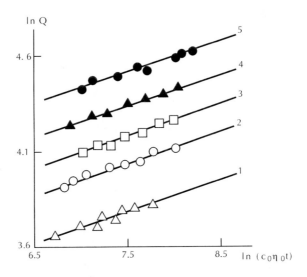

FIGURE 35   The dependences of conversion degree Q on parameter $c_0\eta_0 t$ value for DMDAACh in double logarithmic coordinates at initial monomer concentration c_0: 0.646 (1), 0.729 (2), 0.808 (3), 0.829 (4) and 0.888 (5) kg/l (final polymerization section).

The common linear dependence for autoacceleration section (Fig. 34) is due to small values Q_0. Nevertheless, in this case the points for higher concentrations c0 are located systematically above in respect of points for smaller c_0 as well (Fig. 34).

The second distinction of the Fig. 35 plots is a much smaller value of a straight lines slope in comparison with the Fig. 34 plot, equal to approx. 0.2. This means, that the value D_f in final polymerization section is equal to approx. 2.60, i.e., it corresponds to aggregation mechanism particle–cluster [62]. This result involves two important conclusions. Firstly, at the transition from autoacceleration section to final polymerization section the change of aggregates universality class occurs — the mechanism cluster–cluster on autoacceleration section is replaced by the mechanism particle–cluster on final polymerization section with the corresponding change of fractal dimension D_f. Secondly, such universality class change assumes, that the point of transition between autoacceleration and final polymerization sections is the gelation point for DMDAACh [26].

Now, using the obtained data, the kinetic curve Q_{-t} can be described quantitatively with the aid of the Eqs. (62) and (63), excluding the initial section. The theoretical and experimental kinetic curves Q_{-t} for c_0 = 0.646, 0.729 and 0.888 kg/l comparison is adduced in Fig. 36. As one can see, the offered scaling approach gives the exact description of the experimental curves Q_{-t} (the discrepancy of theory and experiment makes up average 6%).

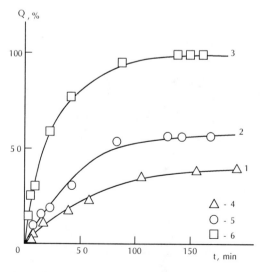

FIGURE 36   Comparison of the experimental (1–3) and calculated according to the Eqs. (62) and (63) kinetic curves for DMDAACh at initial monomers concentration c_0: 0.646 (1, 4), 0.808 (2, 5) and 0.888 (3, 6) kg/l.

The times of transition from initial section to autoacceleration section t_1 and from autoacceleration section to final polymerization section t_2 are also not arbitrary values. In Fig. 37, the dependences of t_1 and t_2 on η_0 are adduced, which turn out to be linear and showed transition time reduction at monomer initial solution growth.

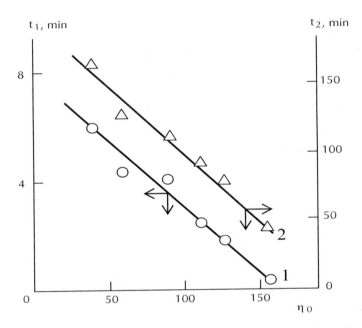

FIGURE 37   The dependences of transition times t_1 (1) and t_2 (2) on relative viscosity η_0 of initial monomers solution for DMDAACh.

Hence, the dynamical scaling of DMDAACh radical polymerization allows to describe quantitatively the kinetic curve Q_{-1} and can be used for its prediction. In this approach base the key physical principles and models (scaling, universality classes, irreversible aggregation models, fractal analysis) are placed. The three key process properties, characterized reactive centers concentration (c_0), diffusive characteristics of reactive medium (η) and "accessibility" degree of reactive centers (D_f), are used for the kinetic curves description. It is supposed, that the offered approach will be valid for the description of radical polymerization process of any polymer.

The authors [63] studied the possibility of static scaling application for conversion degree description in a radical polymerization process on the example of three polymers — polystyrene (PS), poly(methyl methacrylate) (PMMA) and

DMDAACh. For obtaining static scaling of conversion degree Q as a function of the initial monomers concentration c_0 and monomer characteristics the authors [63] used the chemical dimension (or spreading dimension) d_1 notion [11, 64]:

$$A_N \sim n_w^{d_l} ,$$

(64)

where A_N is accessible for reaction sites number, nw is the number of walks steps on the given structure from coordinates origin.

As it has been shown in work [64], the parameter A_N can be considered as particles number reacting in process course, and nw as "chemical distance", i.e., as the shortest way by cluster particles between two its points.

In such treatment the value A_N is proportional to conversion degree Q and initial monomer (particles) concentration c_0 [63]:

$$A_N \sim c_0 Q .$$

(65)

A particles (monomer links) number can be determined, proceeding from the value c_0 and monomer link value V_m, which for PS and PMMA are adduced in work [65] and the value V_m for DMDAACh can be calculated according to the Eq. (55) of Chapter 1. The value D_f for the considered polymers can be calculated according to the Eq. (4) of Chapter 1 accounting for the fact, that the exponent a in Mark–Kuhn–Houwink equation is equal to 0.73 for PS and PMMA [66, 67] and for DMDAACh a=0.82 [68], that gives D_f = 1.73 for the first two polymers and D_f = 1.65 for DMDAACh. Such values D_f allow to attribute the considered polymers macromolecular coils to aggregates, formed by mechanism cluster–cluster [21]. In its turn, this means that for such clusters the chemical dimension dl is equal to 1.42 [69]. Thus, estimating the necessary for calculation parameters (AN, n_w and d_l), the Eq. (64) correctness can be checked for the considered polymers. This is made in Fig. 38 in the form of the plot $Qc_0(n_w^{d_l})$, from which linear and passing through coordinates origin plot, corresponding to the Eq. (64). Thus, the dependence $Qc_0(n_w^{d_l})$ allows to predict the value Q, reached in radical polymerization process, proceeding from three parameters: initial monomer concentration c_0, monomer link structure and aggregation process type. Let us note that final results do not depend on polymerization process kinetics.

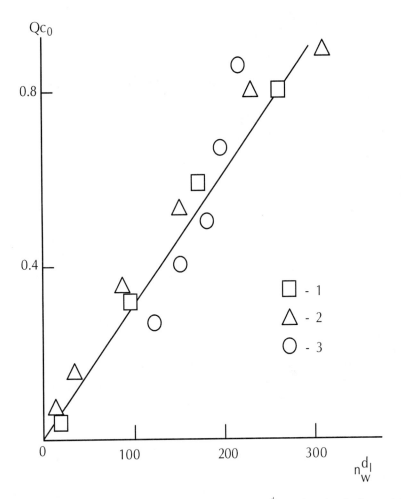

FIGURE 38   The relation between parameters Qc_0 and $n_w^{d_l}$ (explanation in the text) for PMMA (1), PS (2) and DMDAACh (3).

In Fig. 39 the kinetic curves Q_{-t} for the three considered polymers are adduced at approximately the same value c0 (approx. 0.8 kg/l), taken from works [1, 66, 67]. As one can see, despite differing (both qualitatively and quantitatively) kinetic curves, static scaling in the Eq. (64) form gives the final result (the polymer limiting conversion degree Q) correct description.

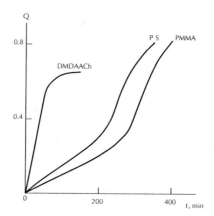

FIGURE 39 The kinetic curves conversion degree — polymerization duration Q_{-t} for PMMA (1), PS (2) and DMDAACh (3).

The authors [70] studied the limiting molecular weight MM variation in DM-DAACh radical polymerization process. In Fig. 40 the dependence of average viscous molecular weight MM on the initial monomers concentration c_0 is shown for DMDAACh (the experimental data are shown by a solid line). As it follows from the plot of Fig. 40, the dependence $MM(c_0)$ has the extreme character, reaching maximum at $c_0 \approx 4$ mol/l, after that MM rapid reduction about twice is observed within narrow enough range c_0 (4.5–5.5 mol/l).

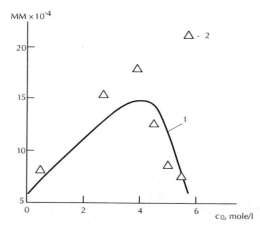

FIGURE 40 The comparison of experimental (1) and calculated according to the Eq. (28) of Chapter 1 (2) dependences of average viscous molecular weight MM on initial monomers concentration c_0 for DMDAACh.

For the dependence $MM(c_0)$ theoretical explanation the authors [70] used the Eq. (28) of Chapter 1. The dependence of reactive medium initial viscosity η_0 on c_0 for DMDAACh is adduced in [1]. At small c_0 (up to approx. 4.0 mol/l) η_0 weak increasing at c0 growth is observed and then within the range of $c_0 = 4.5–5.5$ mol/l rapid (in about 3 times) η_0 enhancement occurs. Therefore, adduced in the expression (1.28) the ratio c_0/η_0 passes through maximum, corresponding on axis c_0 to MM maximum. In Fig. 40 the experimental and calculated according to the Eq. (28) of Chapter 1 dependences $MM(c_0)$ comparison, is adduced showing both qualitative and quantitative correspondence to irreversible aggregation model cluster–cluster (the Eq. (70) of Chapter 1) and experiment. The proportionality coefficient in the Eq. (28) of Chapter 1 was obtained by the matching method. Hence, experimentally observed MM of DMDAACh essential reduction within the monomer concentration range of $c_0 = 4.5–5.5$ mol/l is due to sharp enhancement of monomer solution viscosity 0 within this range of c_0 [70].

The second variant of the dependence $MM(c_0)$ description within the frameworks of irreversible aggregation cluster–cluster model foresees the scaling using Eqs. (66), (26) and (27) of Chapter 1. The estimations according to the indicated relationships have shown, that the value of exponent α in Eq. (66) of Chapter 1, characterizing polymer-solvent system diffusive properties, changes similarly to MM and also passes through maximum at $c_0 = 4.5$ mol/l. It is natural to expect some kind of dependence of α on c_0 and η_0. However, the two last parameters do not have maximum (unlike α) and therefore the authors [70] obtained the dependences of α on parameters c_0 and η_0 combination in two variants: as their product $c_0\eta_0$ and ratio c_0/η_0 to half power. The dependence of α on $(c_0/\eta_0)1/2$ for DMDAACh, adduced in Fig. 41, turns out to be linear and passing through coordinates origin. It is obvious enough, that the initial reactive solution viscosity enhancement should reduce diffusible particles mobility, which is expressed by dependence, adduced in Fig. 41. However, the same dependence assumes its increasing at c_0 growth and this means a strong enough attraction among DMDAACh macromolecular coils availability.

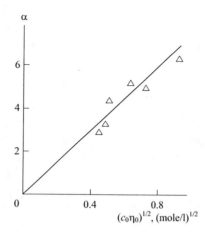

FIGURE 41   The dependence of exponent α (in relative units) on the parameter $(c_0/\eta_0)1/2$ value for DMDAACh.

In Fig. 42 the dependence of the exponent α on product $c_0\eta_0$ is adduced, which confirms the made above suppositions. On the first section of the curve $\alpha(c_0\eta_0)$, corresponding to $c_0 \leq 4.5$ mol/l, the value rises in virtue of c_0 increasing, since η_0 growth on this section is not so significant. At $c_0 > 4.5$ mol/l η_0 sharp enhancement results to essential reduction of diffusible macromolecular coils mobility, characterized by the exponent α.

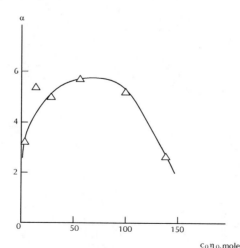

FIGURE 42   The dependence of exponent α (in relative units) on parameter $c_0\eta_0$ value for DMDAACh.

Let us return now to macromolecular coil fractal dimension value D_f for DM-DAACh, which, as it has been noted above, is equal to approx. 1.65. D_f small value assumes, that in DMDAACh radical polymerization process, simulated by diffusion-limited aggregation cluster–cluster, the following conditions should be performed [70]:

- The probability of clusters (particles) reaction at their first contact is very high and close to one [71];
- Attractive forces among particles is their interaction prevalent factor [72];
- The availability of diffusible particles motion Brownian trajectory in reaction process [73];
- Disaggregation and restructurization absence in reaction course [74].

Hence, purely physical model of irreversible diffusion-limited aggregation cluster–cluster can be used successfully for chemical processes description (including a quantitative one) in the considered case — radical polymerization. Let us note, that in paper [75] the Eq. (70) of Chapter 1 was used for gold colloidal particles aggregation, i.e., purely physical process, description.

2.4 A GELATION IN THE RADICAL POLYMERIZATION PROCESS

Large attention was always given to the study gelation during polymers synthesis. However, for radical polymerization processes there are several points of view on definition and place of a position of this phenomenon on a kinetic curve Q_{-t}. So, there is a gel point definition as a point, as the one in which the solution begins to display some properties of a solid body (elasticity, brittleness etc.) [38]. During radical polymerization processes as a gel-effect point the transition from the initial section of curve Q_{-t} to the autoacceleration section is called [31]. In physics of polymers there is a more exact gelation point definition as the point, as the one in which the cluster subtending a system (spreading from one edge of a system up to another) is formed [26, 76]. Different approaches and definitions of this effect can cause uncertainty in its explanation. Therefore, the authors [77, 78] performed the determination of a gelation point position on a kinetic curve Q_{-t} in the terms of definition [26, 76] and with the help of irreversible aggregation models on the example of DMDAACh radical polymerization.

The gelation effect treated as formation of a cluster, subtending the system, results to essential changes of parameters describing this system, in a gelation point [26]. So, the clusters diffusion rate (macromolecular coils rate) with mass m is given by the Eq. (66) of Chapter 1. In a gelation point the exponent α changes the sign from negative on positive one. This means, that up to a gelation point the greater mobility has the smaller clusters, and behind a gelation point the greatest mobility will have a subtending cluster, since its weight is much greater than the weight of any macromolecular coil [26, 76].

The similar sharp change of the exponent ω in Smoluchovskii scaling relationship in a gelation point happens within the frameworks of the kinetic gelation conception [26]. At ω < 0.5 for all distribution of the clusters sizes is valid the common scaling and average cluster size regularly grows with time, and behind a gelation point (ω > 0.5) this scaling is broken: the greatest cluster fastest grows [26].

And at last, in a gelation point the static geometrical characteristics of a system are changed. Theoretically [26] and experimentally [27] it is shown, that the value D_f changes from values about approx. 1.5–2.1 typical for a macromolecular coil in solution [45], up to approx. 2.5, characterizing Witten–Sander cluster [26].

The determined according to the Eq. (4) of Chapter 1 value D_f for DMDAACh is equal to approx. 1.65. Such value D_f simply defines the formation mechanism of a macromolecular coil DMDAACh in solution as diffusion-limited aggregation of a cluster–cluster mechanism [21]. Within the frameworks of this model the dependence of a coil gyration radius R_g on a reaction duration t is determined by the Eq. (26) and the exponent z in the indicated expression linked with D_f and α by the Eq. (27). Within the frameworks of the fractal analysis the relation between Rg and polymerization degree N (or molecular weight MM) is given by the Eq. (1) of Chapter 1. And at last, the exponents and ω are linked among themselves by the Eq. (28).

Thus, using the dependences $MM_{(t)}$, plotted according to the data of paper [32] and shown in Fig. 43, and the Eqs. (1), (26)–(28), of Chapter 1 can determine the values of exponents z, α and ω and, on the basis of the reviewed above their boundary values, define whether gelation happens or does not happen on the given section of a kinetic curve Q_{-t}. One from the methods of gelation point determination and corresponding to it the time tg is a graphic one, where the indicated point can be obtained as a point of interception of the tangents to autoacceleration and finish polymerization sections (Fig. 13).

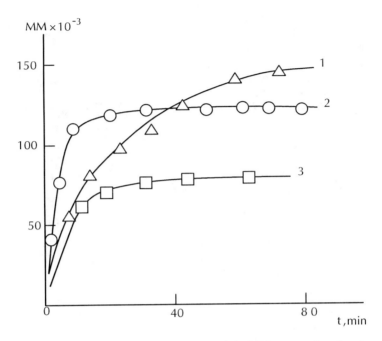

FIGURE 43 The dependences of molecular weight MM on reaction duration t for DMDAACh polymerization at the initial monomers concentration c_0: 4.5 (1), 5.0 (2) and 5.5 (3) mol/l.

In the Table 2, the values of exponents z, α and ω for an autoacceleration and finish polymerization sections are listed. As it follows from the data of the Table 2, on the autoacceleration section $\alpha < 0$ and $\omega < 0.5$. This means, that on the whole autoacceleration section gel formation does not happen and solution is in a sol-state. The opposite picture is observed for exponents α and ω on the finish polymerization section. In this case $\alpha > 0$ and $\omega < 0.5$, that means gel formation on this section. Let us note, that on the finish polymerization section the value of MM reaches asymptotic values, i.e., the condition MM = const is reached. Therefore, for time corresponding to finish polymerization section beginning, value R_g ~ MM1/1.65 and for the time, corresponding to this stage termination — R_g ~ MM1/2.45. Or else, just on this section a transition from $D_f = 1.65$ (mechanism a cluster–cluster) to $D_f \approx 2.5$ (mechanism a particle–cluster) occurs. Hence, the adduced above gelation time tg definition is quite justified.

TABLE 2   Values of exponents of an irreversible aggregation process on autoacceleration and finish polymerization sections of kinetic curve Q_{-t} and relative mobility of clusters on an autoacceleration section for DMDAACh.

c_0, mol/l	Autoacceleration section				Finish polymerization		
	z	α	ω	ϑ	z	α	ω
4.5	3.10	−1.48	−0.435	33.1	−0.392	0.757	0.682
5.0	5.95	−3.60	−1.497	0.25	−0.458	0.783	0.695
5.5	8.63	−4.84	−2.120	0.015	−0.598	0.839	0.723

In the Table 2, values of clusters rates ϑ in solution (in relative units) calculated according to the Eq. (66) of Chapter 1 are also listed. It is possible to see, that c_0 increasing results to very sharp reduction ϑ. At c_0 increasing from 4.5 up to 5.5 mol/l value ϑ decreases in about 2300 times. Besides from the data of the Table 2 it follows, that c_0 increasing does not mean an approaching of polymerized solution to a gelation point: the larger c_0, the less ω.

Thus, within the frameworks of an irreversible aggregation models on the example of DMDAACh it is demonstrated, that the gelation happens on a transient section from an autoacceleration section up to a finish polymerization section. On an autoacceleration section the polymer is in a sol-state. Therefore, the usage of the term "gel-effect" with reference to the beginning of the autoacceleration section does not look not quite justified.

The authors [79, 80] performed the gelation time estimation within the frameworks of an irreversible aggregation models. According to the indicated treatment a macromolecular coil critical gyration radius is determined by the Eq. (48) of Chapter 1 and R_g growth kinetics is described by the Eq. (70) of Chapter 1. The indicated relationships combination allows to obtain the equation for a gelation time tg estimation accounting for the fact, that in the Eq. (70) of Chapter 1 parameters k, T and m_0 are constant ones [79]:

$$t_g = K_1 \frac{\eta_0}{C_0^{\,d/(d-Df)}},$$

(66)

where K_1 is certain constant, which is correct for DMDAACh synthesis concrete conditions.

The values η_0 were taken according to the data of work [1] and the value D_p as was discussed above, was accepted equal to approx. 2.5. In Fig. 44, comparison of the experimental t_g and calculated according to the Eq. (66) t_g^r values of the gelation time is adduced for DMDAACh radical polymerization with five different initial monomer concentrations within the range of 4.0–5.5 mol/l.

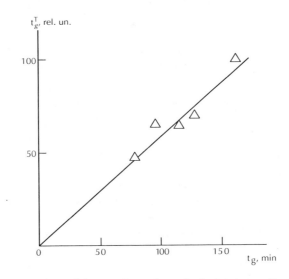

FIGURE 44 Comparison of the experimental t_g and calculated according to the Eq. (66) t_g^T values of gelation time for DMDAACh.

As one can see, between parameters tg and t_g^T the good linear correlation, passing through coordinates origin, is observed. This confirms the Eq. (66) correctness and allows to estimate the value $K_1=1.82$. Hence, the value t_g is defined (in case of the same polymer) by the values of the initial solution viscosity η_0 and the initial monomer concentration c_0 at the condition T = const. It is easy to see, that tg reduction will be observed at T enhancement. For different polymers monomer link mass m_0 increasing will result to t_g corresponding growth.

Hence, the offered fractal model allows to perform the quantitative estimation of gelation time in case of DMDAACh radical polymerization. Since in the model any specific suppositions about polymer structure or synthesis mode are absent, then it is expected, that it will have general correctness at the similar phenomena description. The theory and experiment agreement confirms the offered fractal model correctness, at any rate, in DMDAACh case.

2.5 THE ANALYSIS OF MOLECULAR WEIGHT DISTRIBUTION

The authors [81–83] studied the general laws of molecular weight distribution (MWD) of DMDAACh within the frameworks of the dynamical distribution function for a mechanism cluster–cluster. In works [76, 84, 85] the dynamical scaling description of clusters size distribution in the model of diffusion-limited aggregation cluster–cluster was offered. On this basis for the description of MWD for DMDAACh the dynamical scaling function can be used like this [84]:

$$n_s(t) \sim S^{-2} t^z \,, \tag{67}$$

where t is the reaction duration, accepted equal to the gelation time tg, z is an exponent.

Let us consider the estimation methods of the parameters, including in the Eq. (67). The values $n_s(t_g)$ are determined from the experimental curves of MWD $W_i(N)$ (Fig. 45) as follows. It was assumed that N=S and then the fraction of the polymer corresponding to S is equal to summary weight fraction of the polymer W_i. By dividing W_i on S_i and multiplying by maximum conversion degree Q the value ns(tg) or N_i for S_i can be obtained. The exponent z is determined according to the Eq. (27) and the macromolecular coil diffusivity D_d is calculated according to the Eq. (43). In its turn, the value D_d within the frameworks of an irreversible aggregation models can be determined as follows [37, 84]:

$$D_d \sim S^\alpha \,. \tag{68}$$

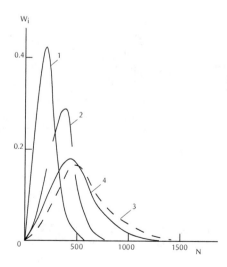

FIGURE 45   MWD curves for DMDAACh, synthesized at initial monomers concentration c_0: 1.0 (1), 2.5 (2), 4.0 (3) and 5.0 (4) mol/l.

From the Eqs. (43) and (68) comparison it is easy to obtain [83]:

$$\alpha \sim -1n\,\eta_0. \tag{69}$$

The values η_0 for monomer solutions, used at DMDAACh synthesis, are accepted in work [1].

As it was indicated in works [37, 84], the value α defines a distribution function shape. For $\alpha < -0.5$ the "bell-shaped" distribution curves, having maximum, are obtained. For $\alpha < -0.5$ the monotonously decreasing curves were obtained. As it follows from the data of Fig. 45, the experimental MWD curves have a bell-shaped form. Proceeding from absolute values η_0, the authors [83] supposed in the Eq. (69) the equality sign and in this case the values α vary within the limits of approx. $-1/-3$ (the values η_0 are given in relative units) and this range corresponds completely to MWD experimental curves shape (Fig. 45). Besides, as it has been shown in work [86], MWD curve maximum position Nmax is determined by the Eq. (67) of Chapter 1. It is easy to see, that c_0 increasing, resulting to η_0 growth, defines the absolute value α growth and Nmax enhancement, that agrees with the data of Fig. 45.

In Fig. 46, the dependences $S^2 n_s(St^z)$ in double logarithmic coordinates are shown for DMDAACh, synthesized at different c_0. As it follows from the data of this figure, all four MWD curves, shown in Fig. 45, are described by a sole generalized curve. This is the most important result, confirming an irreversible aggregation models using correctness for polymerization process description.

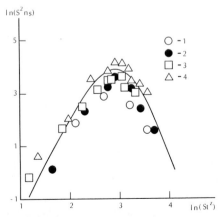

FIGURE 46 The generalized MWD curve in the form of function $S^2 n_s(St^z)$ in double logarithmic coordinates for DMDAACh, synthesized at initial monomers concentration c_0: 1.0 (1), 2.5 (2), 4.0 (3) and 5.0 (4) mol/l.

Using the generalized curve, shown in Fig. 46, MWD change as a function of time can be predicted theoretically. For this purpose at first we set the value $S = N$ and determine the value St^z at arbitrary t. Then from the plot of Fig. 46 the value S^2n_s is found corresponding to it and, according to the described above procedure Wi is determined. Then the same calculation is iterated for another S and so on. In Fig. 47, as the example comparison of MWD curves for $c_0 = 4.0$ mol/l is given at t $= t_g = 108$ min and $t = 53$ min. As one can see from the comparison of curves 1 and 3 in Fig. 47, t decrease ($t < t_g$) results to maximum of MWD curve displacement to lower molecular weights values and MWD narrowing. Besides, in Fig. 47 MWD curve for $c_0 = 1.0$ mol/l at $t - t_g$ (curve 3) is adduced and from the comparison of all three curves it follows, that t and c_0 decrease gives a similar effect.

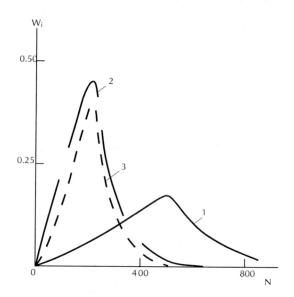

FIGURE 47 The experimental MWD curves at gelation time t_g for DMDAACh, synthesized at initial monomers concentration c_0: 4.0 (1) and 1.0 (3) mol/l. The calculated MWD curve at time 53 min for DMDAACh, synthesized at $c_0 = 4.0$ (2) mol/l.

Hence, the stated above results have shown the correctness of the description of MWD curves for polymers on the example of DMDAACh within the frameworks of the dynamical distribution function of an irreversible aggregation cluster–cluster model. The obtaining of the generalized distribution curve (Fig. 46) confirms the possibility of polymerization process description within the frameworks of the indicated models and allows to predict MWD change kinetics as a function of the initial monomers concentration c_0 and reaction duration t.

The authors [87–89] used for DMDAACh MWD curves theoretical description and determination of the factors influencing their shape, the model [48], successfully applied above for the similar curves description in case of polyarylates, obtained by different polycondensation modes (see section 1.2). As it has been noted above, since in the Eq. (64) of Chapter 1 the value λ does not influence on the distribution $P_s(N)$, then the indicated relationship assumes three main parameters, influencing on the distribution $P_s(N)$: a', b and σ_2. Each from the indicated parameters expresses the certain feature of polymerization process. So, the exponent a' is defined as a matter of fact by macromolecular coil structure, that follows directly from the Eq. (63) of Chapter 1. The value b characterizes a type and intensity of destruction processes. The parameter σ_2 is defined by the stochastic contribution in polymerization process and the dependence of σ_2 on c_0 can be assumed — the larger the initial monomers concentration is, the higher their random collisions probability is. All experimental MWD curves for DMDAACh have unimodal shape (Fig. 45), that assumes coils low mobility in solution or $\sigma_2 < 2/3$ [48].

Let us consider first of all the general aspects of influence of the three indicated parameters on MWD curves shape. For this purpose theoretical dependences $P_s(N)$ have been constructed with a serial variation when only one of the indicated parameters varies, where the other two are constant, that is adduced in Eq. (29)–31 of Chapter 1. Such simulation results were stated in Section 1.2.

The comparison of experimental and theoretical MWD curves has shown that these curves for DMDAACh, synthesized at $c_0 = 4.0$ and 5.0 mol/l, are simulated directly at the following parameters of a $P_s(N)$ curve: $D_f = 1.65$ (a'≈ 0.182), $\sigma_2 = 0.25$ and b = 1. Such simulation for $c_0 = 4.0$ mol/l is adduced in Fig. 48 (curve 6), from which a good theory and experiment agreement may be observed. For MWD curves at $c_0 = 1.0$ and 2.5 mol/l such direct simulation is cannot be obtained, since for them the position of maximum on MWD curve corresponds to $N_{max} = 175$ and 375, but for the theoretical curves such low values of N_{max} are not attained (Eq. (29)–31 of Chapter 1). Nevertheless, to achieve agreement between the theory and the experiment is possible by use of renormalization for these curves, i.e., by the Eq. (65) of Chapter 1 usage. The indicated renormalization was performed by N0 determination at the condition of $\tilde{N}$ equality to the theoretical value N_{max} and N — the experimental value of this parameter.

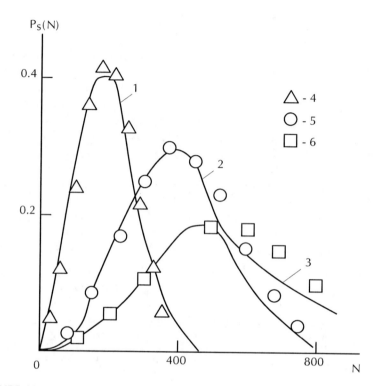

FIGURE 48 Comparison of the experimental (1–3) and calculated according to the Eq. (64) of Chapter 1 (Eqs. 4–6 of Chapter 1) MWD curves for DMDAACh at initial monomers concentration c_0: 1.0 (1, 4), 2.5 (2, 5) and 4.0 (3, 6) mol/l.

The shown in Fig. 48 excellent conformity of experimental and theoretical MWD curves for DMDAACh, synthesized at $c_0 = 1.0$ and 2.5 mol/l, confirms the performed renormalization correctness. In its turn, the indicated renormalization expresses one from the main properties of fractals — their automodelity [90].

As it is known [86], MWD curve in case of aggregation cluster–cluster depends on the diffusive characteristics of a system in many respects, namely on clusters mobility expressed by their rate ϑ, which can be estimated according to the Eq. (66) of Chapter 1. In its turn, the exponent of α value in this relation defines MWD curve maximum position according to the Eq. (67) of Chapter 1. One should expect that the clusters mobility ϑ will be the higher, the smaller initial monomers solution viscosity η_0 is. This supposition is confirmed by the dependence ϑ (η_0^{-2}), adduced in Fig. 49, which is linear and passes through coordinates origin.

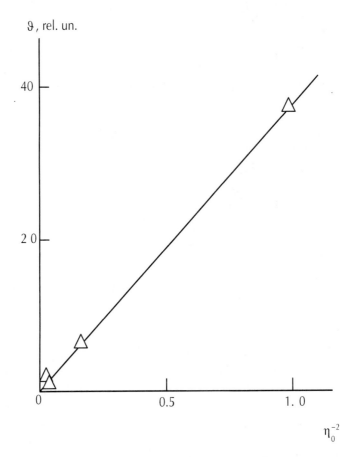

FIGURE 49   The relation between macromolecular coil mobility ϑ and initial monomer solution viscosity η_0 for DMDAACh.

As it has been mentioned above, one should expect an increase of the stochastic contribution in polymerization intensity or white noise σ_2 at the initial monomers concentration c_0 growth. This supposition is confirmed by the dependence $\sigma_2(\bar{n}_0^{1/2})$ shown in Fig. 50, which is also linear and passes through coordinates origin. The data of Figs. 49 and 50 demonstrate that the parameters of the Eq. (64) of Chapter 1 can be expressed through technological characteristics of the polymerization process, for example, c_0 and 0.

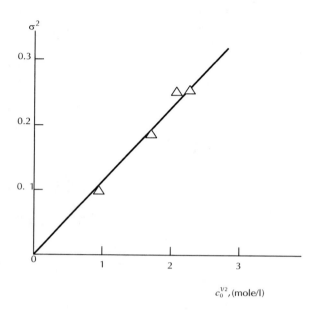

FIGURE 50 The relation between polymerization intensity white noise σ_2 and initial monomer concentration c_0 for DMDAACh.

Hence, the stated above results have shown that DMDAACh molecular weight distribution can be simulated and predicted within the frameworks of an irreversible aggregation model cluster–cluster. The shape and position of MWD curve are controlled by a number of factors, such as macromolecular coil structure, stochastic contribution of a coil environment to a polymerization intensity, and the level of destruction of a coil during its synthesis process. These factors can be linked by simple relationships to technological characteristics of the polymerization process, for example, c_0 and $_{\eta 0}$, that is essentially important for practical applications of the considered theoretical model [89] [23].

REFERENCES

1. Topchiev, D. A.; Malkanduev, Yu. A. The cationic polyelectrolytes of poly-*N,N*-dialkyl-N,N-diallylammonium halogenides: *Features of formation processes, properties and application.* Nal'chik, Kabardino–Balkarian State University, 1997, 181.
2. Gribel, T .; Kulicke, W.-M.; Hansemzaden A. Characterization of water-soluble polyelectrolytes as examined by poly(acrylamide-co-trimethylammonium-acrylate chloride) and establishment of structure-property relationships. *Colloid Polymer Sci.,* 1991, **269**(1), 113–120.

3. Novikov, V. U.; Kozlov, G. V. The fractal analysis of macromolecules. *Successes of Chemistry*, 2000, **69**(4), 378–399.
4. Esmurziev, A. M.; Malkanduev, Yu. A., Kozlov, G. V. *The fractal analysis of dimethyldiallylammonium chloride flocculating ability*. Herald of Kabardino–Balkarian State University, series chemical sciences, 2001, 4,101–104.
5. Kozlov, G. V.; Malkanduev, Yu. A.; Ulybina, A. S. *The Analysis of hydrodynamical properties of copolymers of acrylamide with trimethylammonium methylmethacrylatechloride solutions within the frameworks of fractal analysis*. Herald of Kabardino–Balkarian State University, series chemical sciences, 2003, 5, 175–181.
6. Family, F. Fractal dimension and grand universality of critical phenomena. *J. Stat. Phys.*, 1984, **36**(5/6), 881–896.
7. Meakin, P.; Stanley, H. E. Novel dimension-independent behavior for diffusive annihilation on percolation fractals. *J. Phys. A*, 1984, **17**(2), L173–L177.
8. Shogenov, V. N.; Kozlov, G. V. *The fractal clusters in physics-chemistry of polymers*. Nal'chik, Polygraphservice and T, 2002, 268.
9. Alexander, S.; Orbach, R. Density of states on fractals: "fractons". *J. Phys. Lett.* (Paris), 1982, **43**(17), L625–L631.
10. Kozlov, G. V.; Malkanduev, Yu. A.; Burmistr, M.,N.; Korenyako, V. A. The fractal model of reactive medium for dimethyldiallylammonium chloride radical polymerization. *Problems of Chemistry and Chemical Technology*, 2004, (4), 101–105.
11. Rammal, R.; Toulouse, G. Random walks on fractal structures and percolation clusters. *J. Phys. Lett.* (Paris), 1983, **44**(1), L13–L22.
12. Vilgis, T. A. Flory theory of polymeric fractals — intersection, saturation and condensation. *Physica A*, 1988, **153**(2), 341–354.
13. Novikov, V. U.; Kozlov, G.V. The principles of fractal approach to polymers structure. The non–Euclidean physics of polymers. In collection: *Applied synergetics, fractals and computer structure simulation*. Ed. Oksogoev A.A., Tomsk, TSU, 2002, 268–302.
14. Vannimenus, J. Phase transitions for polymer on fractal lattices. *Physica D*, 1989, **38**(2), 351–355.
15. Alexander, S.; Laermans, C.; Orbach, R.; Rosenberg, H. M. Fracton interpretation of vibrational properties of cross-linked polymers, glasses and irradiabed quartz. *Phys. Rev. B*, 1983, **28**(8), 4615–4619.
16. Gefen, Yu.; Aharony, A.; Alexander, S. Anomalous Diffusion on percolating clusters. *Phys. Rev. Lett.*, 1983, **50**(1),77–80.
17. Kozlov, G. V.; Shustov, G. B.; Zaikov, G.E. The fractal physics of the polycondensation processes. *J. Balkan Tribological Association*, 2003, **9**(4), 467–514.
18. Kozlov, G. V.; Shustov, G. B.; Zaikov, G. E. The fractal physics of the polycondensation processes. In book: *Handbook of Polymer Research, 20*. Ed. Pethrick, A.; Zaikov, G. New York, Nova Science Publishers, Inc., 2006, 37–83.
19. Grassberger, P.; Procaccia, I. The long time properties of diffusion in a medium with static traps. *J. Chem. Phys.*, 1982, **77**(12), 6281–6284.
20. Gladyshev, G. P.; Popov, V. A. Radical polymerization at karge conversion degrees. Moscow, *Science*, 1974, 263.

21. Kokorevich, A. G.; Gravitis, Ya. A.; Ozol–Kalnin, V. G. The scaling approach development at study of lignine supramolecular structure. *Chemistry of wood*, 1989, (1),3–24.

22. Kozlov, G. V.; Malkanduev, Yu. A.; Zaikov, G. E. Fractal kinetics of radical polymerization of dimethyl diallyl ammonium chloride. In book: Fractal Analysis of Polymers: From *Synthesis to Composites*. Ed. Kozlov, G.; Zaikov, G.; Novikov, V. New York, Nova Science Publishers, Inc., 2003, 79–87.

23. Kozlov, G. V.; Malkanduev, Yu. A.; Zaikov, G.E. Fractal kinetics of radical polymerization of dimethyl diallyl ammonium chloride. *J. Balkan Tribological Association*, 2003, **9**(3), 435–441.

24. Kozlov, G. V.; Malkanduev, Yu. A.; Zaikov, G. E. Fractal kinetics of radical polymerization of dimethyl diallyl ammonium chloride. *J. Appl. Polymer Sci.*, 2004, **91**(5), 3140–3143.

25. Muthukumar, M. Dynamics of polymeric fractals. *J. Chem. Phys.*, 1985, **83**(6), 3161–3168.

26. Botet, R.; Yullien, R.; Kolb, M. Gelation in kinetic growth models. *Phys. Rev. A*, 1984, **30**(4), 2150–2152.

27. Kobayashi, M.; Yoshioka, T.; Imai, M.; Iton, Y. Structural ordering on physical gelation of sindiotactic polystyrene dispersed in chloroform studied by time-solved measurements of small angle neutron scattering (SANS) and infrared spectroscopy. *Macromolecules*, 1995,28 22,7376–7385.

28. Kozlov, G. V.; Temiraev, K. B.; Shustov, G. B.; Mashukov, N. I. Modeling of solid state polymer properties at the stage of synthesis: fractal analysis. *J. Appl. Polymer Sci.*, 2002, **85**(6), 1137–1140.

29. Lachinov, M. B.; Dreval, V. E.; Kasaikin, V. A.; Simonyan, R. A.; Shipulina N. I.; Zubov V. P.; Kabanov, V. A. The intercommunication of structure formation and kinetics of methylmethacrylate radical polymerization at $ZnCl_2$ presence at large conversion degrees. *High–Molecular Compounds, A*, 1977, **19**(4), 741–749.

30. Lachinov, M. B.; Simonyan, R. A.; Georgieva, T. G.; Zubov, V. P.; Kabanov, V. A. Nature of gel effect in radical polymerization. *J. Polymer Sci.: Polymer Chem. Ed.*, 1979, **17**(2), 613–628.

31. Bityurin, N. M.; Genkin, V. N.; Zubov, V. P.; Lachinov, M. B. About gel-effect at radical polymerization. *High–Molecular Compounds. A*, 1981, **23**(8), 1702–1709.

32. Yanovskii, Yu. G.; Topchiev, D. A.; Barancheeva, V. V.; Malkanduev, Yu. A.; Kabanov, V. A. Study of features of dimethyldiallylammoniumchloride profound polymerization in water solutions with the aid of the dynamical mechanical spectroscopy method. *High–Molecular Compounds. A*, 1988, **30**(6), 1226–1233.

33. Topchiev, D. A.; Malkanduev, Yu. A.; Korshak, Yu. N.; Mikitaev, A. K.; Kabanov, V. A. Kinetics of N,N-dimethyl–N,N-diallylammoniumchloride in concentrated water solutions. *Acta Polymerica*, 1985, **36**(7), 372–374.

34. Kozlov, G. V.; Malkanduev, Yu. A.; Burya, A. I.; Sverdlikovskaya, O. S. Kinetics of dimethyldiallylammoniumchloride initial polymerization. *Problems of Chemistry and Chemical Technology*, 2003, 2, 73–77.

35. Forsman, W. C. Effect of segment-segment association on chain dimension. *Macromolecules*, 1982, **15**(4), 1032–1040.

36. Meakin, P. Formation of fractal clusters and networks by irreversible diffusion-limited aggregation. *Phys. Rev. Lett.*, 1983, **51**(13), 1119–1122.

37. Kolb, M. Unified description of static and dynamic scaling for kinetic cluster formation. *Phys. Rev. Lett.*, 1984, **53**(17),1653–1656.

38. Encyclopaedia of Polymers. Ed. Kargin, V. Moscow, *Soviet Encyclopaedia*,1, 1972, 1223

39. Hentschel, H. G. E.; Deutch, I. M,; Meakin, P. Dynamical scaling and the growth of diffusion-limited aggregates. *J. Chem. Phys.*, 1984, **81**(5), 2496–2502.

40. Chen, Z. −Y.; Deutch, I. M.; Meakin P. Translational friction coefficient of diffusion limited aggregates. *J. Chem. Phys.*, 1984, **80**(6), 2982–2983.

41. Hentschel, H. G. E.; Deutch, I. M. Flory-type approximation for the fractal dimension of cluster–cluster aggregates. *Phys. Rev. A*, 1984, **29**(3), 1609–1611.

42. Kozlov, G. V.; Malkanduev, Yu. A.; Burmistr, M. V.; Korenyako, V. A. Autoacceleration in radical polymerization process: the fractal analysis. *Problems of Chemistry and Chemical Technology*, 2002, 1, 76–79.

43. Kozlov, G. V.; Ozden, S.; Malkanduev, Yu. A.; Zaikov, G. E. Autoacceleration in the process of radical polymerization: fractal analysis. In book: *Fractals and Local Order in Polymeric Materials*. Ed. Kozlov, G.; Zaikov, G. New York, Nova Science Publishers, Inc., 2001, 11–19.

44. Kozlov, G. V.; Ozden, S.; Malkanduev, Yu. A.; Zaikov, G. E. Autoacceleration in the process of radical polymerization: fractal analysis. *Russian Polymer News*, 2002, **7**(3), 38–44.

45. Baranov, V. G.; Frenkel, S. Ya.; Brestkin, Yu. V. Dimensionality of a linear macromolecule different states. *Reports of Academy of Sciences of SSSR*, 1986, **290**(2), 369–372.

46. Djordjevič, Z. B. The observation of scaling in reaction with traps. In book: *Fractals in Physics*. Ed. Pietronero, L.; Tosatti, E. Amsterdam, Oxford, New York, Tokyo, North–Holland, 1986, 581–585.

47. Kozlov, G. V.; Malkanduev, Yu. A.; Mirzoeva, A. A. The observation of the scaling dependences in dimethyldiallylammoniumchloride radical polymerization process. *Reports of Adygskoi (Cherkesskoi) International Academy of Sciences*, 2002, **6**(1), 82–87.

48. Shiyan, A. A. The stationary distributions of macromolecular fractals mass in diluted polymer solutions. *High–Molecular Compounds, B*, 1995, **3**(9), 1578–1580.

49. Happel, J.; Brenner, G. Hydrodynamics at small Reynolds numbers. Moscow, *World*, 1976, 386.

50. Blumen, A.; Klafter, J.; Zumofen, G. Reactions in fractal models of disordered systems. In book: *Fractals in Physics*. Ed. Pietronero, L.; Tosatti, E. Amsterdam, Oxford, New York, Tokyo, North–Holland, 1986, 561–574.

51. Kozlov, G. V.; Malkanduev, Yu. A.; Filippova, Yu. A. *The influence of large-scale spatial fluctuations on parameters of dimethyldiallylammoniumchloride radical polymerization process*. Herald of Kabardino–Balkarian State University, series chemical sciences, 2003, 5, 57–62.

52. Jullien, R.; Kolb, M. Hierarchical method for chemically limited cluster–cluster aggregation. *J. Phys. A*, 1984, **17**(12), L639–L643.

53. Kozlov, G. V.; Shustov, G. B.; Temiraev, K. B. The dependence of molecular weight of aromatic copolyethersulphoneformal on weight of an elementary link: the fractal

analysis. In book: *Fractals and Local Order in Polymeric Materials*. Ed. Kozlov, G.; Zaikov, G. New York, Nova Science Publishers, Inc., 2001, 29–35.

54. Budtov, V. P. Physical Chemistry of Polymeric Solutions. Sanct–Peterburg, *Chemistry*, 1992, 384.

55. Kozlov, G. V.; Malkanduev, Yu. A.; Burya, A. I.; Burmistr, E. M. Fractal treatment of the dependences of macromolecular coils self-diffusion coefficient on their structure. *Problems of Chemistry and Chemical Technology*, 2003, 2, 69–72.

56. Budtov, V. P. About concentration dependence of the chain molecules self-diffusion coefficient. *High–Molecular Compounds, A*, 1986, **28**(12), 2575–2581.

57. Burlatskii, S. F Oshanin, G. S.; Likhachev, V. N. Diffusion-controlled reactions with polymer chains participation. *Chemical Physics*, 1988, **7**(7), 970–978.

58. Barns, F. S. The electromagnetic fields influence on chemical reaction rate. *Biophysics*, 1996, **41**(4), 790–802.

59. Ivanova, V. S. Synergetics. Strength and Fracture of Metal Materials. Moscow, *Science*, 1992, 160.

60. Kozlov, G. V.; Malkanduev, Yu. A.; Novikov, V. U. Synergetics of dimethyldiallylammoniumchloride radical polymerization. Proceedings of International interdisciplinary seminar *Farctals and Applied Synergetics*. Moscow, Publishers MSOU, 2003, 114–117.

61. Kozlov, G. V.; Malkanduev, Yu. A. Dynamical scaling for dimethyldiallylammonium-chloride radical polymerization. *Chemical Industry Today*, 2004, 2, 50–56.

62. Meakin, P. Diffusion-controlled cluster formation in 2–6-dimensional space. *Phys. Rev. A*, 1983, **27**(3), 1495–1507.

63. Malkanduev, Yu. A.; Kozlov, G. V. *Static scaling of conversion for radical polymerization*. Herald of Kabardino–Balkarian State University, series chemical sciences, 1999, 3, 31–33.

64. Vannimenus, J.; Nadal, J. R.; Martin, H. On the spreading dimension of percolation and directed percolation clusters. *J. Phys. A*, 1984, **17**(6), L351–L356.

65. Bershtein, V. A.; Egorov, V. M. Differential Scanning Calorimetry in Physics–Chemistry of Polymers. Leningrad, *Chemistry*, 1990, 256.

66. Bagdasar'yan, Kh. S. Radical Polymerization. Moscow, *Science*, 1966, 312.

67. Bemford, K.; Bar, B. U.; Jenkins, A.; Onione, P. *Kinetics of Radical Polymerization of Vinyl Monomers*. Moscow, Foreign Literature, 1961, 289.

68. Timofeieva, G. J.; Pavlova, S. A.; Wandrey, C.; Jaeger, W.; Hahn, M.; Linow, K. –J.; Görnitz, E. On the determination of the molecular weight distribution of poly(dimethyl diallyl ammonium chloride) by ultracentrifugation. *Acta Polymerica*, 1990, **41**(9), 479–483.

69. Meakin, P.; Majid, I.; Havlin, S.; Stanley, H. E. Topological properties of diffusion limited aggregation and cluster–cluster aggregation. *J. Phys. A*, 1984, **17**(18), L975–L981.

70. Malkanduev, Yu. A.; Kozlov, G. V.; Khashirova, S. Y. *The molecular weight regulation in dimethyldiallylammoniumchloride radical polymerization process*. Proceedings of Higher Educational Institutions, North–Caucasus region, natural sciences, 2000, 1, 87–90.

71. Kolb, M.; Jullien, R. Chemically limited versus diffusion limited aggregation. *J. Phys. Lett.* (Paris), 1984, **45**(20), L977–L981.

72. Meakin, P. Diffusion-controlled flocculation: The effects of attractive and repulsive interactions. *J. Chem. Phys.*, 1983, **79**(5), 2426–2429.

73. Meakin, P. Effect of cluster trajectories on cluster–cluster aggregation: A comparison of linear and Brownian trajectories in two– and three-dimensional simulations. *Ohys. Rev. A*, 1984, **29**(2), 997–999.

74. Kolb, M. Reversible diffusion-limited cluster aggregation. *J. Phys. A*, 1986, **19**(3), L263–L268.

75. Weitz, D. A.; Huang, J. S.; Lin, M. Y.; Sung, J. Dynamics of diffusion-limited kinetic aggregation. *Phys. Rev. Lett.*, 1984, **53**(17), 1657–1660.

76. Kolb, M.; Herrmann, H. J. The sol-gel transition modeled by irreversible aggregation of clusters. *J. Phys. A*, 1985, **18**(8), L435–L441.

77. Kozlov, G. V.; Malkanduev, Yu. A.; Zaikov, G. E. Gelation in the radical polymerization of dimethyl diallyl ammonium chloride. *J. Appl. Polymer Sci.*, 2004, **93**(3), 1394–1396.

78. Kozlov, G. V.; Malkanduev, Yu. A.; Zaikov, G. E. A gelation on the process of dimethyldiallylammoniumchloride radical polymerization. In book: *Physical Chemistry of Low and High Molecular Compounds*. Ed. Zaikov, G.; Dalinkevich, A. New York, Nova Science Publishers, Inc., 2004, 127–132.

79. Kozlov, G. V.; Malkanduev, Yu. A. Gel formation time in radical polymerization of dimethyl diallyl ammonium chloride: fractal analysis. *Russian Polymer News*, 2001, **6**(1), 32–34.

80. Kozlov, G. V.; Malkanduev, Yu .A.; Afaunova, Z. I.; Zaikov, G. E. The time of gelation in radical polymerization of dimethyl diallyl ammonium chloride: fractal analysis. *J. Balkan Tribological Association*, 2002, **8**(1), 35–39.

81. Kozlov, G. V.; Malkanduev, Yu. A.; Zaikov.; G. E. A fractal analysis of polymers molecular weight distribution: dynamic scaling. In book: *New Perspectives in Chemistry and Biochemistry*. Ed. Zaikov, G. New York, Nova Science Publishers, Inc., 2002, 35–40.

82. Kozlov, G. V.; Malkanduev, Yu. A.; Zaikov, G. E. Fractal analysis of polymers molecular mass distribution. Dynamic scaling. *J. Balkan Tribological Association*, 2003, **9**(2), 252–256.

83. Kozlov, G. V.; Malkanduev, Yu. A.; Zaikov, G. E. Fractal analysis of polymer molecular weight distribution: Dynamic scaling. *J. Appl. Polymer Sci.*, 2003, **89**(9), 2382–2384.

84. Meakin, P.; Vicsek, T.; Family, F. Dynamic cluster-size distribution in cluster–cluster aggregation. Effects of cluster diffusivity. *Phys. Rev. B*, 1984, **31**(1), 564–569.

85. Vicsek, T.; Family, F. Dynamic scaling for aggregation of clusters. *Phys. Rev. Lett.*, 1984, **52**(19), 1669–1672.

86. Botet, R.; Jullien, R. Size distribution of clusters in irreversible kinetic aggregation. *J. Phys. A*, 1984, **17**(12), 2517–2530.

87. Kozlov, G. V.; Malkanduev, Yu. A.; Zaikov G. E. Molecular weight distribution of poly(dimethyl diallyl ammonium chloride): analysis within the frameworks of irreversible aggregation models. In book: *Fractal Analysis of Polymers: From Synthesis to Composites*. Ed. Kozlov, G.; Zaikov, G.; Novikov, V. U. New York, Nova Science Publishers, Inc., 2003, 131–139.

88. Kozlov, G. V.; Malkanduev, Yu. A.; Zaikov, G. E. Molecular weight distribution of poly(dimethyl diallyl ammonium chloride). Analysis within the frameworks of irreversible aggregation models. *J. Balkan Tribological Association*, 2003, **9**(3), 442–448.

89. Kozlov, G. V.; Malkanduev, Yu. A.; Zaikov G. E. Molecular weight distribution of poly(dimethyl diallyl ammonium chloride). Analysis within the frameworks of irreversible aggregation models. J. Appl. Polymer Sci., 2004, 91(5), 3144–3147.
90. Feder, F. Fractals. New York, Plenum Press, 1990, 239.

CHAPTER 3

THE SYNTHESIS OF BRANCHED POLYMERS

CONTENTS

3.1 THE INFLUENCE OF A MACROMOLECULAR COIL STRUCTURE ON SYNTHESIS PROCESSES

In the work [1], the dependences of PHE synthesis main characteristics, namely, reduced viscosity ηred and conversion degree Q, on synthesis temperature T were studied. It was found out, that T rising up to the definite limits influences favorably on the indicated process: a reaction rate rises, η_{red} and Q increase. This effect can be observed in the narrow enough range of T=333–348 K. At T lower than 333 K PHE formation process decelerates sharply, that is due to insufficient activity of epoxy groups at low temperatures. At T > 353 K cross-linking processes proceed, which are due to the activity enhancement of secondary hydroxyls in polymer chain [1]. It is also supposed [1], that at the indicated temperatures of synthesis PHE branched chains formation is possible.

As it is known [2], the macromolecular coil, which is the main structural unit at polymers synthesis in solution, represents a fractal and its structure (coil elements distribution in space) can be described by the fractal dimension Df. Proceeding from this, the authors [1] used the fractal analysis methods for the description of T effect on PHE synthesis course and its main characteristics.

For the description of synthesis processes a general fractal Eq. (27) of Chapter 1 was used. If the indicated relationship is expressed in a diagram form in double logarithmic coordinates, then from its slope in case of such plot linearity the exponent in the indicated relationship and, hence, the value D_f, can be determined. The calculated by the indicated mode according to the curves values D_f are adduced in Table 1. As one can see, the reduction of D_f at T growth is observed.

TABLE 1   The characteristics of branched PHE [1].

The synthesis temperature, K	D_f, the Eq. (27) of Chapter 1	D_f, the Eq. (8)	ds	g
333	1.98	1.93	1.29	0.394
338	1.89	1.88	1.24	0.478
343	1.69	1.67	1.03	0.774
348	1.56	1.54	1.0	1.0

Within the frameworks of fractal analysis fractal (macromolecular coil) branching degree is characterized by spectral (fracton) dimension d_s, which is an object connectivity degree characteristic [3]. For linear polymer $d_s = 1.0$, for a statistically branched one $d_s = 1.33$ [3]. For macromolecular coil with arbitrary branching degree the value ds varies within the limits of 1.0–1.33. Between the dimensions D_f and d_s the Eq. (39) of Chapter 1 exists, which takes into consideration the excluded volume effects. The Eq. (39) of Chapter 1 allows to estimate the values ds according to the known magnitudes D_f (Table 1). As it follows from the data, T increase results to ds reduction, i.e., polymer chain branching degree decrease and at $T = 348$ K PHE polymer chain is a linear one ($d_s \approx 1.0$).

A number of traditional estimations methods of polymer chain branching exists as well [4–6]. So the branching factor g is defined as follows [6]:

$$g = \frac{R_\theta^2}{R_{l,\theta}^2},$$ (1)

where R_θ and $R_{l,\theta}$ are mean gyration radii of a branched polymer and its linear analog in θ-solvent at the same values of molecular weight MM and Kuhn segment of size A, which characterizes chain thermodynamical rigidity.

Within the frameworks of fractal analysis the relationship between coil gyration radius and molecular weight is given as follows [7]:

$$R_\theta \sim MM^{1/D_f^\theta},$$ (2)

$$R_{l,\theta} \sim MM^{1/D_f^{l,\theta}},$$ (3)

where D_θ and $D_{l,\theta}$ are macromolecular coil fractal dimensions of branched polymer and its linear analog in θ-solvent, accordingly.

The fractal equation can be obtained from the Eqs. (1)–(3) combination for g estimation [1]:

$$g = MM^{2\left[\left(D_f^{l,\theta} - D_f^\theta\right)/D_f^{l,\theta} D_f^\theta\right]}.$$ (4)

For a linear macromolecule the dimension $D_f^{l,\theta}$ is always equal to 2.0 [2]. For its branched analog D_f^{θ} determination is more difficult and this dimension value will depend on the branching degree. For statistically branched coil the dimension D_f^{θ} is given by the Eq. (8) of Chapter 1. If to assume, that D_f^{θ} changes proportionally to ds and to use the boundary conditions $D_f^{\theta}=D_f^{l,\theta}=2.0$ at $d_s = 1.0$ and D_f^{θ} =2.286 at $d_s = 1.33$, then according to the obtained above ds values (Table 1) the corresponding dimension D_f^{θ} can be calculated according to the formula [1]:

$$D_f^{\theta} = 2 + 0.858(d_s - 1). \tag{5}$$

The values g, calculated according to the Eqs. (4) and (5), are cited in Table 1. As it was expected, the branching degree growth (g decrease) at T reduction is observed. The cited in Table 1 values g were calculated for MM = 3×10^4.

In Fig. 1, the dependences D_f (g) for PHE and also for polyphenylxalines (PPX) [5] and bromide-containing aromatic copolyethersulfones (B–PES) [8] were shown. As one can see, for all adduced in Fig. 1 polymers D_f similar growth at g reduction is observed, that indicates on the observed effect community. The different D_f values for linear analogs at g = 1.0 are due to different solvents usage, i.e., to different level of interactions polymer-solvent [9].

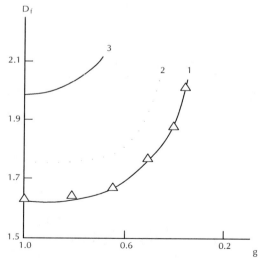

FIGURE 1   The dependences of macromolecular coil fractal dimension D_f on branching factor g for PHE (1), B–PES (2) and PPX (3).

The Eq. (4) supposes chain branching degree increase (g reduction) at MM growth. In Fig. 2 the dependence g (MM), calculated according to the Eq. (4), is adduced, which illustrates this law. The calculation g was fulfilled at $D_f^{l,\theta} = 2.0$ and $D_f^{\theta} = 2.249$ [1].

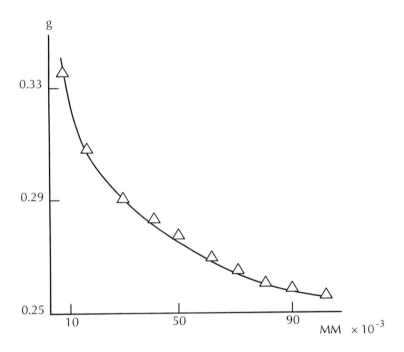

FIGURE 2   The dependence of branching factor g on molecular weight MM for PHE (at $D_f^{l,\theta} = 2$ and $D_f^{\theta} = 2.249$).

An interactions level between macromolecular coil elements and interactions polymer-solvent level can be characterized with the aid of the parameter ε, which is determined with the help of the Eq. (31) of Chapter 1. In its turn, the Eq. (32) of Chapter 1 establishes the intercommunication between ε and D_f. ε positive values characterize repulsion forces between coil elements, the negative ones – attraction forces. In Fig. 3 the dependence of parameter ε on the branching factor g for PHE is adduced. As it follows from the adduced plot, the dependence $\varepsilon(g)$ is linear and can be described by the following empirical equation [1]:

$$\varepsilon = 0.45g - 0.15 . \tag{6}$$

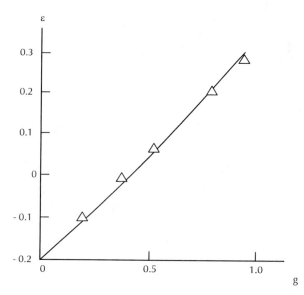

FIGURE 3   The dependence of parameter ε on branching parameter g for PHE.

Therefore, from the Eq. (6) it follows, that macromolecule branching degree increase (g decrease) results to attraction forces growthbetween coil elements and, as consequence, to coil compactization (D_f growth). Let us note, that the entire variation range ε, corresponding to the same variation range $D_f = 1$–3 [10], makes up 1/3–1.0 [11]. From the Eq. (6) the variation range ε for PHE can be obtained at the condition g=0–1 [4]: $\varepsilon = -0.15 - 0.30$. This corresponds to variation of $D_f = 1.54$–2.35 according to the Eq. (32) of Chapter 1. Hence, PHE macromolecule can assume different structural states within the range of leaking coil-coil in θ-solvent [2].

g change results to the exponent a change in Kuhn–Mark–Houwink equation, describing relation between intrinsic viscosity [η] and MM (Eq. (57) of Chapter 1). In its turn, the intercommunication between D_f and a exists, which is defined by the Eq. (4) of Chapter 1. The Eqs. (4), (6), and (32) of Chapter 1 combination allows to obtain the dependence of a on g for PHE [1]:

$$a = 0.275 + 0.68g \quad . \tag{7}$$

Hence, a chain branching degree increase (g reduction) results to a decrease. This conclusion is confirmed experimentally in paper [4] on the example of two polyarylates D_{-1}, received by equilibrium and interphase polycondensation. For the first from them (the branched one) the value a at the same condition is small-er, than for the second (linear) one. Besides, for θ-conditions in case of linear polyarylate a = 0.50, that was expected [2], whereas for branched polyarylate at the same conditions a = 0.36. According to the equation (1.4) this corresponds to $D_f^\theta \approx 2.206$, that corresponds well to the adduced above Family estimation [9] (the equation (1.8)) and, according to the Eq. (5), corresponds to $d_s \approx 1.24$, i.e., to a branched polymer.

Using the Eqs. (6) and (32) combination, the relationship between D_f and g can be obtained as follows [1]:

$$D_f = \frac{2}{0.85 + 0.45g}.$$
(8)

Calculated according to the Eq. (8) D_f values are also adduced in Table 1, from which their good correspondence to D_f values, estimated according to ki-netic curves $Q_{(t)}$ with the aid of the Eq. (27) of Chapter 1, follows (the mean discrepancy of value D_f, received by two indicated methods, makes up ~ 1.4 %).

And in conclusion let us consider the physical significance of PHE polymer chain branching degree decrease at synthesis temperature growth. The branching centers average number per one macromolecule m can be determined according to the equation [5]:

$$g = \left[\left(1 + \frac{m}{7}\right)^{1/2} + \frac{4m}{9\pi} \right]^{-1/2}.$$
(9)

As it is well-known [12], fractal objects are characterized by strong screening of internal regions by fractal surface. Therefore accessible for reaction (in our case – for branching formation) sites are either on fractal (macromolecular coil) surface, or near it. Such sites number N_u is scaled with coil gyration radius R_g as follows [12]:

$$N_u \sim R_g^{d_u},$$
(10)

where d_u is dimension of the unscreened (accessible for reaction) surface, which is determined according to the equation [12]:

$$d_u = \left(D_f - 1\right) + \frac{\left(d - D_f\right)}{d_w}, \qquad (11)$$

where d is dimension of Euclidean space, in which a fractal is considered (it is obvious, in our case d = 3), dw is dimension of random walk on the fractal, which is estimated according to Aarony–Stauffer rule [13]:

$$d_w = D_f + 1 . \qquad (12)$$

As it has been noted above, in case of chemical reactions of various kinds course, including reactions at polymers synthesis, the so-called steric factor p (p ≤ 1) plays an essential role, which shows, that not all collisions of reacting molecules occur with the proper for chemical bond formation these molecules orientation [14]. The value p is connected with R_g as follows [15]:

$$p \sim \frac{1}{R_g} . \qquad (13)$$

Therefore, it can be assumed, that the number of accessible for branching formation sites of macromolecular coil m will be proportional to the product pN_u or [1]:

$$m \sim \frac{R_g^{d_u}}{R_g} \sim R_g^{d_u - 1} . \qquad (14)$$

For this problem solution it is necessary to determine R_g variation at T change. This can be done as follows. Using experimentally determined ηred values, [η] can be calculated according to Shultze–Braschke empirical equation (Eq. (6) of Chapter 1) and then to determine the coefficient K_η in Kuhn–Mark–Houwink equation according to the Eq. (58) of Chapter 1. Further the value MM is determined according to the Eq. (57) of Chapter 1 and polymerization degree N is calculated as follows:

$$N = \frac{MM}{m_0} \ ,$$

(15)

where m_0 is monomer link molecular weight (for PHE $m_0 = 284$).

And at last the value R_g is calculated according to the following fractal relationship [11]:

$$R_g = 37.5 N^{1/D_f} \ , \text{Å}.$$

(16)

In Fig. 4, the dependence of m on $R_g^{d_u-1}$ is adduced, which proves to be linear. Since the value m increases at $R_g^{d_u-1} \sim pN_u$ growth, then this supposes the offered above treatment identity.

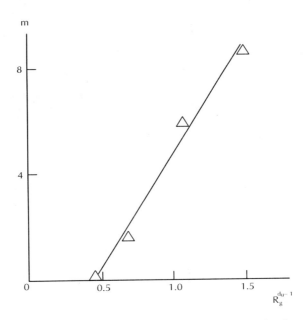

FIGURE 4   The dependence of branching number per one macromolecule m on parameter $R_g^{d_u-1}$ for PHE.

Hence, the stated above results showed, that fractal analysis notions allowed to give principally new treatment of phenomena, occurring at PHE synthesis at synthesis temperature variation. The notion of macromolecular coil structure, characterized by its fractal dimension D_f, forms the basis of this treatment. This structure coil elements interactions among each other and interactions polymer-solution, characterized by parameter ε in the Eq. (31) of Chapter 1, are defined. In its turn, polymer chain branching degree, characterized by branching factor g, is an unequivocal function of D_f according to the Eq. (8). The Eq. (14) gives the physical sense of this correlation. Besides, the macromolecular coil structure defines kinetic curves course according to the Eq. (27) of Chapter 1. It is important to note that fractal analysis methods allow the indicated effects quantitative treatment [1]. Proceeding from these general concepts, the authors [16] gave the description of PHE macromolecular coil structure influence on its synthesis rate at four temperatures in the indicated above range of temperatures T.

As it follows from the data of Fig. 5, synthesis temperature T increase results to PHE synthesis reaction rate enhancement. Within the frameworks of fractal analysis the reaction rate constant kr can be determined according to the Eq. (18) of Chapter 2.

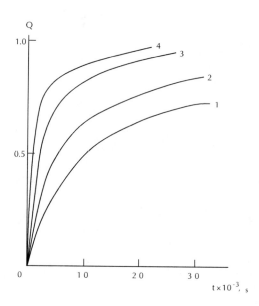

FIGURE 5   The kinetic curves of conversion degree – reaction duration (Q_4) for PHE at synthesis temperatures 333 (1), 338 (2), 343 (3) and 348 K (4).

Since in paper [16] the main studied parameters are given in relative units, then the value c_1 was accepted equal to one. The estimated at t = 3600 s k_r values are adduced in Table 2.

TABLE 2   The characteristics of macromolecular coil of PHE, produced at different synthesis temperatures.

T, K	k_r, relative units	N	R_g, Å	p
333	0.027	54.4	282	0.0181
338	0.057	97.4	427	0.0262
343	0.205	162.6	764	0.0593
348	0.505	227.4	1215	0.1010

The calculation of polymerization degree N and macromolecular coil gyration radius R_g according to the Eqs. (15) and (16), respectively, shows a wide enough variation of the indicated parameters. In paper [17], the following fractal Eq. (76) of Chapter 1 for steric factor p estimation was adduced, in which the constant is also accepted equal to one according to the mentioned above reasons. The values p for PHE at four magnitudes T are given in Table 2.

In Fig. 6, the relationship $k_r(p)$ for PHE is adduced, which has an expected character – p growth results to k_r increase. However, this dependence is not directly proportional. The authors [16] supposed, that for the obtaining of a more general correlation reaction a rate-macromolecular coil structure sites number on coil surface N_u, accessible for reaction (unscreened), should be taken into consideration. In other words, in such treatment the dependence of k_r on complex characteristic pN_u should be plotted, which is shown in Fig. 7. From this Figure plot it follows, that the linear correlation $k_r(pN_u)$ is now obtained, i.e., k_r growth at T increase is defined by PHE macromolecular coil structure change.

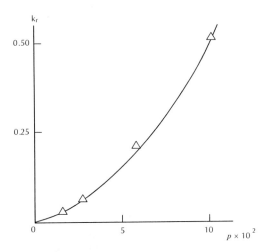

FIGURE 6   The relationship between reaction rate constant k_r and steric factor p for PHE.

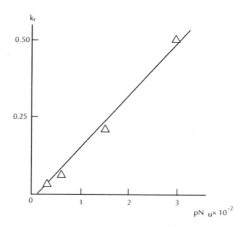

FIGURE 7   The dependence of reaction rate constant k_r on complex characteristic pN_u for PHE.

Let us note, that $k_r = 0$ is achieved at small, but finite quantity pNu. The calculation according to the Eqs. (76) of Chapter 1, Eqs.18 of Chapter 2 andand Eqs. (11) shows, that $D_f \approx 2.20$, i.e., fractal dimension of a branched chain in θ-conditions, corresponds to this value pN_u [9]. After this dimension achievement reaction rate decreases sharply.

In Fig. .8, the dependence of k_r on the branching factor g, calculated according to the Eq. (4) (see Table 1), is adduced in the form of $k_r (g_3)$. As one can see, this dependence is linear and passes through coordinates origin. Hence, the chain branching degree increase, characterized by g reduction, defines PHE synthesis rate sharp decrease (a cubic dependence).

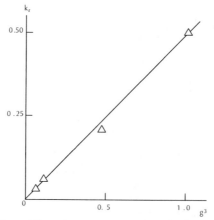

FIGURE 8   The dependence of reaction rate constant k_r on branching factor g for PHE.

Hence, the stated above results showed definite influence of macromolecular coil structure, characterized by its fractal dimension D_f, on PHE synthesis rate. D_f increase, i.e., coil compactization, decreases sharply reaction rate constant kr the value in virtue of two factors influence: the decrease of accessible for reaction sites number N_u and the steric factor p reduction. As a matter of fact, D_f growth defines transition from PHE synthesis diffusive regime to kinetic one [18]. In its turn, polymer chain branching degree growth rises D_f and decreases sharply k_r [16].

In paper [1], strong dependence of the reduced viscosity ηred on reactionary medium components relation at PHE synthesis in mixture water-isopropanol was found. Besides it was revealed, that the process of PHE synthesis can be divided into three modes depending on the contents of isopropanol cis in a mixture. At the small contents of isopropanol c_{is} <15 vol. % the synthesis process practically does not take place and it turns into low-molecular oligomer. At c_{is} = 15–60 vol. % the linear PHE was received and its viscosity η (or molecular weight MM) grows at c_{is} increasing. And at last, at c_{is} > 60 vol. % the cross-linked polymer is formed. Earlier the indicated modes existence was explained only from the chemical point of view as follows. PHE was synthesized by one-step method, namely, by a direct interaction of epichlorohydrin and bisphenol according to the following scheme:

The first regime (c_{is} ≤ 15 vol. %) is explained by the poor ability for mixing of epichlorohydrin with water, that complicates its access to bisphenolate, being present in water. Besides in water formed products drop out as viscous resin and the chain growth is either strongly slowed down, or stops.

As it is known [19], the increase of reactionary ability of hydroxyl group will promote OH-group introduction in hydrogen bond as a donor of protons owing to the displacement of electronic density under the scheme:

It allows to assume, that at c_{is}= 15–60 vol. % there occurs an increase of bisphenols reactionary ability according to the above mentioned mechanism and this results to the increase of PHE production reaction rate and to higher values η_{red}.

And at last, the production of the cross-linked PHE at c_{is} > 60 vol. % is explained by the participation of isopropanol molecules in the reaction, that was observed earlier [20].

The cited above analysis does not take into account the features of macromolecular coil structure, which is a basic element at synthesis of polymers in a solution. It is obvious, that the reactionary medium change should result to the changes of interactions of the second group and, as consequence, to a variation D_f. Therefore the authors [21] proposed the alternative explanation of three modes existence during PHE synthesis process with the fractal analysis representations [22, 23] participation.

In Fig. 9 the dependence of average value of the reduced viscosity η_{red} for PHE on the isopropanol contents cis in reactionary medium water-isopropanol is adduced and the boundaries of the mentioned above synthesis three modes are also indicated (shaded vertical lines). From the data of Fig. 9 it follows, that the first regime (synthesis practical absence) is completed at c_{is} = 15 vol. %.

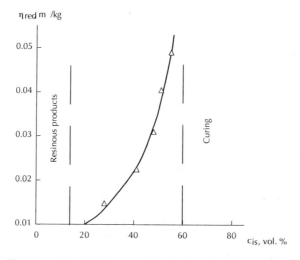

FIGURE 9 The dependence of mean reduced viscosity η_{red} on isopropanol volume contents c_{is} in reactionary mixture water-isopropanol at PHE synthesis. The vertical shaded lines indicate regimes boundary.

The Eq. (27) of Chapter 1allows to obtain synthesis cessation condition. At D_f = d = 3 (d is the dimension of Euclidean space, in which a process is considered) Q = const. Besides the initial conditions of synthesis will be Q = 0 at t = 0. Thus, at D_f = d Q=const the reaction does not take place as well. In other words,

PHE synthesis reaction will be realized in case of fractal reacting objects (macro-molecular coils) only.

The value D_f can be determined according to the Eq. (12) of Chapter 1 and included in this equation Flory–Huggins interaction parameter χ_1, characterizing interactions polymer-solvent level, is determined according to the Eq.10 of Chapter 1. In this case the value of solvent solubility parameter δ_s is the parameter of water-isopropanol solubility mixture and its determination as a function of cis is performed according to the technique [24]. Within the frameworks of the indicated technique the value δ_s is determined by the Eq. (15) of Chapter 1, in which the solubility parameter component δ_f includes the energy of dispersive interactions and the energy of dipole bonds interaction and the component δ_c includes the energy of hydrogen bonds and the energy of interaction between an atom with electrons deficiency of one molecule (acceptor) and an atom with electrons abundance of an other molecule (donor), which requires the certain orientation of these two molecules. For water $\delta_f = 8.12$ (cal/cm³)1/2 and $\delta_c = 22.08$ (cal/cm³)1/2, for isopropanol $\delta_f = 7.70$ (cal/cm³)1/2 and $\delta_c = 5.25$ (cal/cm³)1/2 [24].

The value δ_f^m (δ_c^m) for a solvents mixture can be determined according to the mixtures rule [24]:

$$\delta_f^m = \sum_{i=1}^{n} \phi_i \delta_{f_i},$$

(17)

$$\delta_c^m = \sum_{i=1}^{n} \phi_i \delta_{ci}$$

(18)

where the index "m" designates mixture, the index i designates i-th mixture component, the complete number of which is equal to n and ϕ_i is that a component volumetric fraction.

The empirical constant χ_s in the Eq. (10) of Chapter 1 can be estimated according to the following considerations, but at first the authors [21] showed such estimation necessity. The value D_f calculation for cis = 30 vol. % ($\delta_p = 9.15$ (cal/cm³)1/2, $\delta_s^m = 18.8$ (cal/cm³)1/2) gives the value $D_f = 3.15$, that is physically impossible, since $D_f < d = 3$ [25]. This means, that there exists significant entropic contribution to value $\chi 1$ and, hence, D_f. In Fig. 5 a number of kinetic curves $Q_{(t)}$ for PHE was adduced. Using kinetic curve Q(t) at T = 348 K plotting in double logarithmic coordinates (Fig. 10), the exponent in the Eq. (27) of Chapter 1 can be estimated and, hence, the value D_f, which is equal to ~ 2.60. Then according to the

Eq. (12) of Chapter 1 at the known E_{ev} (the values E_{ev}^m for mixtures were estimated according to the mixtures rule by analogy with the Eqs. (17) and (18)), δ_s^m and δ_p can estimate the parameter χ_s, which is equal to ~ 1.23. In further calculations χ_s=const is accepted [21].

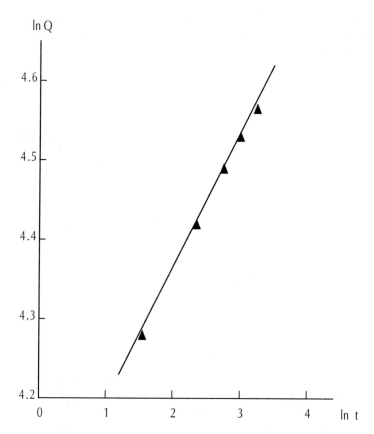

FIGURE 10 The dependence of conversion degree Q on reaction duration t in double logarithmic coordinates for PHE synthesis at T = 348 K.

In Fig. 11, the dependence $D_f (c_{is})$ calculated by the indicated method, is adduced. At c_{is} = 15 vol. % D_f= 2.967, i.e., macromolecular coil structure, as indicated above, does not allow synthesis reaction proceeding. At c_{is} < 15 vol. % $D_f \approx$ 3 = const in virtue of the mentioned above condition $D_f \leq d$.

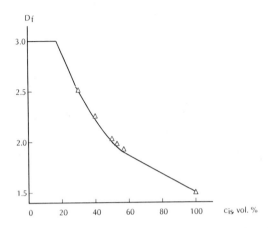

FIGURE 11 The dependence of macromolecular coil fractal dimension D_f on isopropanol volume contents cis in reactionary mixture water-isopropanol for PHE.

As it is known [26], polymers cross-linking macroscopic process provides joining cluster formation (from one end of a reactionary bath up to another one). This is gelation process for cross-linking polymer and the cluster dimension at this point is equal to ~ 2.5 [27, 28]. The gelation process proceeds in medium with cross-linked macromolecular coils (so-called microgels) plenty, that results to reactionary medium viscosity essential enhancement and D_f increasing. According to work [29] the fractal dimension value $\overline{D_f}$ for "dense" solution is linked with value D_f in case of diluted solution by the Eq. (23) of Chapter 2, according to which for $\overline{D_f}$ =2.5 we shall receive Df = 1.786. At $\overline{D_f}$ =3, i.e., a compact globule, for which any chemical reaction, including cross-linking process, is impossible, we shall receive D_f = 1.875. Hence, the cross-linking process must begin within the range of D_f = 1.786–1.875. According to the Eqs. (10), (12), (15) of Chapter 1, (17) and (18) for this value D_f we shall obtain $\delta_s^m \approx 14.8$ (cal/cm³)1/2, that corresponds completely to the experimental data, adduced in Fig. 9.

Hence, the stated above results showed, that macromolecular coil structure, characterized by its fractal dimension D_f, could be a critical factor in polymers synthesis regimes definition, in particular, PHE. Compact coil (D_f = d) does not allow reaction proceeding and this condition defines the first regime-resin-like products formation. D_f decreasing defines polycondensation reaction proceeding possibility in diluted solutions, and for the reaction proceeding on the gelation stage one needs even smaller D_f value, defining $\overline{D_f}$ magnitude for "dense" solution. In the considered case the value Df is controlled by interactions polymer-solvent change.

3.2 THEORETICAL ANALYSIS OF MOLECULAR WEIGHT CHANGE IN SYNTHESIS PROCESS

As it has been noted above, the synthesis temperature increase within the range of T = 333–348 K results to final characteristics enhancement of this process for PHE – conversion degree Q and reduced viscosity ηred, in the first approximation characterizing polymer molecular weight MM. In Fig. 12 the dependences of η_{red} on synthesis duration t are adduced for PHE at four different T. As it follows from these plots, at first ηred sharp increase at t growth is observed and then values η_{red} achieve asymptotic branch. Thus, one can suppose, that at some t values η_{red} (or MM) magnitudes achieve their limit, depending on T.

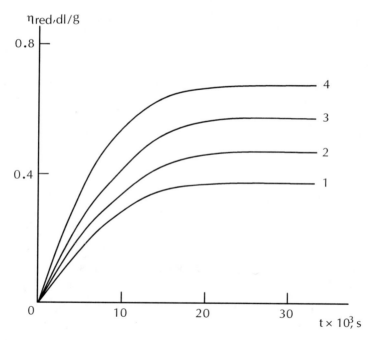

FIGURE 12   The dependences of reduced viscosity η_{red} on synthesis duration t for PHE at synthesis temperature T: 333 (1), 338 (2), 343 (3) and 348 K (4).

This effect can be explained within the frameworks of irreversible aggregation models [15], in which a macromolecular coil in solution is considered as a fractal and its structure is characterized by the dimension D_f [2]. The greatest attainable radius R_c of the coil can be estimated according to the scaling Eq. (48) of

Chapter 1. The authors [30] estimated applicability of the indicated relationship for the limiting values MM determination and elucidated the macromolecular coil structure, characterized by the dimension D_f, as far as T changes at PHE synthesis.

The range of D_f values (see Table 1) at PHE synthesis assumes, that this process proceeds according to the mechanism cluster-cluster, i.e., a large macromolecular coil is formed by merging of smaller ones [15]. This circumstance allows to estimate R_c value according to the Eq. (48) of Chapter 1. The value c_0 choice does not influence on the dependence R_c (D_f) course, but it can change R_c growth rate at D_f reduction. Proceeding from these considerations, in paper [30] the value $c_0 = 25$ was chosen. The experimental values of molecular weight MMe for PHE can be estimated according to Kuhn–Mark–Houwink equation Eq. (57) of Chapter 1, which after determination of constants Kη and a for PHE acquires the following form [30]:

$$[\eta] = 2.84 \times 10^{-4} \left(MM^e \right)^{0.714} .$$

(19)

Theoretical limiting value of molecular weight MMT was determined according to the following scaling relationship [7]:

$$MM^T \sim R_c^{D_f} .$$

(20)

In Fig. 13, the comparison of the dependences MM_e and MMT on synthesis temperature T for PHE is adduced. The constant coefficient in the Eq. (20) was determined by method of experimental and theoretical values MM superposition. As one can see, a good correspondence of theory and experiment is obtained, that confirms application correctness of the Eq. (48) of Chapter 1 for limiting values MM estimation at PHE synthesis.

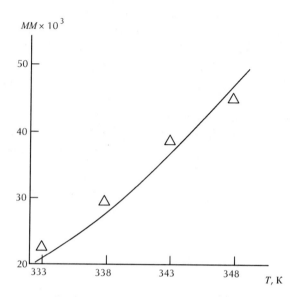

FIGURE 13 The comparison of theoretical (solid line) and experimental (points) dependences of limiting molecular weight MM on synthesis temperature T for PHE.

Hence, the stated above results showed that limiting values of molecular weight, attainable in PHE synthesis process at different T, could be described within the frameworks of irreversible aggregation cluster-cluster model by the usage of the Eq. (48) of Chapter 1. MM indicated limiting value is controlled by macromolecular coil structure, characterized by its fractal dimension D_f.

At present it is known [31, 32], that polymer branching degree influences essentially on molecular weight change kinetics of polymer in its formation process. This process simulation for branched polymers by Monte–Karlo method shows correctness of the following scaling relationship [32]:

$$\frac{dQ}{dt} = k_r (1-Q)(1+cQ),$$ (21)

where t is reaction duration.

For branched chains, growing in critical conditions, the following relationship was obtained [32]:

$$\gamma_t^{-1} = 1 - g',$$ (22)

where g' is the branching degree, determined as the exponent in the scaling relationship between branching centers number per one macromolecule m and MM [32]:

$$m \sim MM^{g'} .$$ (23)

Hence, the Eq. (22) supposes the exponent γ_t growth at polymer branching degree g' increase. The authors [33] obtained the relationship between branching degree and molecular weight in case of real polymer synthesis on the example of PHE.

The experimental values MM for PHE were determined according to the Eq. (19) and the constants K_η and a – according to the Eqs. (4) and (58) of Chapter 1, respectively. The values K_η and a, obtained by the indicated mode, are adduced in Table 3. It is necessary to note, that for the same polymer (PHE) different values of constants in Kuhn–Mark–Houwink were obtained, that is due to different structure of PHE macromolecular coils, received at different T.

TABLE 3   The scaling parameters of PHE, synthesized at different temperatures.

T, K	a	K_η	γ_t^e	γ_t^T
333	0.515	1.4×10^{-3}	0.75	0.606
338	0.587	6.0×10^{-4}	0.54	0.522
343	0.775	7.0×10^{-5}	0.40	0.226
348	0.923	1.3×10^{-5}	0.33	0

Further the dependences $MM_{(t)}$ in double logarithmic coordinates, corresponding to the Eq. (21), can be plotted for determination of the exponent γ_t (γ_t) experimental values in the indicated relationship. The plotted by the indicated mode dependences $MM_{(t)}$ for four T are shown in Fig. 14 and the values γ_t^e are adduced in Table 3.

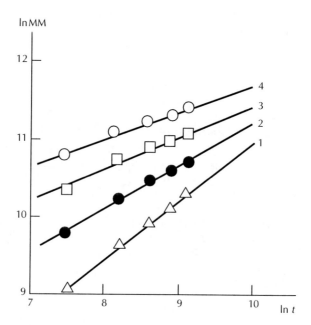

FIGURE 14   The dependences of molecular weight MM on reaction duration t in double logarithmic coordinates for PHE at synthesis temperatures T: 333 (1), 338 (2), 343 (3) and 348 K (4).

As it follows from these data, the value γ_t^e grows at D_f increase (see Table 1), i.e., at polymer chain branching degree enhancement. This situation corresponds completely to papers [31, 32] conclusions. However, the Eq. (22) usage for theoretical value $\gamma t \, (\gamma_t^T)$ estimation in case of PHE is impossible. Since $\gamma_t^T < 1$ (see Table 3), then this means negative values g', that does not have physical significance. Therefore for the estimation of PHE polymer chain branching degree the authors [33] used an other parameter – the branching factor g, which is determined according to the Eq. (1). Let us note, the principal difference between parameters g' and g, which follows from the Eqs. (1) and (22) comparison. Polymer branching increase is characterized by g' increase within the range of 0–1 and g decrease within the same range. The values $g' = 0$ and g = 1.0 correspond to linear polymer [4, 6].

Proceeding from the stated above and also from absolute values γ_t^e, the following form of the dependence γ_t^T (g) can be supposed [33]:

$$\gamma_t^T = 1 - g \, .$$
(24)

The comparison of γ_t^e and γ_t^T is adduced in Table 3, from which their satisfactory correspondence follows.

In Fig. 15, the comparison of the experimental (calculated according to the Eq. (19)) and theoretical (calculated according to the Eq. (21) at $\gamma_t = \gamma_t^T$ and for linear PHE at $\gamma_t = \gamma_t^e$) dependences $MM_{(t)}$ is adduced. As one can see, the theory and experiment good correspondence was obtained.

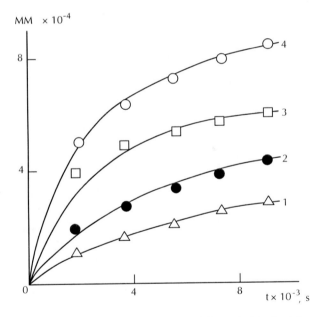

FIGURE 15   The comparison of experimental (lines) and theoretical (points) dependences of molecular weight MM on reaction duration t for PHE at synthesis temperatures T: 333 (1), 338 (2), 343 (3) and 348 K (4).

Thus, the stated above results showed the correspondence of computer simulation of branched polymers formation and PHE synthesis in the aspect, that the scaling Eq. (21) describes correctly molecular weight change kinetics in both cases. However, there exists principal difference in the exponent γ_t determination in the indicated relationship, although in both cases the value γ_t increases at polymer branching degree growth, whichever parameter it is characterized by. The indicated discrepancy can be explained by different conditions of branched chains formation.

In paper [1], it has been shown, that at PHE synthesis reduced viscosity ηred (or polymer molecular weight MM) at reagents initial concentration c_0 growth occurs. Let us note, that similar dependences were also observed at another polymers number synthesis [34]. However, there was no quantitative description of this effect. Therefore the authors [35] performed the dependence $\eta_{red}(c_0)$ or MM(c_0) for PHE with the usage of irreversible aggregation models and fractal analysis.

In Fig. 16, the experimental dependence $\eta_{red}(c_0)$ for PHE (solid line) is adduced, from which monotone growth ηred from 0.23 up to 0.62 dl/g follows in the indicated range c_0. At $c_0 > 0.7$ mole/l polymer cross-linking process begins, that restricts an upper concentration limit of reaction proceeding [21].

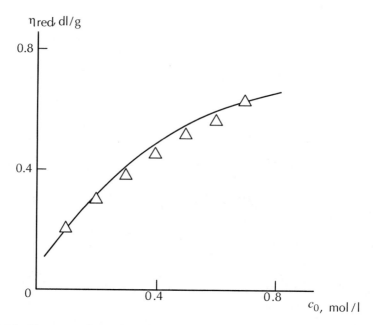

FIGURE 16 The comparison of experimental (solid curve) and theoretical (points) dependences of reduced viscosity ηred on initial reagents concentration c_0 for PHE.

For theoretical description of the dependence $\eta_{red}(c_0)$ kinetic scaling for diffusion – limited aggregation at the condition $R_g \ll \xi$ (R_g is gyration radius, ξ is a scale of aggregation processes universality classes transition), described by the Eq. (38) of Chapter 2, was used. Since theoretical molecular weight MMT can be written as m0N, where m0 is molecular weight of a polymer repeating link (for PHE $m_0 = 284$) and N is polymerization degree, then proportionality coefficient in the Eq. (38) of Chapter 2 can be obtained by the curves $MM_e(c_0)$ and $m_0N(c_0)$ superposition and estimated values MMT. Then according to these values MMT using the Eq. (6) of Chapter 1 and (19) theoretical values ηred, corresponding to the model [36], are calculated. In Fig. 16 the comparison of the theoretical (points) and experimental (solid line) dependences red(c_0) for PHE is adduced, from which the theory and experiment good correspondence follows.

Hence, the model of irreversible diffusion-limited aggregation allows correct quantitative description of the reduced viscosity ηred (or molecular weight MM) change with variation of reagents initial concentration c_0. At the condition of value Df know ledge and polymer chemical constitution Kuhn–Mark–Houwink equation for it can be received theoretically, that allows to avoid labour measurements. The Eq. (38) of Chapter 2 predicts, that value MM at the fixed t depend not only on c_0, but also on macromolecular coil structure [35].

It has been shown above, that characteristics of PHE, synthesized in reactionary medium water-isopropanol, depend to a considerable extent on the indicated components ratio. The increase of isopropanol contents in mixture from 30 up to 55 vol. % results to the adduced viscosity ηred growth from ~ 0.16 up to 0.50 dl/g. This effect was explained within the frameworks of Gammet model, who supposed, that isopropyl spirit addition decreased interaction between molecules of water and bisphenol, resulting to enhancement of dioxicompaund nucleophyl reactionary ability at the expence of hydrogen bond formation [37].

The adduced above analysis does not take into consideration features of macromolecular coil structure, which is the main element at polymers synthesis in solution. Therefore the authors [38] explained the considered above effect from the positions of irreversible aggregation models and fractal analysis.

From the Eq. (38) of Chapter 2 it follows, that value of polymerization degree N (and, hence, molecular weight MM) at the fixed c_0 and t is determined only by the fractal dimension D_f: the smaller D_f, the larger exponent in the Eq. (38) of Chapter 2 and the higher N (or MM). Knowing the value D_f (see Fig. 11) and PHE chemical constitution, Kuhn–Mark–Houwink equation can be received and the experimental values MM (MM_e) according to the experimentally determined η_{red} magnitudes can be calculated. Then the proportionality coefficient in the Eq. (38) of Chapter 2 can be determined by N and MMe superposition and thus to obtain theoretical values MM (MMT), predicted by the model of irreversible aggregation [36]. In Fig. 17 the comparison of the dependences $MM_e(c_{is})$ and $MMT_{(cis)}$ for PHE is adduced (in case of MM_e the error limits are adduced), from which theory and experiment good correspondence follows. This indicates to PHE synthesis process description equivalency within the frameworks of irreversible aggregation models.

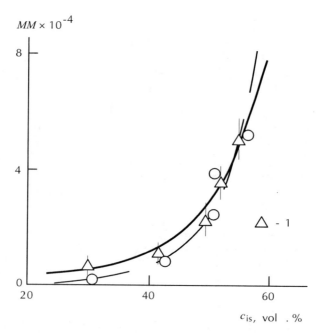

FIGURE 17   The comparison of experimental (1) and calculated according to the Eq. (38) of Chapter 2 at using different (2) and one (3) Kuhn–Mark–Houwink equations dependences of molecular weight MM on isopropanol contents cis in reactionary mixture for PHE synthesis.

Since PHE, produced in different mixtures water-isopropanol, has different values Df (i.e., different structure), then from the Eqs. (4) and (58) of Chapter 1 it follows, that for them values a and K_η will be different, i.e., the relation between [η] and MM for the same polymer will be defined by different Kuhn–Mark–Houwink equations. Therefore, strictly speaking, polymer viscosity should be determined in the same solvent, in which it was synthesized. This rule is confirmed by a well-known fact [4], that synthesized by different polycondensation modes polyarylates, having the same chemical constitution, have different values a (and, hence, D_f according to the Eq. (4) of Chapter 1) and K_η and also distinguishing properties. In Fig. 17, the dependence MM_c(cis), calculated according to the same Kuhn–Mark–Houwink, is adduced.

As it follows from the data of Fig. 17, MM calculation accounting for coil structure and without it gives close enough results at large MM, but at small MM (<104) the discrepancy can be even quintuple one.

Hence, the stated above results showed that PHE molecular weight changes at water-isopropanol ratio variation in reactionary mixture could be described quantitatively within the frameworks of irreversible aggregation models and fractal analysis. The indicated ratio change defines interactions polymer-solvent change that results to macromolecular coil structure variations. An interactions polymer-solvent weakness, characterized by Flory–Huggins interaction parameter χ_1 reduction, results to D_f decreasing, that makes macromolecular coil more accessible for synthesis reaction proceeding.

3.3   THE BRANCHING DEGREE AND MACROMOLECULAR COIL STRUCTURE

As it has been noted above, the branching degree of polymer chain can be characterized by several parameters. One of them is a number of branching centers per one macromolecule m. The branching degree g', determined from the scaling Eq. (23), serves as another parameter. As a rule, the value $g'<1$ [6] while this means that the number of branching centers is not proportional to the length of macromolecule or its polymerization degree, N. From the chemical point of view, such effect is difficult to explain, since each monomer link in macromolecule has the same probability of branch formation, and then one can expect m~N. However, in the real conditions of polymers synthesis there is a number of causes which in principle can cause the decrease of the ratio m/N. One of such reasons can be the fact that the branching reactive centers, formed in the initial stages of synthesis, are proved to be "buried" inside a macromolecular coil and, consequently, are less accessible [31]. Such situation defines the necessity of the macromolecular coil structure allowance, that can be fulfilled with the aid of its fractal dimension D_f. Therefore the authors [39] carried out the description of the macromolecular coil structure influence on a number of branching centers accessible for reaction at polymer molecular weight change. This description is given within the frameworks of fractal analysis on the example of PHE.

Besides the characteristics indicated above one more parameter, the branching factor g, can be used for the estimation of polymer branching degree, which is determined according to the Eq. (1). Within the frameworks of fractal analysis the formulae. (4) allows to determine the value g, which supposes the dependence of g on molecular weight. The parameters g and m are connected by the Eq. (9).

In Fig. 18 the dependence m(MM), where the value m was calculated according to the Eqs. (4) and (9), is adduced. As one can see, this dependence is a nonlinear one, i.e., the value m grows much weaker than MM.

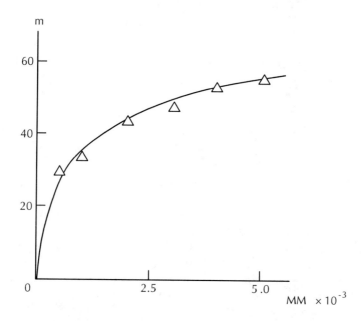

FIGURE 18   The dependence of branching center number per one macromolecule m on molecular weight MM for PHE synthesized at T = 333 K.

In Fig. 19 the same dependence is adduced in double logarithmic coordinates, which proved to be linear, and from its slope the value $g' \approx 0.272$ can be estimated. The small value g' supposes strong influence of macromolecular coil structure on m value and this effect can be estimated quantitatively as follows. As it is known [12], one of the main features of the fractal object structure is the strong screening of its internal regions by the surface. Therefore, the accessible for chain branching reaction macromolecular coil sites are disposed either on its surface or near it. The number of such sites N_u is determined according to the scaling Eq. (10).

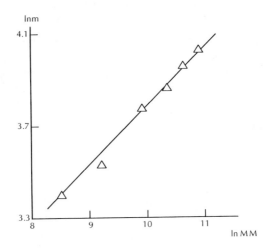

FIGURE 19   The dependence of branching center number per one macromolecule m on molecular weight MM in double logarithmic coordinates for PHE synthesized at T = 333 K.

The dependence m($R_g^{d_u-1}$), corresponding to the Eq. (14), is shown in Fig. 4. As one can see, it is linear and has an expected character: m is increased with $R_g^{d_u-1}$ enhancement. The extrapolation of this plot to m=0 gives $R_g^{d_u-1} \approx 2.2$, which corresponds to the smallest size of PHE macromolecule for the beginning of the branching. This size is equal to ~ 9.9 Å. The volume of the repeating link of PHE V0 can be estimated according to the Eq. (55) of Chapter 1. Then, believing that the cross-sectional area of PHE macromolecule is equal to 30.7 Å2 [40] and using the volume V0 ≈ 410 Å3, calculated according to the Eq. (55) of Chapter 1, the monomer link length for PHE, l0, can be estimated as equal to ~ 13.4 Å. The comparison of the smallest value Rg and l0 indicates that PHE chain branching process begins already at the initial synthesis stage.

Let us note, that the dependences m(MM) with fractional exponent g<1 are typical for other polymers as well. So, in work [4] the constants of Kuhn–Mark–Houwink equation for -solvent in case of the branched polyarylate D$_{-1}$ and its linear analog have been adduced, that allows to calculate intrinsic viscosities [η] θ and [η]$_{l,0}$, accordingly, for arbitrary MM. Then, the value g can be estimated according to the relationship [4]:

$$\frac{[\eta]_\theta}{[\eta]_{l,\theta}} = g^{2-a},$$

(25)

where a is an exponent in Kuhn–Mark–Houwink equation for a linear analog in θ-solvent (a = 0.5 [4]).

Knowing the values g, m magnitudes can be calculated according to the Eq. (9). The estimations have shown, that for polyarylate D_{-1} at MM increase from 5×10^4 up to 10×10^4 m growth from 2.0 up to 2.8 is observed, which corresponds to g ≈ 0.66 [39].

Taking into consideration that gyration radius, R_g, scales to MM according to the Eqs. (2) and (3), from the Eq. (14) one obtains [39]:

$$m \sim MM^{(d_u-1)/D_f} . \tag{26}$$

The comparison of the Eqs. (23) and (26) allows to receive the following equation [39]:

$$g = c_{ch}\left(\frac{d_u-1}{D_f}\right) . \tag{27}$$

The proportionality coefficient, c_{ch}, in the Eq. (27) has a clear physical significance: it defines the greatest density of "chemical" branching centers per one macromolecule. Parameter $(d_u - 1)/D_f$ is defined by this density decrease with macromolecular coil structural features. For PHE the value $c_{ch} \approx 1.41$. The values g, $(d_u-1)/D_f$ and $c_{ch}[(d_u-1)/D_f]$ for PHE at three synthesis temperatures T are listed in Table 4, from which a good correspondence of the first and the third parameters from the indicated ones follows.

TABLE 4   The experimental and theoretical characteristics of PHE chain branching at different synthesis temperatures.

T, K	g	$\dfrac{d_u-1}{D_f}$	$c_{ch}\left(\dfrac{d_u-1}{D_f}\right)$
333	0.272	0.193	0.270
338	0.229	0.145	0.203
343	0.130	0.105	0.147

For branched polyarylate D_{-1} D_f value can be determined according to the equation (1.4) and further, according to the Eqs. (11) and (12), by using a = 0.36 [4] and $D_f \approx 2.20$, du can be determined and cch can be calculated for D_{-1} according to the Eq. (27). In this case $c_{ch} \approx 3.22$, i.e., the greatest "chemical" branching centers density for D_{-1} is much higher than for PHE.

And in conclusion the integral dependence of m on chemical and physical factors can be written [39]:

$$m \sim MM^{c_{ch}\left[(d_u-1)/D_f\right]}. \qquad (28)$$

The dependence corresponding to the Eq. (28) is shown in Fig. 20. As one can see, now this correlation is linear and passes through the coordinates origin. This allows to assert that all factors controlling m value are taken into consideration [39].

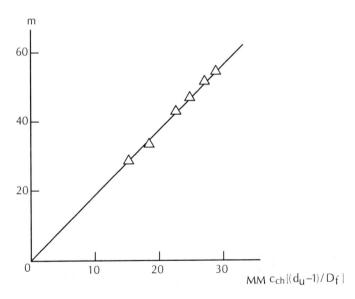

FIGURE 20 The dependence of branching center number per one macromolecule m on complex parameter $MM^{c_{ch}[(d_u-1)/D_f]}$ for PHE synthesized at T = 333 K.

Hence, the fractal analysis methods are efficient for clear structural identification of both chemical and physical factors, controlling a chain branching degree.

The number of effective branching centers per one macromolecule m is controlled by four factors: polymer molecular weight MM, maximum "chemical" density of reactive centers cch, dimension of unscreened surface du of macromolecular coil and its fractal dimension D_f. The Eq. (27) allows to determine the critical value D_f (D_f^{cr}), below of which g=0 (i.e., branching does not occur): $D_f^{cr} = 1.10$ [39].

REFERENCES

1. Kozlov, G. V.; Beeva, D. A.; Mikitaev, A. K. The fractal physics of branched polymers synthesis: polyhydroxyether. *The New in Polymers and Polymer Composites*, 2012, 1, 68–99.
2. Baranov, V. G.; Frenkel, S. Ya.; Brestkin, Yu. V. Dimensionality of linear macromolecule different states. *Reports of Academy of Sciences of SSSR*, 1986, **290**(2), 369–372.
3. Alexander, S,; Orbach, R. Density of states on fractals: "fractons." *J. Phys. Lett.* (Paris), 1982, **43**(17), L625–L631.
4. Askadskii, A. A. The Physics–Chemistry of Polyarylates. Moscow, *Chemistry*, 1968, 216 .
5. Korshak, V. V.; Pavlova, S.-S. A.; Timofeeva, G. I.; Kroyan, S. A.; Krongauz, E. S.; Travnikova, A. P.; Raubah, H.; Shultz, G.; Gnauk, R. Hydrodynamic properties of randomly branched polyphenylquinoxalines with low degrees of branching. *High–Molecular Compounds. A*, 1984, **24**(9), 1868–1876.
6. Budtov, V. P. Physical Chemistry of Polymer Solutions. Sankt–Peterburg, *Chemistry*, 1992, 384 .
7. Vilgis, T. A. Swollen and condensed states of polymeric fractals. *Phys. Rev. A*, 1987, **36**(3), 1506–1508.
8. Kozlov, G. V.; Mikitaev, A. K.; Zaikov, G. E. The side groups influence on polymer limiting mechanical characteristics: fractal analysis. *Polymer Research J.*, 2008, **2**(4), 381–388.
9. Family, F. Fractal dimension and grand universality of critical phenomena. *J. Stat. Phys.*, 1984, **36**(5/6), 881–896.
10. Kozlov, G. V.; Afaunova, Z. I.; Zaikov, G. E. Methods of describing oxidation reactions in fractal spaces. *Polymer International*, 2005, **54**(4), 1275–1279.
11. Kozlov, G. V.; Dolbin, I. V.; Zaikov, G. E. Fractal physical chemistry of polymer solutions. *J. Balkan Tribological Association*, 2005, **11**(3), 335–373.
12. Meakin, P.; Coniglio, A.; Stanley, E. H.; Witten, T. A. Scaling properties for the surfaces of fractal and nonfractal objects: an infinite hierarchy of critical exponents. *Phys. Rev. A*, 1986, **34**(4), 3325–3340.
13. Sahimi, M.; McKarnin, M.; Nordahl, T.; Tirrell, M. Transport and reaction of diffusion-limited aggregates. *Phys. Rev. A*, 1985, **32**(1), 590–595.
14. Barns, F. S. An electromagnetic fields influence on chemical reactions rate. *Biophysics*, 1996, **41**(4), 790–802.

15. Kokorevich, A. G.; Gravitis, Ya. A.; Ozol–Kalnin, V. G. The scaling approach development at lignin supramolecular structure investigation. Chemistry of Wood, 1989, 1, 3–24.

16. Kozlov, G. V.; Shustov, G. B.; Mikitaev, A. K. A macromolecular coil structure effect on polyhydroxyether synthesis rate. Electronic Journal *Studied in Russia*, 010, 78–80, 2009, http://zhurnal.ape.relarn.ru/articles/2009/010.pdf.

17. Kozlov, G. V.; Shustov, G. B.; Zaikov, G. E. The fractal physics of the polycondensation processes. *J. Balkan Tribological Association*, 2003, 9(4), 467–514.

18. Jullien, R.; Kolb, M. Hierarchical method for chemically limited cluster-cluster aggregation. *J. Phys. A*, 1984, 17(12), p. L639–L643.

19. Sorokin, M. F.; Chebotareva, M. A. The esterification reaction of pepargone acid by α-oxyethylether of benzene acid. *Plastics*, 1985, (5), 8–10.

20. Shvets, V. F.; Lebedev, N. N. *About kinetics and mechanism of ethylene oxide reaction with sulfamides and phenols*. Proceedings of MKhTI, 1963,(43), 72–77.

21. Kozlov, G. V.; Zaikov, G. E. Models of synthesis of polyhydroxyether. The fractal analysis. *J. Balkan Tribologic. Association*, 2003, 9(2), 196–202.

22. Kozlov, G.V.; Zaikov, G. E. *Fractal Analysis and Synergetics of Catalysis in Nanosystems*. New York, Nova Biomedical Books, 2008, 163 .

23. Naphadzokova, L. Kh.; Kozlov, G. V. *Fractal Analysis and Synergetics of Catalysis in Nanosystems*. Moscow, Publishers of Academy of Natural Sciences, 2009, 230.

24. Wiehe, I. A. Polygon mapping with two-dimensional solubility parameters. *Ind. Engng. Chem. Res.*, 1995, 34(2), 661–673.

25. Mandelbrot, B. B. *The Fractal Geometry of Nature*. San–Francisco, W.H. Freeman and Company, 1982, 459.

26. Hess, W.; Vilgis, T. A.; Winter, H. H. Dynamic critical behavior during chemical gelation and vulcanization. *Macromolecules*, 1988, 21(8), 2536–2542.

27. Botet, R.; Jullien, R.; Kolb, M. Gelation in kinetic growth models. *Phys. Rev. A*, 1984, 30(4), 2150–2152.

28. Kobayashi, M.; Yoshioka, T.; Imai, M.; Itoh, Y. Structural ordering on physiccal gelation of syndiotactic polystyrene dispersed in chloroform studied by time-resolved measurements of small angle neutron scattering (SANS) and infrared spectroscopy. *Macromolecules*, 1995, 28(22), 7376–7385.

29. Muthukumar, M. Dynamics of polymeric fractals. *J. Chem. Phys.*, 1985, 83(6), 3161–3168.

30. Kozlov, G. V.; Grineva, L. G.; Mikitaev, A. K. The scaling analysis of limiting molecular weight at polyhydroxyether synthesis. Mater. of VI International Sci.–Pract. Conf. *New Polymer Composite Materials*. Nal'chik, KBSU, 2010, 200–204.

31. Alexandrowich, Z. Kinetics of formation and mean shape of branched polymers. *Phys. Rev. Lett.*, 1985, 54(13), 1420–1423.

32. Alexandrowich, Z. Fractals dimension and branched polymers synthesis. In book: *Fractals in Physics*. Ed. Pietronero, L.; Tosatti, E. Amsterdam, Oxford, New York, Tokyo, North–Holland, 1986, 172–178.

33. Kozlov, G. V.; Shustov, G. B.; Mikitaev, A. K. The fractal analysis of molecular weight change kinetics in polyhydroxyether synthesis process. Mater. of V International Sci.–Pract. Conf. *New Polymer Composite Materials*. Nal'chik, KBSU, 2009, 117–123.

34. Korshak, V. V.; Vinogradova, S. V. A Nonequilibrium Polycondensation. Moscow, *Science*, 1972, 696 .

35. Kozlov, G. V.; Mikitaev, A. K. The fractal treatment of the dependence of molecular weight on reagents concentration at polyhydroxyether synthesis. Mater. of VI International Sci.–Pract. Conf. *New Polymer Composite Materials*. Nal'chik, KBSU, 2010, 205–210.

36. Hentschel, H. G. E.; Deutch, J. M.; Meakin, P. Dynamical scaling and the growth of diffusion-limited aggregates. *J. Chem. Phys.*, 1984, **81**(5), 2496–2502.

37. Gammet, L. *The Principles of Physical Organic Chemistry*. Moscow, World, 1972, 326.

38. Kozlov, G. V.; Mikitaev, A. K. The dependence of polyhydroxyether molecular weight on water and isopropanol ratio in reactionary medium. Mater. of VI International Sci.–Pract. Conf. *New Polymer Composite Materials*. Nal'chik, KBSU, 2010, 192–199.

39. Kozlov, G. V.; Burya, A. I.; Shustov, G. B. The dependence of the chain branching degree on molecular weight: fractal analysis. *Chemical Industry and Chemical Engineering* Quarterly, 2008, **14**(3), 181–184.

40. Aharoni, S. M. Correlations between chain parameters and failure characteristics of polymers below their glass transition temperature. *Macromolecules*, 1985, **18**(12), 2624–2630.

CHAPTER 4

THE CROSS-LINKED POLYMERS CURING

CONTENTS

4.1   TWO TYPES OF FRACTAL REACTIONS AT CURING CROSS-LINKED POLYMERS

The application of irreversible aggregation models and fractal analysis for the description of the curing kinetics of two haloid-containing epoxy polymers was considered in paper [1]. For this purpose let us describe briefly experimental methods used for the considered in the present Chapter epoxy polymers study. Firstly, the curing kinetics of haloid-containing oligomer, having chemical structure [1] was studied. This oligomer (conditional designation EPS-4) was cured by 4,4'-diaminodiphenylmethane (DDM) at stoichiometric ratio DDM: EPS-4. The system EPS-4/DDM curing kinetics was studied by method of inversed gas chromatography (IGC) [1]. The basic parameter received from processing of the experimental data was the constant of the reaction rate kr determined for an interval of reaction conversion degree Q = 0.1–0.7 of kinetic curve Q_{-t} part, where t is curing reaction duration. Ketones (methyl ethyl ketone, 1,4-dioxane, cyclohexanone) were chosen as the standard substances for the retention time determination and argon as the gas-carrier [1]. The system EPS-4/DDM curing was studied at three curing temperatures T_{cur}: 383, 393 and 403 K [1, 2].

$$\left(H_2C - CH - CH_2 - O - \bigcirc - \overset{\overset{CH_3}{|}}{\underset{\underset{CH_3}{|}}{C}} - \bigcirc - O \right) \left. \right]_2 \bigcirc_{Cl_4}$$

Besides, curing kinetics of haloid-containing oligomer on the basis of diphenylpropane and hexachloroethane (2DPP+HCE) was studied. This oligomer was also cured by DDM at the stoichimetric ratio DDM: 2DPP+HCE. The study of the system 2DPP+HCE/DDM curing kinetics was carried out by method of IR-spectroscopy using spectrometer Perkin-Elmer. In order to avoid the dependence on the thickness of an oligomer layer put on a substrate the method of internal standard was applied. The optical density of analytical band of 920 cm^{-1} was not accepted as the epoxy groups contents measure, but as its ratio to the optical density of a standard in the capacity of which the IR-band of skeletal vibrations for aromatic ring 1510 cm^{-1} are used, as this concentration is constant during the curing process. The optical densities of an analytical band and the bands of the standard were determined by a baseline method. The system 2DPP+HCE/DDM curing studies were carried out at T_{cur} values: 333, 353, 373, 393 and 513 K [1, 2].

In Figs. 1 and 2, the kinetic curves $Q_{(t)}$ for the systems 2DPP+HCE/DDM and EPS-4/DDM are given correspondingly.

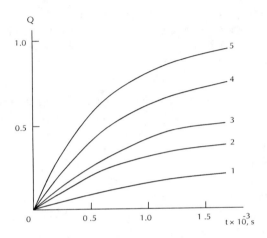

FIGURE 1   The dependences of curing reaction conversion Q on reaction duration t for system 2DPP+HCE/DDM at curing temperatures: 333 (1), 353 (2), 373 (3), 393 (4) and 513 K (5).

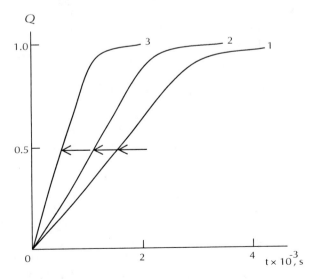

FIGURE 2   The dependences of curing reaction conversion Q on reaction duration t for system EPS-4/DDM at curing temperatures: 383 (1), 393 (2) and 403 K (3).

Two main distinctions of curves $Q_{(t)}$ for the indicated systems attract our attention. Firstly, for system 2DPP+HCE/DDM the smooth decrease of a slope of curves $Q_{(t)}$ is observed in the process of increase t, whereas for system EPS-4/DDM the linear dependence $Q_{(t)}$ up to the large (about 0.8) values Q is observed. Secondly, if the limiting conversion degree of curing reaction for the system 2DPP+HCE/DDM is the function of T_{cur} (the more T_{cur} the more this degree), then for the system EPS-4/DDM such dependence is not present and at all used T_{cur} maximum value Q is observed, close to one.

Let us consider the reasons of the mentioned distinctions, involving the representations of the irreversible aggregation models and of fractal analysis. Within the frameworks of fractal analysis polymerization kinetics is described according to the general Eq. (27) of Chapter 1. The curing formal kinetics within the frameworks of traditional approaches is formulated by the Eq. (16) of Chapter 2, which differentiating on t, gives the Eq. (17) of Chapter 2. The Eqs. (16) and (17) of Chapter 2 allow to received the Eq. (18) of Chapter 2 for the estimation of dimension D_f according to the curing process kinetic parameters [1].

Now it is possible to calculate value D_f, using the reaction rate constants k_r, obtained according to the described above experimental methods as function of t or Q. The calculation has shown the principal distinction of behavior D, which is the microgels structure characteristic, for the systems 2DPP+HCE/DDM and EPS-4/DDM. For the first from the indicated systems the value D_f does not depend on t at the initial part of the curve $Q_{(t)}$, but is the function of Tcur. So, at $T_{cur} =$ 353 and 373 K $D_f \approx 1.5$, at $T_{cur} = 393$ $D_f \approx 1.7$ and at $T_{cur} = 513$ K $D_f \approx 2.3$. It means that the increase of curing temperature determines the formation of more compact microgels during smaller intervals of t. For the system EPS-4/DDM the similar (but weaker) dependence of D_f on T_{cur} is observed, but simultaneously there appears clearly expressed dependence of D_f on t. So, in a curing temporal interval 300–3600 s the value D_f varies as follows: at $T_{cur} = 383$ $D_f = 1.60$–2.38 and at $T_{cur} = 403$ K $D_f = 1.51$–2.42 [1].

In Fig. 3, the modeling calculations of a curve $Q_{(t)}$ are given at $T_{cur} = 383$ K for different situations. So, the curve 1 is an experimental curve $Q_{(t)}$. The calculation according to the Eq. (27) of Chapter 1 under condition $D_f = $ const $= 1.76$ gives curve 2, which does not agree with the experimental curve, but it is qualitatively very similar to the curves $Q_{(t)}$ for the system 2DPP+HCE/DDM (Fig. 1), that was expected, proceeding from the condition $D_f = $ const. The calculation according to the 27 of Chapter 1 at the condition $D_f = $ const $= 2.24$ gives curve 3, which does not agree once again with the experimental curve, but it is very similar to curve $Q_{(t)}$ for the same system at $T_{cur} = 403$ K (Fig. 2). This fact is proved to be true at the comparison of curve 3 with the experimental points 4 for the indicated experimental curve $Q_{(t)}$. In other words, this comparison shows that the curve $Q_{(t)}$ form at the initial parts of curing reaction is controlled by forming microgels structure, which is characterized in the present Chapter by its fractal dimension D_f. And at last, the calculation according to the Eq. (27) of Chapter 1, but with the variable value D_f, determined according to the Eq. (18) of Chapter 2, gives an excellent correspondence with the experiment (points 5). Adduced in Fig. 3 the modeling curves $Q_{(t)}$ and their comparison with the corresponding experimental curve confirm the made above assumption about the reason of the kinetic curves different form for the systems 2DPP+HCE/DDM and EPS-4/DDM.

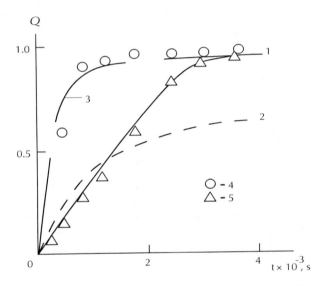

FIGURE 3 The comparison of experimental (1, 4) and modeling (2, 3, 5) curves $Q_{(t)}$ for system EPS-4/DDM. 1 – experimental curve for T_{cur} = 383 K; 2 – calculation according to the Eq. (27) of Chapter 1 under the condition of D_f = const = 1.76; 3 – calculation according to the Eq. (27) of Chapter 1 under the condition of D_f = const = 2.24; 4 – experimental data for T_{cur} = 403 K; 5 – calculation according to the Eq. (27) of Chapter 1 with D_f, calculated according to the Eq. (18) of Chapter 2.

Let us consider the different dependences $D_f(t)$ reasons for the studied systems. As it has been pointed out above, general Df variation for both systems as the function of t and T_{cur} makes up D_f = 1.51–2.42. This interval D_f corresponds approximately to the cluster–cluster aggregation mechanism [3]. It was shown in paper [4] that the dimension Df of the cluster, formed by joining two clusters with dimensions D_{f_1} and D_{f_2} ($D_{f_1} \geq D_{f_2}$) in case of the mentioned aggregation mechanism was determined according to the Eq. (68) of Chapter 1, from which it follows directly that for realization of the condition D_f = const irrespectively from t the fulfillment of the criterion is needed Eq. (1):

$$D_{f_1} = D_{f_2} = D_f = \text{const} \cdot \qquad (1)$$

For the increase of D_f realization with t raising it is required that at a curing previous stage there should be clusters (microgels) corresponding to the condition $D_{f_1} \neq D_{f_2}$ or, in other words clusters dimension distribution is required. So, in order to receive an average dimension $D_f = 1.64$ at the previous stage an interval $D_{f_1} \div D_{f_2} = 1.70\text{–}1.58$, i.e., $\Delta D_f = D_{f_1} - D_{f_2} = 0.12$ is needed. The obtainment of average dimension Df=2.24 already needs an interval $D_{f_1} \div D_{f_2} = 2.35\text{–}2.12$, i.e., $\Delta D_f = 0.23$. Hence, for systems, which are similar to EPS-4/DDM, in process t increasing not only microgels average value of dimension D_f increasing occurs, but also their distribution width raising happens. Proceeding from the obtained results, the authors [1] determined curing kinetics, which is similar to the observed one for the system 2DPP+HCE/DDM (Fig. 1) and corresponding to the condition D_f= const as a homogeneous and corresponding to the condition D_f = variant as a heterogeneous one. One of the probable reasons determining the distinction of curing kinetics mentioned types is a different level of density fluctuations in these systems that will be considered in details below.

Consequently, the stated above results showed the benefit of irreversible aggregation models and fractal analysis notions for haloid-containing epoxy polymers curing kinetics description. There are two different curing regimes (homogeneous and heterogeneous ones) corresponding to the criteria D_f = const and D_f = variant as reaction duration function. The first condition corresponds to the same dimension of forming in curing process microgels and the second one – to these dimensions distribution. The authors [1] showed the influence of microgels structure, characterized by their fractal dimension D_f, on epoxy polymers curing kinetic curves shape and parameters.

The authors [5–7] studied epoxy polymers curing kinetic curves mentioned obvious difference reasons with attraction of fractal analysis [8–10] and scaling approach [11] methods. One of the probable methods of the kinetic curves $Q_{(t)}$ is the general fractal Eq. (26) of Chapter 1 (more complete analogue of the Eq. (27) of Chapter 1). The Eq. (26) of Chapter 1 presented itself in a good light at the description of curves $Q_{(t)}$ for linear polymers synthesis in case of both radical polymerization [12] and polycondensation [13, 14]. D_f value can be determined according to the slope of linear plot Q as a function t in double logarithmic coordinates as follows from the Eq. (26) of Chapter 1and complex constant $K_1 c_0 \eta_0$ by a fitting method. In Fig. 4 such simulation was shown for the system 2DPP+HCE/DDM as points at the following conditions: $K_1 c_0 \eta_0 = \text{const} = 8.06 \times 10^{-3}$ and $D_f = \text{const} = 1.78$. As one can see, a good enough correspondence of experimental and theoretical curves was obtained up to t = 2400s, where system universality class change occurs owing to gel formation and dimension D_f corresponding increase from 1.78 up to ~ 2.5 [15, 16]. To obtain an analogous description for the system EPS–4/DDM failed, since for this system the value D_f is a function of the reaction duration t. In principle curve 2 in Fig. 4 can be described with variable D_f and η_0 using, but such approach is a formal one, since the equation with two variables describes practically any smooth monotonous curve [6].

Proceeding from this, the authors [5-7] attempted to describe the curves $Q_{(t)}$, shown in Fig. 4 (see also Figs. 1 and 2) within the frameworks of scaling approaches for low-molecular substances reactions [17], which give the Eqs. (86)–(88) of Chapter 1, in which the parameter ρA is replaced on $c_{(t)}$. In Fig. 5 the dependences of ln (1–Q) on t, corresponding to the Eq. (86) of Chapter 1, are given for both indicated systems. As it follows from the adduced plots, the system 2DPP+HCE/DDM curing kinetics is well described by linear dependence in co-ordinates of Fig. 5, whereas the dependence [ln (1–Q)](t) for system EPS-4/DDM deviates from linearity. This means that the system 2DPP+HCE/DDM curing reaction, described by the Eq. (86) of Chapter 1, at the indicated above condition is a classical reaction of the first order, occurring in reactive medium with small density fluctuations.

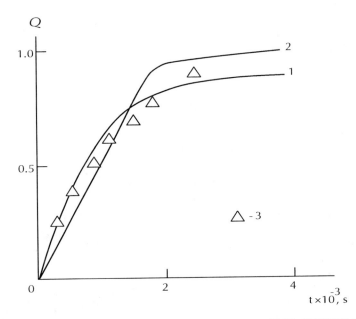

FIGURE 4   The experimental kinetic curves $Q_{(t)}$ for systems 2DPP+HCE/DDM (1) and EPS-4/DDM (2). 3 – calculation according to the equation (1.26).

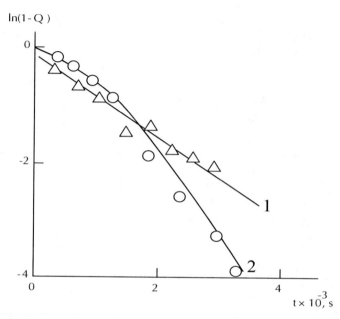

FIGURE 5   The dependences of $(1-Q)$ on reaction duration t in logarithmic coordinates corresponding to the Eq. (86) of Chapter 1 for the systems 2DPP+HCE/DDM (1) and EPS-4/DDM (2).

Attempts of the authors [5–7] to linearize the dependence of $(1-Q)$ on t for the system EPS-4/DDM by the Eqs. (87) and (88) of Chapter 1 were not successful. Therefore the following assumption was made [6]. The Eq. (87) of Chapter 1 describes low-molecular substances reaction kinetics at large density fluctuations in Euclidean space with dimension d, which is equal to 3 in the considered case. If we assume that the fractal clusters (microgels) formation with dimension D_f defines reaction curing course in a fractal space with dimension D_f, the dimension d in the Eq. (87) of Chapter 1 should be replaced by D_f. The dependence of $\ln(1-Q)$ on $t^{D_f/(D_f+2)}$, corresponding to the Eq. (87) of Chapter 1 with the indicated replacement, is adduced in Fig. 6. In such treatment the indicated dependence is linear and this circumstance points out that the system EPS-4/DDM curing reaction proceeds in the conditions of large density fluctuations in fractal space with dimension D_f [5-7].

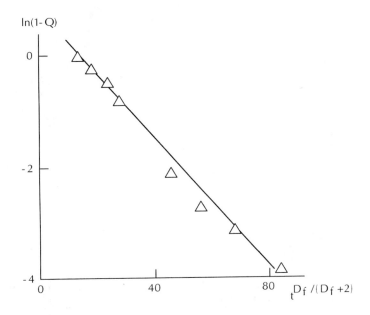

FIGURE 6 The dependence of (1–Q) on parameter $t^{D_f/(D_f+2)}$ in logarithmic coordinates corresponding to the Eq. (87) of Chapter 1 for the system EPS-4/DDM.

Hence, the fractal reactions of polymerization can be divided, as a minimum, into two classes: fractal objects reaction whose kinetics is described similarly to the curve 1 in Fig. 4 and reactions in fractal space whose kinetics is described by the curve 2 in Fig. 4. The second class reactions correspond to structures formation on fractal lattices and the first class – on Euclidean ones [18, 19].

The main principal difference of the indicated reactions classes is the dependence of their proceeding rate on fractal dimension D_f forming in reaction I products (macromolecular coils, microgels). The first class of reactions is well described by the Eq. (26) of Chapter 1 (see Fig. 4). The indicated equation was received on the basis of paper [20] theoretical conclusions, where it is assumed that the smaller D_f is the less compact is the fractal and there are more sites on it, which are accessible to the reaction. This means that D_f decreasing results to reaction rate growth. In Fig. 7 three modeling curves Q(t) corresponding to the Eq. (26) of Chapter 1 at c_0 = const, η_0 = const for D_f = 1.5, 1.8 and 2.1 are adduced. As it follows from the adduced plots, D_f increasing reduces really sharply the reaction rate and decreases Q at t comparable values.

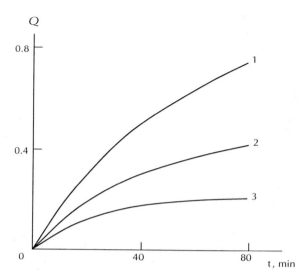

FIGURE 7   The modeling curves $Q_{(t)}$ for fractal objects reactions calculated according to the Eq. (26) of Chapter 1 at $D_f = 1.5$ (1), 1.8 (2) and 2.1 (3).

As to reactions in fractal spaces, here the situation is quite opposite. As it is known [21], if to consider a trajectory of oligomer and curing agent molecules diffusive movement as a random walk trajectory, the sites number $\langle s \rangle$, visited by such walk, is written as follows:

$$\langle s \rangle \sim t^{d_s/2}, \tag{2}$$

where ds is spectral dimension of space, characterizing its connectivity degree [22].

For Euclidean spaces $d_s = 2$ [23], for cross-linked microgels $d_s \approx 1.33$ [22]. From the Eq. (2) it follows, that $\langle s \rangle$ value which can be treated as a reacting macromolecules (microgels) contacts number, is proportional to t in Euclidean and to $t^{D_f/(D_f+2)}$ in fractal spaces. At the same t an indicated contacts greater number in Euclidean space determines the faster curing reaction rate in comparison with fractal space [24].

In this connection let us note an interesting detail. As it was shown in paper [25], for an ideal phantom network the relationship is true Eq. (3):

$$\frac{D_f}{D_f + 2} = \frac{d_s}{2}.$$ (3)

It is easy to see the obvious analogue between the Eq. (87) of Chapter 1 (at replacement of d on D_f) and the Eq. (2) exponents.

In Fig. 8 the kinetic curves $Q_{(t)}$, calculated according to the Eq. (87) of Chapter 1 at B = const for D_f=1.5, 1.8 and 2.1 and also for d = 3, are adduced. It is easy to see, that according to the quoted above treatment reaction rate grows at dimension Df increasing and reaches the largest value in Euclidean space at d=3. It should be noted that in fractal objects reactions according to the Eq. (26) of Chapter 1 at D_f = d = 3, Q = const and accounting for boundary condition Q = 0 at t = 0, Q = const = 0. This means, that such reactions for three-dimensional Euclidean objects do not proceed at all.

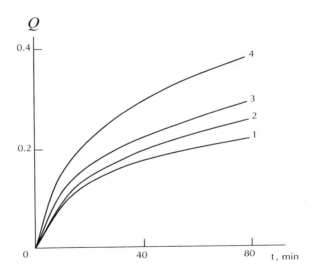

FIGURE 8   The modeling curves $Q_{(t)}$ for reactions in fractal space, calculated according to the Eq. (87) of Chapter 1 at D_f = 1.5 (1), 1.8 (2), 2.1 (3) and d = 3.

Hence, the stated above results have shown that fractal reactions at cross-linked polymers curing can be of two classes: fractal objects reactions and reactions in fractal space. The main distinction of the two indicated reaction classes is the dependence of their rates on fractal dimension D_f of reaction products. Such theoretical dependences knowledge allows to operate the real curing reaction course.

The authors of papers [26–28], using the determined from slopes of corresponding linear plots of Figs. 5 and 6 coefficient A and B values, described theoretically kinetic curves $Q_{(t)}$ with the aid of the Eqs. (86) and (87) of Chapter 1 for the systems 2DPP+HCE/DDM and EPS-4/DDM, respectively. The comparison of the calculated by the indicated method and experimentally received curves $Q_{(t)}$ for both considered systems is adduced in Fig. 9. As it follows from the shown in Fig. 9 plots, the theory and experiment good correspondence was obtained.

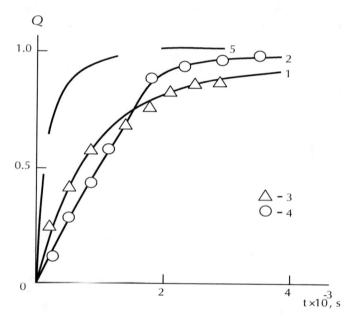

FIGURE 9   The experimental (1, 2) and theoretical (3–5) kinetic curves $Q_{(t)}$ for systems 2DPP+HCE/DDM (1, 3) and EPS-4/DDM (2, 4, 5). The curve 5 corresponds to system EPS-4/DDM curing reaction in Euclidean space.

As to plotting a theoretical curve $Q_{(t)}$ for the system EPS-4/DDM one important remark should be made. As it follows from the Eq. (87) of Chapter 1, at t=0 $c_{(t)} = 1$ or $(1–Q) = 1$ (Eq. (89) of Chapter 1), i.e., at t = 0 Q = 0, that is obvious. However, from the plot of Fig. 6 it follows that the straight line $[\ln(1 - Q)](t^{D_f/(D_f+2)})$ extrapolates to $\ln(1 - Q) = 0$ (or Q = 0) at finite value $t^{D_f/(D_f+2)}$, which is equal to ~1.2. This means that nonhomogeneous reaction of curing for the system EPS-4/DDM has some induction period tin, estimated approximately as equal to 150s. The physical basis of such effect is clear. In curing reaction beginning (at t = 0) fractal clusters (microgels) are absent in reactive mixture and available curing agent and oligomer can be considered as either points (zero-dimensional objects) or short rods (one-dimensional objects). For the formation of microgels, creating reaction fractal space, certain time is necessary, which is an induction period t_{in}. Therefore at theoretical curve α(t) calculation one should use not actually reaction duration t, but the difference $(t-t_{in})$ [28].

As it was noted above, cross-linked polymers formal curing kinetics could be described by the Eq. (16) of Chapter 2. Besides, there are two modified variants of the indicated equation accounting for curing reaction proceeding with auto-acceleration or auto-deceleration [29]:

$$\frac{dQ}{dt} = k_r\left(1-Q\right)\left(1+cQ\right),$$ (4)

$$\frac{dQ}{dt} = k_r\left(1-Q\right)\left(1-\xi Q\right),$$ (5)

where c and ξ are characteristics of auto-acceleration and auto-deceleration effects, respectively.

It was shown above that with fractal analysis and irreversible aggregation models methods using two types of curing reactions exist: homogeneous and heterogeneous ones. The first from the indicated reaction types is characterized by microgels fractal dimension D_f constant value almost up to gel formation point, which is treated as the formation of network tightening the entire system [15] or the transformation of liquid fluid system with cross-linked microgels in elastic polymer [16]. In heterogeneous reaction process on the same part of kinetic curve D monotonous growth is observed [1]. The authors of paper [30] compared two indicated types of curing reaction on the example of two haloid-containing epoxy polymers for elucidation of physical basis of auto-acceleration (auto-deceleration) effect availability in cross-linked polymers curing process.

As it is shown above, polymerization processes can be described by the general fractal Eq. (27) of Chapter 1. Differentiating the indicated relationship on time t and equating the derivative dQ/dt to available ones in the Eqs. (4) and (5), let us receive [30]:

$$t^{(D_f-1)/2} = \frac{c_1}{k_r(1-Q)(1+cQ)}, \tag{6}$$

$$t^{(D_f-1)/2} = \frac{c_1}{k_r(1-Q)(1-\xi Q)}, \tag{7}$$

for curing reactions with auto-acceleration and auto-deceleration, respectively, where c_1 is constant determined from boundary conditions.

In its turn, the dimension D_f value can be determined according to the Eq. (18) of Chapter 2. From the comparison of the Eqs. (18) of Chapter 2, (6) and (7) it is easy to see that in case D_f = const (homogeneous curing reaction) the members in numenator of the Eqs. (6) and (7) [(1+cQ) and (1–ξQ), respectively] are equal to 1. In the case $D_f = D_f(t)$ (heterogeneous curing reaction) two modes of kinetic curves $Q_{(t)}$ description are possible (see Fig. 10 and also Figs. 1, 2 and 6). The first from them, used in paper [1], provides application of function D_f (t), determined by any independent method and then c = 0 and ξ=0. The second method is used below and supposes D_f =const, D_f value can be selected arbitrarily from an interval D_f (t) for the given concrete curing reaction. The comparison of the adduced in Fig. 10 kinetic curves for the systems 2DPP+HCE/DDM and EPS-4/DDM shows that at t < 1200 s curing reaction for the first system proceeds faster than for the second one and for t>1200 s is quite reverse. This means that at curing of the system EPS–4/DDM relatively to 2DPP+HCE/DDM at t<1200 s the auto-deceleration effect is observed and at t>1200 s the auto-acceleration effect is observed. Having calculated the value Df at t=1200 s by the known kr and Q values according to the Eq. (18) of Chapter 2 and believing it as constant, one can estimate parameters c and ξ in the Eqs. (6) and (7), respectively. Strictly speaking, at such approach two indicated equations usage is not quite required: if in the Eq. (6) c < 0, then this means auto-deceleration effect availability and c>0 availability of the auto-acceleration effect. In Fig. 11 the dependence $D_f(t)$ is adduced for the system EPS-4/DDM, which was calculated according to the Eq. (18) of Chapter 2 and D_f value for the system 2DPP+HCE/DDM was shown by a shaded line. From the Fig. 11, data it follows that at t = 1200s D_f values for both systems are equal, at t<1200s Df value for system EPS-4/DDM is lower than the corresponding magnitude for the system 2DPP + HCE/DDM and at t > 1200s it is higher. The comparison of plots in Figs. 10 and 11 allows to make two following conclusions. Firstly, auto-acceleration (auto-deceleration) effect can be realized only in heterogeneous curing reactions. In homogeneous curing reactions (D_f =

const) $c = 0$ and $\xi = 0$. c (or ξ) value is defined by values D_f ratio in homogeneous (D_{hom}) and heterogeneous (Dhet) curing reactions. In case $D_{hom} > D_{het}$ $c < 0$ auto-deceleration effect is just observed; in case $D_{het} > D_{hom}$ $c > 0$ and auto-acceleration effect is also observed. Therefore, in the proposed treatment both an absolute value and a sign of the constant c are defined by the difference$\Delta D_f = D_{hom} - D_{het}$. In Fig. 12, the relation $c(\Delta D_f)$ for the system EPS-4/DDM is adduced, where the value c was calculated according to the Eq. (6) at the condition $D_f = const = 1.75$. And as it was supposed, negative values ΔD_f correspond to negative magnitudes c (auto-deceleration) and positive values D_f correspond to positive magnitudes c (auto-acceleration). Secondly, hence, absolute values ΔD_f and c relation is well approximated by a linear correlation (Fig. 12) [30].

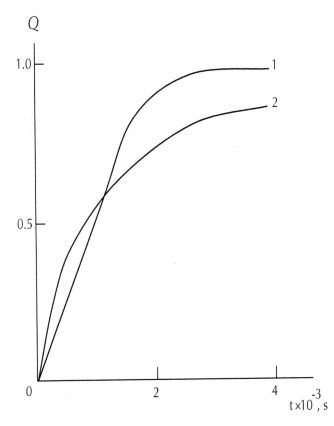

FIGURE 10 The kinetic curves $Q_{(t)}$ of curing process for the systems EPS-4/DDM (1) and 2DPP + HCE/DDM (2).

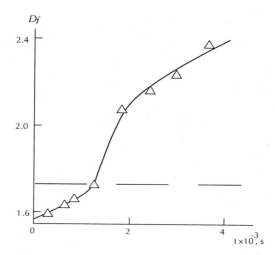

FIGURE 11   The dependence of microgels fractal dimension D_f on curing reaction duration t for the system EPS-4/DDM. The condition D_f = const for the system 2DPP + HCE/DDM was shown by a shaded line.

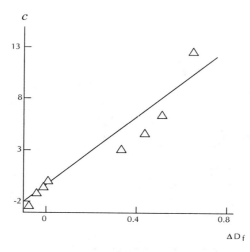

FIGURE 12   The relation between auto-acceleration (auto-deceleration) characteristic c and difference $\Delta D_f = D_{hom} - D_{het}$ for the system EPS-4/DDM.

Hence, the authors of paper [30] have demonstrated that the introduction in the consideration of structure of forming in curing reaction process microgels, characterized by its fractal dimension D_f, and also the introduction of the indicated reaction of two types allowed to determine of physical nature auto-acceleration (auto-deceleration) effect. This effect is realized only for heterogeneous curing reactions and its sign and intensity are defined by values D_f ratio at D_f = const and D_f = variant = $D_f(t)$. In other words, the indicated factors are defined by microgels structure change character in curing process as a reaction duration function.

4.2 THE DESCRIPTION OF CURING REACTIONS WITHIN THE FRAMEWORKS OF SCALING

It has been indicated in the previous section that the system EPS-4/DDM curing kinetics can be described according to the scaling Eq. (87) of Chapter 1 with dimension d replacement on D_f. The authors of paper [31] considered this question in more details. In Figs. 13–15 the dependences of Q on t (Q is given in per cents) are adduced, corresponding to the Eqs. (86), and (87) of Chapter 1 and the Eq. (87) of Chapter 1 with d replacement on Df. As it follows from the adduced plots, the dependences [ln (100–Q)](t) are curvilinear ones and, hence, the system EPS-4/DDM curing kinetics cannot be described according to the Eq. (86) of Chapter 1, therefore curing reaction is not a classical reaction of the first order. This also means large-scale density fluctuations (heterogeneity) availability in reactive medium [17]. The specimens for IGC studies were prepared in such sequence: oligomer and curing agent were dissolved in acetone, the mixture from solutions was piled up on substrate with subsequent drying. Existing in solution mixture heterogeneity was fixed in solvent evaporation process and it was preserved in solid-phase curing reaction. Nevertheless, one should note, that at T_{cur} raising from 383 up to 403 K the shown in Fig. 13 curve linearity is somewhat smoothed out and the curve [ln (100–Q)] (t) for T_{cur} = 403 K can already be approximated by a straight line. Therefore one can assume that temperature raising results to certain diffusive processes intensification [32, 33] both at mixture drying and in solid-state curing course, that smooth out spatial heterogeneity influence [17].

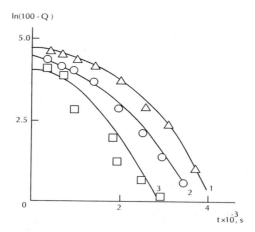

FIGURE 13   The dependences of concentration decay of reacting substances (100–Q) on reaction duration t in logarithmic coordinates for system EPS-4/DDM at Tcur: 383 (1), 393 (2) and 403 K (3).

In Fig. 14, the dependences ln (100–Q) on parameter $t^{d/(d+2)}$ corresponding to the Eq. (87) of Chapter 1 at d=3 are adduced for the same epoxy system. It was unable again to obtain linear correlations and this means that despite the essential density fluctuations (see Fig. 13) the Eq. (87) of Chapter 1 does not also describe the system EPS-4/DDM curing reaction.

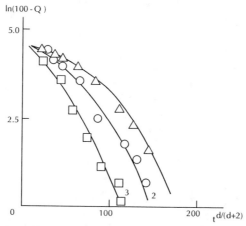

FIGURE 14   The dependences of concentration decay of reacting substances (100–Q) on parameter $t^{d/(d+2)}$ in logarithmic coordinates for system EPS-4/DDM at T_{cur}: 383 (1), 393 (2) and 403 K (3).

   As it has been shown above cross-linked clusters (microgels), forming in curing process on its initial stages, have fractal structure, characterized by fractal
dimension D_f [25]. To determine D_f value with curing process kinetic parameters
application is possible according to the Eq. (18) of Chapter 2. The dimension D_f
calculation has shown that its value is actually a function of t (or Q) and for the
system EPS–4/DDM varies within the limits 1.51–2.38 (see Fig. 11). This means
that aggregation process, defining microgels formation, occurs according to the
mechanism cluster-cluster [3]. As it was noted above, in the polymerization processes case the dimension d in the Eq. (87) of Chapter 1 should be replaced on Df
and then the indicated equation acquires the following form [31]:

$$(100 - Q) \approx \exp\left(-B \ ^{D_f/(D_f+2)}\right). \qquad (8)$$

   In Fig. 15, the dependences of ln (100–Q) on parameter $t^{D_f/(D_f+2)}$ for the studied system at Tcur three values are adduced, from which it follows that in such
treatment the dependences are linear and, hence, the modified Eq. (8) describes
the system EPS–4/DDM curing reaction as a reaction with large spatial fluctuations and reaction fractal products [31].

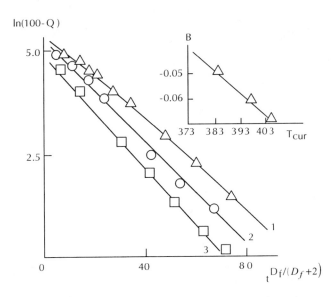

FIGURE 15   The dependences of concentration decay of reacting substances (100–Q)
on parameter $t^{D_f/(D_f+2)}$ in logarithmic coordinates for the system EPS–4/DDM at T_{cur}: 383
(1), 393 (2) and 403 K (3). In insert: the dependence of coefficient B in the Eq. (8) on T_{cur}.

The adduced in Fig. 15 plots linearity allows to determine according to their slope the coefficient B value in the Eq. (8). In insert to Fig. 15 the dependence of B on curing temperature T_{cur} was shown. As it is expected, Tcur raising results to absolute value B growth that causes Q growth at T_{cur} raising at the same values t.

Hence, the stated above results demonstrated once more that in polymerization reactions in general and curing ones in particular it was necessary to take into account the forming products structure. The cross-linked clusters (microgels) structure can be characterized by its fractal dimension D_f. The Eq. (8) modified with accounting for this factor describes well haloid-containing epoxy oligomer EPS-4 curing process at various curing temperatures [31].

As it was noted above, for analysis of polymerization reactions in general and cross-linked polymers curing reactions in particular a number of physical conceptions can be used, from which two of them will be considered below. The indicated conceptions were used as basic relationships of the Eq. (26) and (86) of Chapter 1. In Fig. 16, the dependences of ln (100–Q) on t in logarithmic coordinates, corresponding to the Eq. (86) of Chapter 1, are adduced for the system 2DPP + HCE/DDM at six T_{cur} values, where Q is given in percentage.

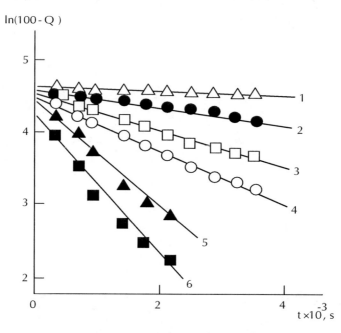

FIGURE 16   The dependences of concentration decay of reacting substances (100–Q) on parameter t in logarithmic coordinates for system 2DPP+HCE/DDM at T_{cur}: 295 (1), 333 (2), 353 (3), 373 (4), 393 (5) and 513 K (6).

   As it follows from the adduced in Fig. 16 plots, the dependences [ln (100–Q)] (t) proved to be linear, i.e., corresponding to the Eq. (86) of Chapter 1. This means that the system 2DPP + HCE/DDM curing reaction at all used T_{cur} can be considered as classical reaction of the first order, proceeding in medium with small density fluctuations [17]. The plots linearity allows to determine the coefficient A value in the Eq. (86) of Chapter 1 from their slope. It is easy to see from the data of Fig. 16 that T_{cur} increase results to A growth. Therefore in Fig. 17, the dependence of $A_{1/2}$ on T_{cur} for the studied system is adduced which is approximated well enough by linear correlation and it can be expressed analytically as follows [34]:

$$A = 2.8 \times 10^{-5} \left(T_{cur} - T_0 \right)^2 .$$
(9)

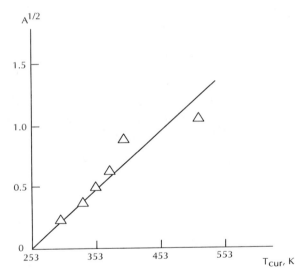

FIGURE 17   The dependence of parameter $A_{1/2}$ on curing temperature T_{cur} for system 2DPP+HCE/DDM.

   The temperature $T_0 = 253$ K in the Eq. (9) corresponds to the condition $A = 0$ (see Fig. 17) and this means that at all $T < T_0$ the system 2DPP+HCE/DDM curing reaction does not occur. The comparison of coefficient A values, determined from linear plots of Fig. 16 A_e and calculated according to the Eq. (9) A_T, was adduced in Table 1, from which their satisfactory correspondence follows.

TABLE 1   The comparison of experimental Ae and theoretical A_T constants in the Eq. (86) of Chapter 1 for the system 2DPP + HCE/DDM.

T_{cur}, K	A^e	A^T
295	0.047	0.049
333	0.147	0.179
353	0.260	0.128
373	0.410	0.400
393	0.740	0.550
513	1.16	1.89

Let us consider further the Eq. (27) of Chapter 1 application for the system 2DPP+HCE/DDM curing kinetics description. In Fig. 18 the dependences $Q_{(t)}$ in double logarithmic coordinates corresponding to the Eq. (27) of Chapter 1, were adduced. As one can see, they are linear, that allows to determine from their slope the value of fractal dimension D_f of microgels, forming in curing process. As the calculations have shown, D_f value grows at T_{cur} increase and varies within limits 1.20–1.95. D_f increasing means that T_{cur} raising results to more compact microgels formation [23]. Besides, the received values D_f interval indicates that microgels formation occurs according to the irreversible aggregation mechanism cluster-cluster or by joining small microgels in much larger ones [35].

For the interval $T_{cur} = 353$–513 K the plots $Q_{(t)}$ in double logarithmic coordinates dissociate into two linear parts with different slope. The part for larger t has a smaller slope and, hence, a higher D_f magnitude. D_f calculation for these parts shows approximately constant D_f value within the range of 2.35–2.50. Such D_f magnitudes correspond to aggregation mechanism particle-cluster [3]. Besides, at present theoretically [15] and experimentally [16] it is shown that polymerizing system transition to dimension ~ 2.5 means gel formation point reaching, which is understood as a network formation tightening the system [15]. Hence, the intersection points of two linear parts in Fig. 18 (at the same T_{cur}) correspond to the system 2DPP + HCE/DDM gelation time in the indicated above sense. Knowing D_f values, a proportionality coefficient in the Eq. (27) of Chapter 1 can be determined. As it has been shown earlier [13, 14], the indicated coefficient consists of three factors: $K_1 c_0 \eta_0$, where K_1 is constant, c_0 is reacting substances initial concentration, η_0 is reacting medium initial viscosity (see the Eq. (61) of Chapter 2). Since for the studied system the value c_0 is constant, then the obtained by the indicated method proportionality coefficient can be considered as η_0 value in relative units. The increase of T_{cur} from 295 up to 513 K results to the growth of η_0 from 0.0053 up to 2.25, i.e., in about 400 times.

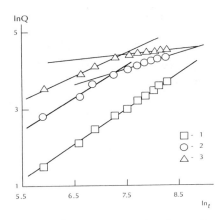

FIGURE 18 The dependences of reaction conversion degree Q on reaction duration t in double logarithmic coordinates for system 2DPP+HCE/DDM at T_{cur} = 333 (1), 373 (2) and 393 K (3).

Using values of constant A and η_0 determined by the indicated methods, theoretical kinetic curves $Q_{(t)}$ for each Tcur can be calculated according to the Eq. (26) and 86 of Chapter 1. The example of experimental and two theoretical curves $Q_{(t)}$ comparison for the system 2DPP+HCE/DDM at T_{cur} = 353 K is adduced in Fig. 19. As one can see, both considered in the present Chapter models simulate well enough the experimental kinetic curve $Q_{(t)}$ [34].

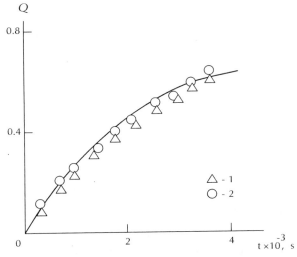

FIGURE 19 The kinetic curves $Q_{(t)}$ for the system 2DPP+HCE/DDM at T_{cur} = 353 K. The line is experimental data, points – calculation according to the Eqs. (86) (1) and (27) (2) of Chapter 1.

Since the Eq. (26) and (86) of Chapter 1 describe well enough an experimental kinetic curve, then one should expect, that between these equations parameters (A in the Eq. (86) of Chapter 1, D_f and η_0 in the Eq. (26)) a certain interconnection should exist. For this supposition checking in Fig. 20 the correlations between D_f, η_0 and A are adduced for the system 2DPP + HCE/DDM at six used T_{cur}. As one can see, the supposed interconnection actually exists: A increasing results to growth of both D_f and η_0 (and vice versa: D_f and η_0 growth results to an increase). The dependences of D_f and η_0 on A can be expressed analytically as follows [34]:

$$D_f = 1 + 0.98 A^{1/2}, \qquad (10)$$

$$\eta_0 = 3A^2 . \qquad (11)$$

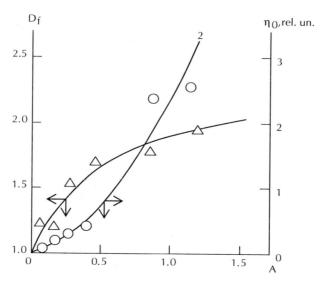

FIGURE 20   The dependences of microgels fractal dimension D_f (1) and reactive medium initial viscosity η_0 (2) on the parameter A value for the system 2DPP+HCE/DDM.

It is easy to see, that the dependence of η_0 on A is much stronger than similar dependence for D_f. In Tables 2 and 3, the comparison of the obtained according to the graphic data values D_f and η_0 and calculated according to the Eqs. (10) and (11) D_f^T and η_0^T these parameters magnitudes, respectively, was adduced. As it follows from the indicated comparison, satisfactory correspondence of parameters D_f and η_0, estimated by both considered methods, is received again.

TABLE 2   The comparison of experimental D_f^e and theoretical D_f^T fractal dimension of microgels for system 2DPP+HCE/DDM.

T_{cur}, K	D_f^e	D_f^T
295	1.22	1.212
333	1.20	1.376
353	1.53	1.500
373	1.68	1.630
393	1.78	1.840
513	1.95	2.050

TABLE 3   The comparison of experimental η_0^e and calculated theoretically η_0^o reactionary medium initial viscosity (in relative units) for the system 2DPP+HCE/DDM.

T_{cur}, K	η_0^e	η_0^o
295	0.0053	0.007
333	0.019	0.065
353	0.161	0.203
373	0.400	0.500
393	2.40	1.64
513	2.25	4.0

The dependences of parameters D_f and η_0 on the same coefficient A, described by the Eqs. (10) and (11), suppose D_f and η_0 interconnection of such kind: D_f increasing should result to a strong growth η_0. Such dependence exists actually and for linearization of correlation $\eta_0(D_f)$ the usage for η_0 power 1/8 (Fig. 21) is required. The relationship between η_0 and D_f can be written analytically as follows [34]:

$$\eta_0^{1/8} = 0.585 D_f \quad .$$ (12)

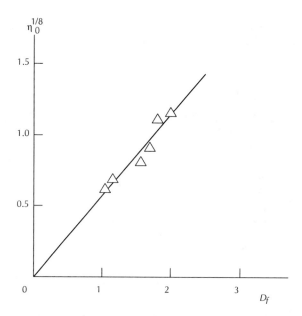

FIGURE 21   The dependence of parameter $\eta_0^{1/8}$ on microgels fractal dimension D_f for the system 2DPP+HCE/DDM.

Thus, the stated above results have demonstrated that both scaling Eq. (86) of Chapter 1 and fractal Eq. (27) of Chapter 1 (or Eq. (61) of Chapter 2) describe well to an equal extent haloid-containing epoxy polymer 2DPP+HCE/DDM curing reaction kinetics at different curing temperatures. In virtue of this circumstance there exists interconnection between parameters included into the indicated equations. The fractal Eq. (61) of Chapter 2 introduces in the kinetics problem consideration reaction products structure (in the given case structure of microgels and condensed state after gel formation point), characterized by its fractal dimension D_f that makes this conception physically more informative [34].

The classical problem in chemical kinetics is a diffusive processes influence on this kinetics [8, 9, 32, 33]. In diffusion-controlled reactions, their rate is defined by diffusion time, which is necessary for reagents to reach one another. Simulation of similar reactions on Euclidean lattices has shown that reactions of the type (Eq. (2) of Chapter 2) and (Eq. (3) of Chapter 2) are described by the (Eq. (43) of Chapter 2) and (Eq. (53) of Chapter 2), respectively. As it is known [36], the change in space type from Euclidean to fractal strongly changes the chemical reaction course. In this case the reactions (Eq. (2) of Chapter 2) and (Eq. (3) of Chapter 2) are described by the (Eq. (6) of Chapter 2) and (Eq. (7) of Chapter 2), respectively.

Let us consider the physical model application described above for the curing process treatment of a haloid-containing epoxy polymer (the system 2DPP+HCE/DDM) [37-39]. It is obvious that in this case the relationships (Eq. (5) of Chapter 2) or (Eq. (7) of Chapter 2), corresponding to the reaction (Eq. (3) of Chapter 2), should be used.

Using relationships of type Eqs. (4)–(7) of Chapter 2 in double logarithmic coordinates, the exponent value for them can be estimated according to the slope of the obtained linear plots. The difference $(1-Q)$ was used as the density ρA of particles "survived" in the curing process (oligomer molecules) [39]. In Fig. 22, the dependences $(1-Q)$ on t for five kinetic curves $Q_{(t)}$ shown in Fig. 1, plotted by the indicated method, are adduced. As it follows from the data of Fig. 22, a change curing temperature results to essential growth in the slope of the linear plots and, hence, the exponent in the Eqs. (5) or (7) of Chapter 2. In its turn, the exponent increasing in these relationships results to growth of the dimension, controlling the studied curing process of epoxy polymers.

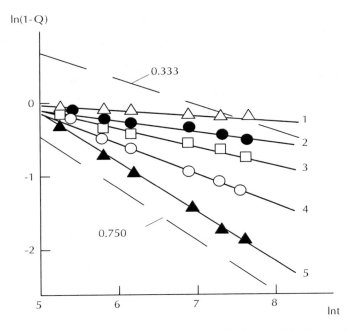

FIGURE 22 The dependences of $(1-Q) = \rho A$ on t in double logarithmic coordinates up to gelation point for the system 2DPP+HCE/DDM at T_{cur}: 333 (1), 353 (2), 373 (3), 393 (4) and 513 K (5). The upper shaded line gives a slope for reaction in fractal space, the lower one in Euclidean space.

In Fig. 22, linear plots were also shown by shaded lines, corresponding to the slope for reactions in fractal (d_s= 1.33, the slope Δ = 0.333 [40]) and Euclidean (d = 3, =0.75 [40]) spaces, obtained by computer simulation method. As one can see, the straight lines obtained for the system 2DPP+HCE/DDM have a slope that does not correspond to the two indicated cases, except for T_{cur} = 513 K, where the curing process is close to reaction in Euclidean space with dimension d = 3. Such behavior of dimensions, which is changed within the range of 0.296–2.964, allows to suppose that it is effective spectral dimension d_s', accounting for availability in real systems of spatial and energetic disorder [41]. This means that Tcur raising increases the effective connectivity of the solvent-oligomer-curing agent system and diffusion of the reacting particles in this case is strange (anomalous) [42]. For diffusivity D_{dif} of particles in liquid matrix, which is Euclidean object with dimension d=3 in supposition of Newtonian rheology its value is determined by Einstein (Eq. (43) of Chapter 2). From the indicated equation it follows that T_{cur} raising, accompanied by η_0 sharp increasing, should result to D_{dif} strong enough decay, that creates premises for static traps formation and, hence, for the Eq. (86) of Chapter 1 application (see Fig. 16).

Using the Eq. (27) of Chapter 1, D_f value can be determined, replotting the indicated relationship in double logarithmic coordinates, after that D_f value is calculated according to the obtained linear plots slope [39]. D_f values calculated by the indicated method for the system 2DPP+HCE/DDM are adduced in Table 4 together with the effective spectral dimension d_s' magnitudes, estimated according to the linear plots slope in Fig. 22. As the authors [43] showed, the dimension d_s' was linked with dimensions of space (reactionary medium) d and microgels D_f according to the following equation Eq. (13):

$$d_s' = \frac{2(3D_f - d)}{d+2} .$$

(13)

TABLE 4   The main dimensions of the system 2DPP+HCE/DDM curing process.

T_{cur}, K	D_f	d_s'	d	d_w	D_f^T
333	1.20	0.296	3.27	20.27	1.27
353	1.53	0.748	2.73	8.02	1.60
373	1.68	1.012	2.91	5.93	1.76
393	1.78	2.0	3.0	3.0	2.40
513	1.95	2.964	3.0	2.02	2.50

Using d_s' and D_f values adduced in Table 4, d magnitude for a reactionary medium, which is the analogue of a lattice in computer simulation chemical reactions, can be calculated. D values are adduced in Table 4, from which it follows that within the limits of an error of estimations their average value is equal to 3. Therefore, a reactionary medium in case of curing the system 2DPP+HCE/DDM represents itself a "virtual" fractal with Hausdorff dimension, which is equal to Euclidean space dimension, but with connectivity degree, typical for fractal objects. This distinguishes real chemical reactions from computer simulation, where the condition $d_s'=d_s=4/3$ for fractal lattices and $d_s'=d_s=d=3$ for Euclidean ones is assumed a priori. The authors [44] obtained the result similar to the described above, in case of the scaling treatment of radical polymerization.

The dimension dw of reagents (oligomer and curing agent molecules) random walk trajectory can be estimated according to the equation [22] Eq. (14):

$$d_w = \frac{2d}{d_s'}.$$

(14)

Further theoretical value of microgels dimension D_f^T can be calculated according to the generalized model of diffusion-limited irreversible aggregation (Eq. (92) of Chapter 1), where η is the parameter, which the authors [45] interpreted as ratio n/m (n, m are whole positive numbers), characterizing chemical reaction of n statistically walking particles with m aggregate perimeter sites. Since in a curing reaction case one curing agent molecule reacts with one oligomer molecule, then as the first approximation it was accepted that n = 1, m = 1 and n/m = 1. In Table 4, the dimensions dw and D_f^T are adduced and the comparison of the latter with estimation of this dimension according to the Eq. (27) of Chapter 1 shows their well good enough correspondence, accounting for the made approximations. Therefore, up to gelation point the curing reaction is controlled completely by the dimension d_s' [39].

In Fig. 23, the dependences of (1–Q) on t in double logarithmic coordinates are adduced for the curing reaction after the gelation point. As one can see, in this case the slope of the obtained linear plots is independent on curing temperature and is equal to about 0.333. Such plots are in agreement with the Eq. (7) of Chapter 2, describing reaction on fractal lattice with dimension ~ 2.5 and superuniversal exponent $d_s/4 \approx 0.333$, which is independent on Euclidean space dimension, in which a fractal is considered [40]. This is explained by the formation in gelation point of cluster with dimension ~ 2.5 tightening the reactionary space [15, 16]. As one can see, the change of the space type, in which curing reaction occurs, from

Euclidean (up to gelation point) to fractal (after it) reduces sharply curing rate of system 2DPP+HCE/DDM. The fact that after the gelation point the plots slope in Fig. 23 is independent on Tcur, indicates unequivocally on spatial (due to diffusion conditions reagents) disorder nature, defining replacement ds on d_s'.

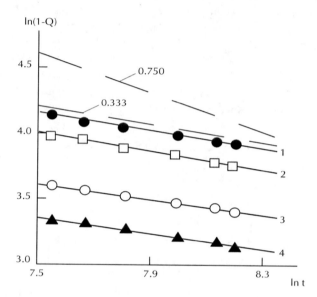

FIGURE 23 The dependences of (1–Q) on t in double logarithmic coordinates after gelation point for the system 2DPP+HCE/DDM at T_{cur}: 353 (1), 373 (2), 393 (3) and 513 K (4). The upper shaded line gives a slope for reaction in Euclidean space, the lower one in fractal space.

Characteristic size r(t) of space region, which was visited by a particle (reagent molecule) up to time moment t can be estimated as follows [42]:

$$r(t) \sim t^{1/d_w} .$$

(15)

In Fig. 24 the dependence of d_s' on r(t) is adduced, which proves to be linear and passing through coordinates origin. This means that at zero level of reagents diffusion zero connectivity of reactionary space (d_s'=0) was reached, which does not allow curing process realization. Figure 25 shows the dependence of experimentally received proportionality coefficient c in the Eq. (7) of Chapter 2, showing that c is also a function of r(t) and is described analytically by the empirical equation [39]:

$$c^{2/3} = 0.8 \ r(t) \cdot \tag{16}$$

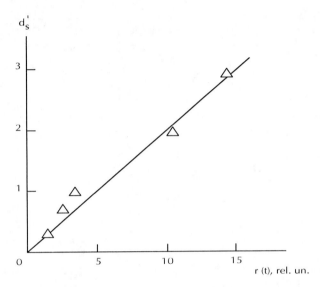

FIGURE 24   The depen dence of reactionary medium effective spectral dimension d_s' on region size r(t), visited by reagents, for the system 2DPP+HCE/DDM.

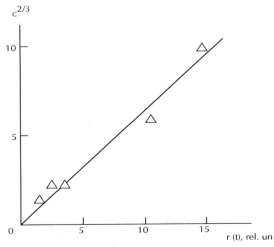

FIGURE 25   The dependence of proportionality coefficient in the Eq. (7) of Chapter 2 on region size r(t), visited by reagents, for the system 2DPP + HCE/DDM.

At r(t) = 0, c = 0 again and curing reaction cannot be realized.

Hence, the stated above results have shown scaling approach correctness at haloid-containing epoxy polymer curing reaction description. The indicated conception application allows to elucidate this process physical aspects and the main distinctions of real chemical reactions from the obtained ones by computer simulation. Up to gelation point spatial disorder, defined by reagents diffusion different intensity at various curing temperatures, controls completely curing reaction course. After a gelation point, tightening cluster formation levels these distinctions [39].

Let us consider physical significance of particles strange (anomalous) diffusion at epoxy polymers curing [8, 9, 42]. As it is known [42], transport strange processes are described according to the equation:

$$\left\langle \overline{r^2(t)} \right\rangle = 2D' + t^\mu, \tag{17}$$

where $\left\langle \overline{r^2(t)} \right\rangle$ is mean-square removing of particle from its movement start, D' is generalized transport coefficient and power exponent at t is equal [42]:

$$\mu = \frac{d_s}{d_f} = \frac{2}{2+\theta}, \tag{18}$$

where d_f is fractal (Hausdorff) dimension of system, θ is this system connectivity index.

At $0 \leq \mu < 1$ it is said about subdiffusive transport processes, at $1 < \leq 2$ – about superdiffusive ones and $\mu = 1$ corresponds to classical (Gaussian) diffusion. In its turn, the exponent μ is connected with Hurst exponent H by the equation [42]:

$$\mu = 2H. \tag{19}$$

The value H defines particles trajectory fractal dimension dw [42]:

$$d_w = \frac{1}{H}, d_w 1. \tag{20}$$

The dimension dw can also be expressed through system connectivity index θ [42]:

$$d_w = 2 + \theta .$$ (21)

Assuming ds = d'_s and d_f = d, one can calculate μ and θ values according to the Eq. (18). The exponent μ changes within the limits of 0.099–0.988, the index θ – within the limits of 18.3–0.04 at Tcur change from 333 up to 513 K. This means that the dimension d_w varies within the interval of 20.27–2.02 (see Table 4) for the system 2DPP+HCE/DDM at the same conditions [46]. Let us note that high dw values at small T_{cur} suppose repeated returning of nonreacted particles to their movement start. Particle removement from its movement start can be described with the aid of the Eq. (15) and also in the following way [42]:

$$r(t) \sim t^{1/(2+\theta)} .$$ (22)

For t = 1700s (see Fig. 1) the calculation according to the Eqs. (15) and (22) shows that r(t) value increases in about 28 times at T_{cur} growth within the range of 333–513 K. It is obvious, the smaller r(t) is the fewer reagent molecules can be diffused up to curing process realization and the more regions in the system are, where oligomer and curing agent molecules (jointly) do not get to that decreases curing rate and conversion degree limiting value Q_{lim}.

The quoted above values μ = 0.099–0.988 suppose that at the system 2DPP+HCE/DDM curing only slow diffusion (subdiffusive transport) is realized and reaching at T_{cur} = 513 K the condition $d'_s = d_s = d$ means that in this case classical Gaussian diffusion (μ=1) is an upper limit and superdiffusive transport realization is impossible. In Fig. 26, the dependence $Q_{lim}(\mu)$ for the system 2DPP+HCE/DDM is adduced, from which the expected linear relationship between the indicated parameters follows. This relationship confirms that the system 2DPP+HCE/DDM curing process is controlled by diffusion [47].

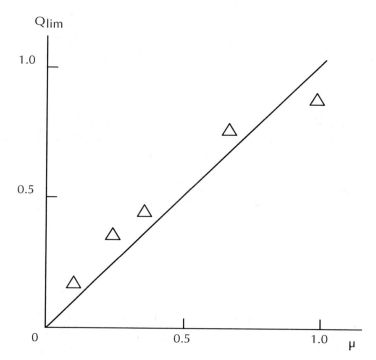

FIGURE 26 The dependence of conversion limiting degree Q_{lim} of curing reaction on exponent for the system 2DPP+HCE/DDM.

4.3 MICROGELS FORMATION IN THE CURING PROCESS

The first theory of cross-linked polymers gelation, elaborated by Carothers and Flory, considered a gel formation point as chemical nodes infinite network formation [48]. Since this theory does not always agree with experimental data then "gelation period" concept was proposed. According to the indicated concept two gelation points exist. The first from them corresponds to an appearance moment in reactionary medium of cross-linked clusters (microgels), characterized by non-fusibility and non-solubility. The second gelation point corresponds to an essentially later stage of reaction – transformation of liquid fluid system with cross-linked clusters in an elastic polymer [49]. The authors of papers [50, 51] considered theoretical conditions of the first gelation point realization with fractal analysis and percolation theory methods using on the example of system 2DPP + HCE/DDM.

In Fig. 27 the kinetic curves $Q_{(t)}$ for the system 2DPP+HCE/DDM at five curing temperatures Tcur are adduced. As it follows from this figure data the reaction rate grows at T_{cur} increase. The curves $Q_{(t)}$ can be described theoretically within the frameworks of the general fractal Eq. (26) of Chapter 1. As it was noted above, if the dependence $Q_{(t)}$ is plotted in double logarithmic coordinates, then from these plots in their linearity case microgels fractal dimension D_f value could be estimated. In Fig. 28 such dependences for the system 2DPP+HCE/DDM at three curing temperatures are adduced. As it follows from these plots, T_{cur} growth is accompanied by the slope reduction of linear plots $Q_{(t)}$ in double logarithmic coordinates, or D_f increasing. Within the range T_{cur} = 295–513 K D_f increasing from 1.20 up to 1.95 is observed (see Table 4). For the highest T_{cur} = 513 K the plot slope discrete change is observed, which corresponds to Df growth from 1.95 up to ~2.68. Such transition within the frameworks of fractal analysis corresponds to the second gel formation point [15, 16], i.e., to network formation, which tightens the whole sample. From the Fig. 28 data it also follows that the second gelation point for T_{cur} = 353 and 373 K in t scale of Fig. 27 is not reached.

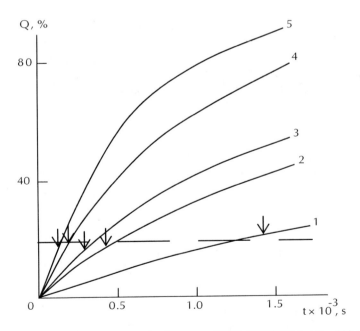

FIGURE 27   The kinetic curves $Q_{(t)}$ for the system 2DPP+HCE/DDM at T_{cur}: 333 (1), 353 (2), 373 (3), 393 (4) and 513 K (5). The vertical arrows indicate the first gelation point.

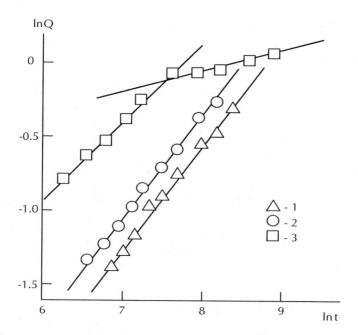

FIGURE 28 The dependences of curing reaction conversion degree Q on reaction duration t in double logarithmic coordinates, corresponding to the Eq. (27) of Chapter 1, for the system 2DPP+HCE/DDM at T_{cur}: 353 (1), 373 (2) and 513 K (3).

In the Table 5, the experimental values of gelation time t_g^e for the first gelation point, determined by IR-spectroscopy method, are adduced. These values t_g^e in Fig. 27 are indicated by vertical arrows. It is interesting to note that the indicated gelation point is reached for all Tcur at approximately the same Q value, which is equal to ~19%. For this observation explanation the authors [50, 51] used the percolation theory [52] and irreversible aggregation model [53], corresponding to many clusters simultaneous growth that agrees with real situation at epoxy polymers curing. According to the model [53], such clusters growth ceases in their contact case. Therefore it is possible to consider the first gelation point, characterized by time t_g^e as the point, in which many spherical microgels contact is realized. According to the percolation theory [52], such spheres volume fraction f can be determined according to the relationship Eq. (23):

$$fx_c \approx 0.15 \ , \qquad\qquad (23)$$

where xc is percolation threshold.

TABLE 5 The dependence of the first gelation point time t_g^e reaching on curing temperature T_{cur} for the system 2DPP+HCE/DDM.

T_{cur}, K	$t_g^e \times 10^{-3}$, s
295	5.04
333	1.44
353	0.36
373	0.30
393	0.18
513	0.12

If to suppose that time t_g^e corresponds to percolation threshold of spherical microgels, closely filling reactionary space, then $x_c = 0.19$ and $f = 0.79$. Such value f corresponds actually to close packing of spheres of approximately equal diameter [54]. From this it follows that the first gelation point is characterized by microgels growth stopping, closely filling reactionary medium at their contact.

From the Eq. (26) of Chapter 1 under the condition K_1 = const and c_0 = const it can be written [51]:

$$t_g^T \sim \left(\frac{Q_1}{\eta_0}\right)^{2/(3-D_f)}, \qquad (24)$$

where t_g^T is theoretical magnitude of the first gelation point reaching time, Q_1 is Q value, corresponding to t_g^e and equal to ~19%.

The combination of Eqs. (12) and (24) allows to estimate the value t_g^T. The comparison of experimental t_g^e and theoretical t_g^T dependences of gelation time for its first point on microgels fractal dimension D_f for the system 2DPP+HCE/DDM is adduced in Fig. 29. As it follows from this comparison, a theory and experiment good correspondence is received (the usage of logarithmic scale for t_g was made from convenience considerations).

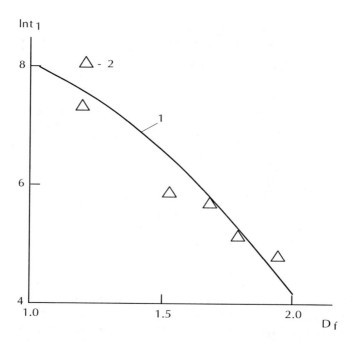

FIGURE 29   The comparison of experimental (1) and theoretical (2) dependences of the first gelation point time t_g^c on microgels fractal dimension Df in logarithmic coordinates for the system 2DPP+HCE/DDM.

Thus, the application of fractal analysis and percolation theory allows to elucidate that the first gelation point of cross-linked polymers in model [49] is structural transition, which is realized at reactionary space filling by microgels. The gelation time in the indicated point is controlled by microgels fractal dimension D_f. Between D_f value and reactionary medium viscosity η_0 the correlation exists: D_f increasing causes strong growth in η_0 [51].

It was shown above that curing of epoxy polymers can proceed in both Euclidean (three-dimensional) and fractal spaces. In the last case on the part of kinetic curve conversion degree-reaction duration $Q_{(t)}$ almost up to gelation point continuous change in the microgels structure occurs, which is characterized by its fractal dimension D_f and more precisely, monotonous increase in D_f occurs. At such D_f variation a curve $Q_{(t)}$ has even qualitative distinctions from a similar curve for curing of epoxy polymers in Euclidean space, namely, practically linear Q growth as a t function is observed in the indicated part up to gelation point (Q < 0.8). The authors of paper [55] studied the reasons and mechanism of microgels structure indicated variation on the example of system EPS-4/DDM.

The estimations of microgels dimension D_f according to the methods described above with the use of the Eq. (27) of Chapter 1 showed that D_f value changes within the limits of 1.61–2.38. Such D_f variation range assumes that microgels formation process proceeds according to the cluster-cluster mechanism, i.e., large microgels are formed from smaller ones, but not directly from oligomeric units [3]. In this case the Eq. (68) of Chapter 1 is just for large microgel dimension D_p, which is formed from smaller ones with dimensions D_{f_1} and D_{f_2} ($D_{f_1} \geq D_{f_2}$). From the indicated equation it follows that in case of microgels distribution by their fractal dimensions absence, i.e., in case $D_{f_1} = D_{f_2} = \ldots = D_{f_i}$ a large cluster differs from smaller ones which form it by size only and the Eq. (68) of Chapter 1 gives $D_f = D_{f_1} = D_{f_2} = \ldots = D_{f_i}$. Hence, the distribution of the microgels structure, characterized by dimensions Df distribution, i.e., $D_{f_1} \neq D_{f_2} \neq \ldots \neq D_{f_i}$, is required for D_f variation at curing reaction course (of t growth). Possible change $D_f(t)$ can be estimated theoretically as follows. If we assume that at time moment ti D_f distribution of microgels is equal to D_{f_i}—$D_{f_{i+1}}$, then according to the Eq. (68) of Chapter 1 $D_{f_i} = D_f$ under the condition $D_{f_{i+1}} = D_{f_1}$, $D_{f_i} = D_{f_2}$. The time range $\Delta t = t_i - t_i - 1 = ti + 1 - ti$ is accepted to be equal to 500 s. The corresponding experimental D_f values are accepted as $D_{f_i}, D_{f_{i+1}}$. In Fig. 30 the comparison of experimental and calculated dependences $D_f(t)$ according to the indicated method for the system EPS-4/DDM is adduced. As one can see, the cluster-cluster aggregation mechanism, for which the Eq. (68) of Chapter 1 was obtained [4], explains quantitatively the variation $D_f(t)$ observed experimentally [55].

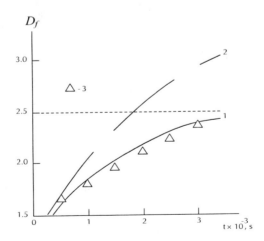

FIGURE 30 The dependences of fractal dimension D_f (D_{sp}) of microgels on reaction duration t for the system EPS-4/DDM. 1 – D_f calculation according to the Eq. (68) of Chapter 1; 2 – D_{sp} calculation according to the Eq. (25); 3 – D_f experimental values.

As it has been noted above, variation $D_f(t)$ reason is curing reaction proceeding in fractal space. This process by its physical significance is similar to the formation of clusters with dimension D_f on lattice with dimension D_{sp} [18]. In paper [44] it was supposed $D_{sp} = D_f$. The relation between D_f and D_{sp} is given by the following equation [18]:

$$D_f = D_{sp} \frac{2 + 2d_l - d_s}{4d_l - d_s} , \tag{25}$$

where dl and ds are chemical and spectral dimensions of cluster (microgel), accordingly.

Between the dimensions D_f and ds the intercommunication is given by the equation (1.39) and ds and dl are linked with one another as follows [56]:

$$d_s = \frac{2d_l}{d_l + 1} , \tag{26}$$

From the Eq. (39) of Chapter 1, (25) and (26) it follows that the condition $D_{sp} = D_f$ is realized only at the following values of the indicated dimensions: $D_f \approx 1.67$, $d_l = 1.0$ and $d_s = 1.0$. The last condition means, that microgel is formed by linear macromolecule [22]. However, at curing process proceeding polymeric chain branching occurs and this results to ds growth [22]. In its turn, as it follows from the Eq. (26), this will cause dl corresponding increase. If as ds to accept this dimension value for branched polymer chain ($d_s = 1.33$ [22]) and as d_l — this dimension value, determined for the formed by mechanism cluster–cluster aggregates in computer simulation ($d_l = 1.42$ [56]), then the Eq. (25) can be written as follows [55]:

$$D_f \approx 0.807 D_{sp} . \tag{27}$$

The Eq. (27) defines the following condition for branched polymers [55]:

$$D_{sp} > D_f . \tag{28}$$

The condition in Eq. (28) allows to suppose that the largest cluster in system is the cluster, forming fractal space in the system EPS-4/DDM curing process

(Fig. 30). As it is known [15], D_f growth is observed at macromolecular coil (microgel) molecular mass increasing. The plots of Fig. 30 reveal precisely such tendency. The tightening cluster, i.e., spreading from one system end up to the other, is such cluster after gel formation point [15]. Such cluster has dimension $D_f \approx 2.5$ [15], which is shown in – by a horizontal shaded line.

In previous sections the curves $Q_{(t)}$ were described within the frameworks of scaling approaches for low-molecular substances reactions [17, 40]. The Eq. (87) of Chapter 1 is used for the description of reaction in medium, having large density fluctuations. In section 1 it was shown that the dependence (1–Q) on parameter $t^{d/(d+2)}$, corresponding to the Eq. (87) of Chapter 1, is non-linear and it can be linearized by microgel dimension D_f usage instead of d in the indicated equation. This served as a reason for the assumption that curing reaction proceeds in fractal space. In Fig. 31 the dependence of ln (1–Q) on parameter $t^{D_{sp}/(D_{sp}+2)}$ is adduced for the system EPS-4/DDM, which also proves to be linear. The data of Fig. 31, together with the assumptions, stated in previous sections, suppose that curing reaction of the system EPS-4/DDM proceeds in fractal space with dimension D_{sp}. Let us note that at $t = 3 \times 10^3$ s the deviation of the adduced in Fig. 31 dependence from linearity is observed. From the comparison with the data of Fig. 30 one can see that this deviation corresponds to $D_{sp} \approx 3$, i.e., to transition to non-fractal behavior at $D_{sp} = d$ that was expected.

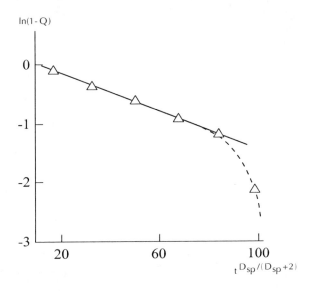

FIGURE 31   The dependence of (1–Q) on parameter $t^{D_{sp}/(D_{sp}+2)}$ in logarithmic coordinates, corresponding to the Eq. (87) of Chapter 1 for the system EPS-4/DDM.

Therefore, the stated above results have confirmed again that D_f values distribution is the main reason of microgels structure variation, characterized by its fractal dimension D_f. D_f change at reaction duration growth is well described quantitatively within the frameworks of aggregation mechanism cluster-cluster. Fractal space, in which curing reaction proceeds, is formed by the structure of the largest cluster in system [55].

In Section 1, the interconnection of reactive medium initial viscosity η_0 and microgels fractal dimension D_f, was shown in the system 2DPP+HCE/DDM case, which was given analytically by the Eq. (12). Let us consider the same question for the system EPS-4/DDM. In Fig. 32, the dependence $D_f(t)$, where D_f value was calculated according to the Eq. (18) of Chapter 2, is adduced for the system EPS-4/DDM. As it was noted above, according to the conception [49] the first gel formation point (theoretical) corresponds to the appearance moment in reactive medium of the first cross-linked microgels, characterized by nonfusibility and nonsolubility. This point corresponds to a slope change at t = 1200 s in the dependence $D_f(t)$ (Fig. 32). The second gelation point corresponds to a reaction considerably later stage – transformation of liquid fluid system with cross-linked particles in elastic polymer. On the dependence $D_f(t)$ this point corresponds to the condition $D_f \approx 2.5$, which is reached at t ≈ 3800 s. The second gelation point corresponds to generally accepted in physics notion about sol-gel transition as network formation tightening sample [15]. Besides, within the frameworks of fractal analysis and irreversible aggregation models it has been shown [15] that this transition is characterized by the system universality class change, namely, by transition from cluster-cluster aggregation (D_f 1.68) at t = 1200 s to particle-cluster aggregation ($D_f \approx 2.5$) at t = 3800 s.

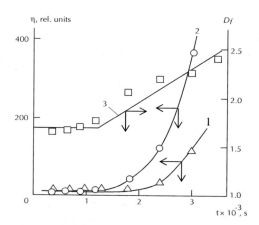

FIGURE 32   The dependences of reactive medium viscosity η (1, 2) and microgels fractal dimension D_f (3) on reaction duration t for the system EPS-4/DDM, cured at T_{cur}=383 (1) and 393 K (2, 3).

The known Q and D_f values allow to estimate parameter $K_1c_0\eta_0$ as t function according to the Eq. (61) of Chapter 2. Since the values K_1 and c_0 in paper [57] experiment conditions are constant, then the indicated parameter can be considered as reactive medium current viscosity η, expressed in relative units. In Fig. 32 the dependences $\eta(t)$ for the system EPS-4/DDM at T_{cur}=383 and 393 K are shown. As one can see, up to microgels formation point (t = 1200 s) very weak η growth is observed and then sharp η increasing (on about two orders) occurs. Such η increasing can be explained theoretically within the frameworks of the model, proposed in paper [25], which uses percolation theory representations. η value is given as follows [25]:

$$\eta \sim |Q_c - Q|^{-m}, \qquad (29)$$

where Q_c is reaction conversion critical degree, corresponding to the second gel formation point or percolation network formation, m is critical index. Q_c value was accepted equal to maximum value Q ($Q_c \approx 0.94$ for T_{cur}=383 and 393 K).

In its turn, complex critical index m is determined as follows [25]:

$$m = \frac{vd_w d_s}{2}, \qquad (30)$$

where v is percolation critical index, which is equal to 0.8 [25], d_w is random walk dimension, ds is spectral dimension.

In Fig. 33, the comparison of experimental (calculation according to the Eq. (61) of Chapter 2) and theoretical (calculation according to the Eq. (29)) η values, which are plotted for convenience in logarithmic scale, is adduced.

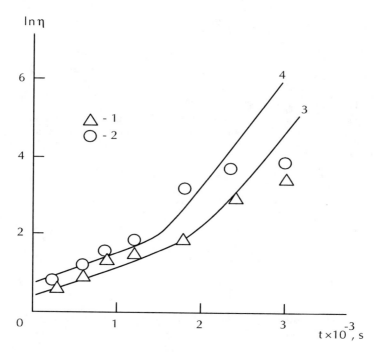

FIGURE 33 The comparison of the dependences of reactionary medium viscosity η on reaction duration t calculated according to the Eq. (61) of Chapter 2 (1, 2) and the Eq. (29) (3, 4) in logarithmic coordinates for the system EPS-4/DDM, cured at T_{cur} = 383 (1, 3) and 393 K (2, 4).

As one can see, the good correspondence of values η, calculated by the indicated methods, was obtained. It is particularly important to note that the Eq. (29) explains η sharp growth at t > 1200s. The experimental and theoretical η values discrepancy is due to a number of the used at their calculation approximations.

The Eq. (29) indicates unequivocally η sharp growth reason at t > 1200s. From this relationship follows that at t growth exponent m increases owing to raising D_f and, hence, d_w at ν = const and d_s = const. This means, that $|Q_c-Q|$ decreasing or curing reaction conversion degree increasing is the only reason of sharp η growth. This postulate can be confirmed within the frameworks of Muthukumar conception [58], describing viscosity of branched polymeric fractals solutions. According to the conception [58], viscosity increment Δη depends on molecular weight MM of such fractal (microgel) as follows:

$$\Delta \eta \sim c_0^{-2/(D_f-d)} MM^{2/D_f},$$ (31)

where d is dimension of Euclidean space, in which a fractal is considered. It is obvious that in our case d=3.

The dependences $MM_{(t)}$ for the system EPS-4/DDM, cured at two T_{cur}, are adduced in Fig. 34. As one can see, weak growth MM is observed up to the first gel formation point, indicated in Fig. 34 by stroke-dotted lines, and in the range of t = 1200–3600s strong (five-fold) MM increasing occurs. This means that MM growth and, hence η is due to curing process of space between microgels and microgels themselves in the indicated range t.

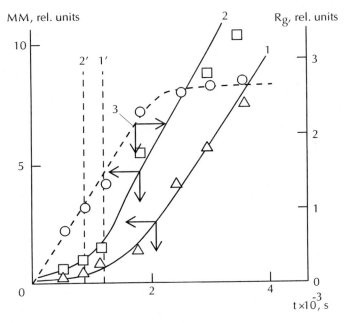

FIGURE 34 The dependences of molecular weight MM (1, 2) and microgels gyration radius R_g (3) on reaction duration t for the system EPS-4/DDM, cured at T_{cur}=383 (1) and 393 K (2, 3). The lines 1' and 2' show the first gelation point at T_{cur}=383 (1') and 393 K (2').

Let us consider the change in curing reaction process of cured microgels gyration radius R_g. R_g value is connected with MM according to the fractal Eq. (1) of Chapter 1. In Fig. 34, the dependence $R_g(t)$ for the system EPS-4/DDM, cured at T_{cur} = 393 K, is adduced. It is interesting to note the absence of similarity between the dependences MM(t) and $R_g(t)$. This observation requires special explanation. D_f value in solution with consideration of the excluded volume interactions can be determined according to the Eq. (39) of Chapter 2. Gelation period (the range of t = 1200–3800s, see Fig. 32) or transition in condensed state is characterized by microgels environment change and now instead of solvent, oligomer and curing agent low-molecular molecules they are in similar microgels environment. This results to fractal dimension change and for the condensed state its value df is determined according to the Eq. (100) of Chapter 1. The Eqs. (39) and (100) of Chapter 1 combination at the indicated above conditions $d_s = 1.33$ and $d = 3$ gives for branched (cross-linked) polymers [57]:

$$d_f \approx 1.6 \ D_f . \tag{32}$$

From the Eq. (32), it follows that at $D_f \approx 1.81$ d_f value will be larger than d. Such microgels growth is restricted since they do not "enter" in three-dimensional space and their density grows with sizes increasing (the effect is similar to blood coagulation) [59]. This results to microgels compactness increasing, expressed by D_f growth (Fig. 32) and the formation from them in the condensed state denser morphological formations (floccules) [60]. Let us note that D_f growth begins at the point $t \approx 1200$ s, where $D_f \approx 1.81$. The Fig. 34 data show that within the range of t = 1800–3600 s at twofold MM increase very weak R_g growth occurs that is confirmed by the conclusion made above.

Hence, the stated above results demonstrated that the reason of reactive medium viscosity sharp increase after microgels formation was curing of space between them and corresponding growth of curing reaction conversion degree. The growth of microgels molecular weight after the mentioned point occurs practically at their constant sizes. These data confirm and work out the details of the conception of "gelation period."

As it is known [61], in case of different chemical reactions proceeding, including cross-linked polymers curing, an essential role is played by the so-called steric factor p (p ≤ 1), showing that not all reacting molecules collisions occur with proper for chemical bond formation these molecules orientation. This factor importance in such treatment is defined by its proportionality to reaction rate constant k_r – the smaller p the smaller k_r and reaction proceeds with less rate.

As it is elucidated within the frameworks of irreversible aggregation processes computer simulation, steric factor p plays in them an important role, defining in essence both aggregation process mechanism and final aggregate structure, characterized by its fractal dimension D_f [62]. So, at larger values p, close to one, diffusion-controlled mechanism of aggregate growth with relatively small values D_f (~1.65) is realized and at small p of order 0.01 mechanism of chemically-limited aggregation with more compact final aggregates ($D_f \approx 2.11$) is realized [3, 63].

Proceeding from the considered above importance of steric factor p, the authors of paper [64] performed the study of change character of the indicated parameter value in epoxy polymers curing process on the system EPS-4/DDM example.

Let us consider the interconnection of steric factor p value with reaction products structure, which can be characterized by its fractal dimension D_f. Using the Eqs. (74) and (75) of Chapter 1 in the supposition that p = 1 is reached at very high cluster "accessibility" degree and estimating this degree by minimal value $D_f = 1.5$ [65] and assuming also that the indicated D_f value is realized at small t of order 100 s [57], the authors [64] obtained the following relationship:

$$p \approx \frac{3.16}{t^{(D_f - 1)/2}} . \tag{33}$$

The other limiting case of the model [62] corresponds to $D_f = 2.11$. Assuming that the indicated D_f value is reached at t = 3000 s [57] (see also Fig. 32) let us obtain from the Eq. (33) that these conditions correspond to $p \approx 0.037$ that is close enough to p = 0.01 according to the model [62].

Using the Eq. (18) of Chapter 2, D_f variation can be calculated as a function of Q or t. In Figs. 35–37 the dependences p(t), calculated according to the Eq. (33) for the system EPS-4/DDM at curing temperatures $T_{cur} = 383$, 393 and 403 K, respectively, are adduced. These figures comparison shows qualitative identity of the adduced in them dependences – at curing reaction proceeding p value decreases, being in small t part this decay is rapid enough and then it decelerates sharply at $p \leq 0.1$ reaching. T_{cur} change has clearly expressed quantitative consequences. So, the initial values p are higher for larger T_{cur}. In other words, the larger Tcur is the more pronounced reaction diffusive regime is expressed. Besides, Tcur increasing results to faster p decay with time t that is due to reaction rate increasing.

In Figs. 35–37 the dependences of chromatographic peak IGC height h (in relative units) as a function of t for corresponding T_{cur} are also shown. It is easy to see, that minimum on curves $h_{(t)}$, designated by the letter A and identified as corresponding to gel formation point, coincides on time scale with a bend in curve $p_{(t)}$ for all three used T_{cur}. As it was noted above, the bend point in curves $p_{(t)}$, corresponding to the condition $p \leq 0.1$, answers to the formation of compact clusters with large fractal dimension $D_f \geq 2.11$ [63].

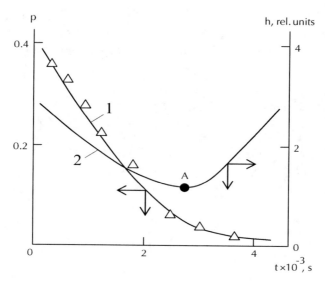

FIGURE 35   The dependences of steric factor p (1) and chromatographic peak height h (2) on reaction duration t for the system EPS-4/DDM at $T_{cur} = 383$ K.

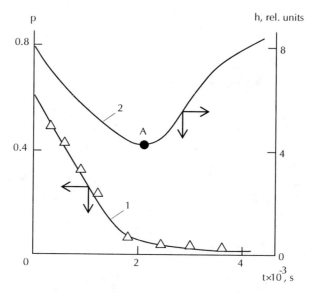

FIGURE 36   The dependences of steric factor p (1) and chromatographic peak height h (2) on reaction duration t for the system EPS-4/DDM at $T_{cur} = 393$ K.

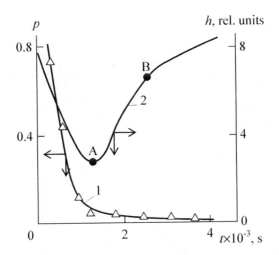

FIGURE 37 The dependences of steric factor p (1) and chromatographic peak height h (2) on reaction duration t for the system EPS-4/DDM at Tcur=403 K.

In Fig. 38, the kinetic curves $Q_{(t)}$ for the system EPS at three used T_{cur} are adduced. The points in the curves $Q_{(t)}$, corresponding to points A in cures $h_{(t)}$ (Figs. 35–37) are indicated by horizontal arrows.

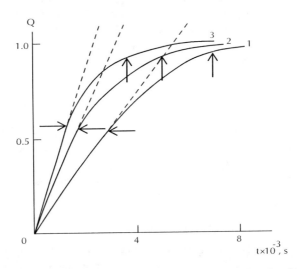

FIGURE 38 The kinetic curves $Q_{(t)}$ for the system EPS-4/DDM at T_{cur}: 383 (1), 393 (2) and 403 K (3). The shaded lines are tangents to the initial parts of curves Q(t). Horizontal arrows indicate completion of microgels formation, vertical arrows indicate spatial network of entire sample formation.

The shaded lines in Fig. 38 show tangents to the initial parts of curves $Q_{(t)}$. And at last, the vertical arrows in Fig. 38 show points in curves $Q_{(t)}$, corresponding to points B in curves $h_{(t)}$ – the so-called glass transition (see Fig. 37). Let us note that the points in curves $Q_{(t)}$, corresponding to A, define slope reduction of tangents to the indicated curves or reaction rate decreasing. The points in curves $Q_{(t)}$, corresponding to B, signify the system fractal dimension change (that is easy to define out of the dependence $Q_{(t)}$ in double logarithmic coordinates, corresponding to the Eq. (27) of Chapter 1) or system universality class change [3].

The data of Figs. 35–38 allow to describe the system EPS-4/DDM curing process within the frameworks of irreversible aggregation models and fractal analysis as follows. The formation of separate cross-linked clusters (microgels) occurs in curing reaction initial stage [48]. The range of these clusters fractal dimensions ($D_f = 1.60$–2.22) indicates that this formation is realized by cluster-cluster mechanism, i.e., small microgels form a larger microgel and so on [3]. This excludes the formation of chemical cross-linking homogeneous network over the entire sample (by the type of "tennis-racket") [66, 67]. When microgels dimension reaches limiting value ($D_f \approx 2.20$ [68]), their growth is ceased. This situation in curves $p_{(t)}$ corresponds to a bend, in curves $h_{(t)}$ – to point A (to minimum), in curves $Q_{(t)}$ to reaction rate reduction. During this stage reactive medium viscosity systematic growth occurs that is reflected in h reduction, which is proportional to square root of diffusivity [64]. The bend point in curves $p_{(t)}$ can be identified as transition from a diffusive regime of curing reaction to a kinetic one and this transition is characterized by the following parameters: $p \approx 0.1$ and $D_f \approx 2.1$. It should be noted that similar classification of curing regimes is highly conditional. It is significant that if the indicated transition corresponds to different t on temporal scale for various T_{cur}, then on the scale Q it is reached at the condition $Q \approx \text{const} \approx 0.57$ (Fig. 38).

After microgels formation curing reaction of space between them begins, which, as it is noted above, occurs according to particle-cluster mechanism. Oligomer and curing agent molecules smaller sizes in comparison with microgels size result to h growth and, hence, to diffusivity increase. In point B in curves $h_{(t)}$ this process is completed and now the system EPS-4/DDM structure is in condensed state and is described within Witten–Sander model with fractal dimension $d_f \approx 2.5$ [69]. Therefore, universality class replacement from cluster-cluster to particle-cluster occurs [3]. Gel formation by similar mode is confirmed experimentally in paper [16] on the example of polystyrene physical gels.

It is obvious that the system EPS-4/DDM proposed curing mechanism corresponds completely to the notions about gel formation phenomena as "gel formation period", but not "gelation point" [48]. This period occupies the temporal range A–B in the curve $h_{(t)}$ (Fig. 37). Strictly speaking, gelation is the critical structural transition and it should be identified as spatial network formation tightening the entire reactive system [15]. Besides, it has been shown both experimentally [16] and theoretically [15] that in the gel point fractal dimension of gel-forming system structure is equal to ~ 2.5. Therefore the gel formation point is identified in such (physically the most strict) treatment as point B in curve $h_{(t)}$.

p reduction at t growth (Figs. 35–37) is due to purely steric reasons, caused by reacting objects structure complexity. The more complex this structure is the more difficult it's to realize such reacting objects mutual orientation, which allows chemical reaction [70]. The most powerful factor in this sense is aggregate internal regions screening by external ones that is typical for fractal objects [53]. In Fig. 39 the dependences $p_{(Q)}$ for the system EPS-4/DDM at the three used values T_{cur} are adduced. These dependences demonstrate clearly p change in curing reaction course: at reaction proceeding p reduces linearly and at Q = 1 p=0 [64].

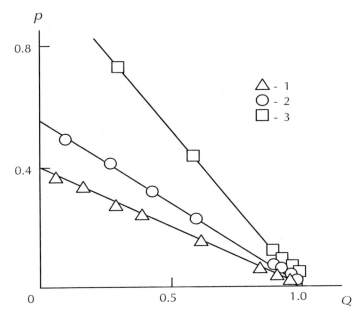

FIGURE 39 The dependences of steric factor p on curing reaction conversion degree Q for the system EPS-4/DDM at T_{cur}: 383 (1), 393 (2) and 403 K (3).

Hence, the stated above results have shown that the change of microgels structure, characterized by its fractal dimension, in the system EPS-4/DDM curing reaction course influences on both steric factor value and curing reaction conversion degree. The irreversible aggregation models and fractal analysis application allows to receive complete description of the system EPS-4/DDM curing reaction and to give physical treatment of gelation process in them.

4.4   SYNERGETICS OF THE CURING PROCESS

As it is known [71], synergetics studies universal rules of spatial structures self-organization in dynamic systems of different nature. This discipline is based on physical essence of systems adaptation process to external influence by way of structures self-organization and is a universal one for animate and inanimate nature systems. Adaptation is the process of structure reforming, lost stability, with new more stable structure self-organization. Fractal (multifractal) structures are formed in reformation process, which is impossible to describe correctly within the frameworks of Euclidean geometry [71].

Everything said above related to a full extent to polymers synthesis processes of various types – in these processes by self-organization way fractal objects (macromolecular coils, microgels) are formed, for which temperature, synthesis (curing) duration and so on are external influence. The authors of paper [72] studied the indicated factors influence on epoxy polymers curing process within the frameworks of synergetics.

The calculated with the aid of the Eq. (27) of Chapter 1 D_f values for the system 2DPP+HCE/DDM are adduced in Table 6. As it follows from this table data, T_{cur} increasing results to microgels fractal dimension within broad enough range (D_f= 1.20–1.95). The spectral dimension ds, characterizing microgels connectivity degree [22], can be determined with the aid of the equation (1.39). The calculated according to the indicated equation ds values for the studied epoxy polymers microgels are also adduced in Table 6. It is significant that for two the least curing temperatures (333 and 353 K) linearly connected microgels are not formed (d_s <1), at T_{cur} = 373 K macromolecular coil is formed ($d_s \approx 1$) and only at T_{cur} =393 and 513 K completely cured microgels are formed ($d_s \rightarrow 1.33$). This means that in T_{cur}=333 and 353 K case epoxy polymers curing is realized on gel formation stage only, i.e., at t >1.5 ×10^3 s [39].

TABLE 6   The dimensions of cross-linked structures for the system 2DPP+HCE/DDM.

T_{cur}, K	Up to gelation point		After gelation point
	D_f	d_s	d_f
333	1.20	0.74	2.30
353	1.53	0.86	2.34
373	1.68	1.01	2.45
393	1.78	1.24	2.79
513	1.95	1.32	2.80

As it is known [71], the adaptivity universal algorithm, described by the relationship (2.58), is realized at the self-organization of structures to transition from the previous point of structure instability to the following one. As a governing parameter at epoxy polymers curing two factors can be used: curing temperature T_{cur} and curing duration t, which reflect external influence on self-organization process of cross-linking spatial structures. Let us consider the first Am estimation with using as a governing parameter t (A_m^t). With this purpose let us accept Z_n and Z_{n+1} equal to Q values at t=500 and 1500s, respectively. In Fig. 40, the dependence D_f of microgels and d_f of the condensed state structure after gel formation point on A_m^t value is adduced.

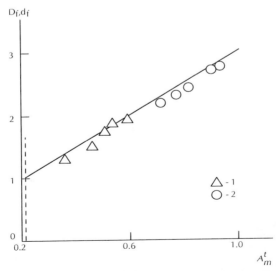

FIGURE 40   The dependence of cross-linked structures fractal dimension up to D_f(1) and after d_f (2) gelation point on adaptability measure A_m^t for the system 2DPP+HCE/DDM.

The data of this figure allow to make the following conclusions. Firstly, for cross-linked structures A_m^t value is necessary, which is larger than minimal adaptability measure A_m^{min}, equal to 0.213 according to the gold proportion rule [71]. Secondly, the maximum value $A_m^t \approx 1.0$ gives Euclidean structures with dimension $d_f = d = 3$. Thirdly, the dependence of fractal dimension of the formed cured structures on A_m^t has the following form [72]:

$$D_f(d_f) = \frac{A_m^t - A_m^{min}}{0.42} + 1 \ .$$ (34)

From the Eq. (34) it follows that in the considered case adaptivity measure is equivalent to the system ability to curing. In its turn, limiting conversion degree Q_{lim} is defined by microgels dimension D_f, formed up to a gelation point that follows directly from the Fig. 41 data. As one should expect, for oligomer molecules $(D_f = 1)$ $Q_{lim} = 0$, i.e., curing reaction does not occur. The dependence $Q_{lim}(D_f)$ can be described according to the following equation [72]:

$$Q_{lim} = 0.79(D_f - 1) \ .$$ (35)

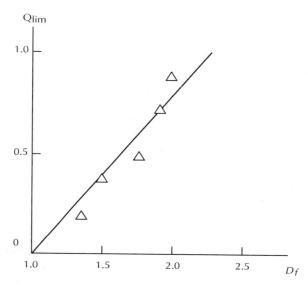

FIGURE 41   The dependence of limiting conversion degree Q_{lim} of curing reaction on microgels fractal dimension D_f for the system 2DPP+HCE/DDM.

Let us note that maximum value $Q_{lim} = 1.0$ is reached at $D_f = 2.27$, i.e., at the greatest possible dimension of macromolecular coil (microgel) of branched (cross-linked) polymer [68].

The Eqs. (34) and (35) combination allows to obtain the dependence of Q_{lim} on A'_m [72]:

$$Q_{lim} = 2.05\left(A'_m - A^{min}_m \right). \qquad (36)$$

It is significant that for maximum value $Q_{lim} = 1.0$ $A'_m = 0.70$ that according to the gold proportion rule corresponds to $\Delta i = 0.232$, m=4 and for the smallest value Q_{lim} of order 0.05 $A'_m = 0.232$, i.e., it corresponds to $\Delta i = 0.232$, m = 1. In other words, limiting conversion degree sharp increase is realized at constant and small microgel stability measure, corresponding to its ability to structure change (curing) owing to possible reformation number m change (increasing) only.

The proposed synergetic model allows to estimate temperature boundaries of curing process realization and to select its optimal temperature regime. For this let us estimate system adaptability measure by curing temperature A^T_m as follows [72]:

$$A^T_m = \frac{T_{cur\,i}}{T^\infty_{cur}}, \qquad (37)$$

where maximum value $T_{cur} = 513$ K was used as T^∞_{cur}.

In Fig. 42, the relation of adaptability measure A'_m and A^T_m for the system 2DPP+HCE/DDM up to and after gel formation point is adduced. As it follows from this figure data, at A^T_m growth A'_m increase is observed and at $A^T_m \approx 0.785$ A'_m value reaches its asymptotic value (~ 0.60 up to gelation point and ~ 0.95 after it). The indicated A^T_m value corresponds to $T_{cur} = 403$ K. At A^T_m, which are equal to 0.60 up to gel formation point and to 0.54 after this point $A'_m = 0$, i.e., curing process is not realized. The indicated A^T_m magnitudes allow to determine T_{cur} lower value as equal to 287 and 318 K, respectively. Let us note that curing after gelation point can proceed at lower T_{cur} than before it (even at room temperature).

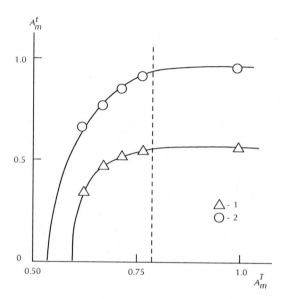

FIGURE 42 The relation of adaptability measures by curing duration A_m^t and curing temperature A_m^T up to (1) and after (2) gelation point. The vertical shaded line indicated A_m^t value, corresponding to curing stationary temperature regime transition for the system 2DPP+HCE/DDM.

From the data of Fig. 42 it follows that at $T_{cur} \geq 403$ K for the system 2DPP+HCE/DDM curing stationary temperature regime is reached, corresponding to the condition A_m^t = const. As it is known [71], for synergetic systems the general law is observed: at external parameter variation system behavior changes from simple to chaotic one. However, the external parameter certain interval exists, in which system behavior is ordered and periodic. The ordering consists in the fact that system behavior is reproduced at each time moment. Doubling number a can be received from the following equation [71]:

$$Z_\infty - Z_n = \delta^{-a},\tag{38}$$

where Z_n is a governing parameter value, at which period doubles in a times, Z_∞ is this parameter limiting value, δ is Feigenbaum's constant ($\delta \approx 4.67$ [71]).

Presenting Z_∞ as $T_{cur}^\infty / T_{cur}^\infty = 1$ and $Z_n = T_{cur} / T_{cur}^\infty$ and assuming also a=1, let us receive [72]:

$$1 - A_m^T = \delta^{-1} \; . \tag{39}$$

From the Eq. (39) the value $\mathrm{T}_{cur} = A_m^T \, T_{cur}^\infty$, corresponding to the transition from simple behavior to chaotic one, can be estimated. This value T_{cur}=403 K, that is excellently agreed with empirically established temperature boundary of transition to curing stationary regime (Fig. 42). Hence, the indicated transition is defined by synergetics laws and is a general effect. So, an analogous phenomenon of transition to chaotic behavior at a = 1 was found out for polymer composites structure [73].

In Fig. 43, the dependence of microgels spectral dimension d_s on A_m^T is adduced for the system 2DPP+HCE/DDM.

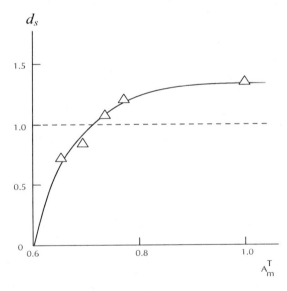

FIGURE 43 The dependence of microgels spectral dimension d_s on adaptability measure A_m^T for the system 2DPP+HCE/DDM. The horizontal shaded line indicates linear connectivity of microgels.

As one can see, the value ds = 1.0 is reached at $\mathrm{T}_{cur} \approx 363$ K, which corresponds to linear macromolecule [22] and at T_{cur}<363 K linearly connected chains in curing process are not formed. At T_{cur}= 403 K, corresponding to the transition to stationary temperature regime ($A_m^T \approx 0.785$), ds value is close to 1.33, i.e., to maximum spectral dimension for tightly cured macromolecules (microgels) [22].

Hence, the stated in the present section results show synergetics principles applicability to epoxy polymers curing processes description. This means, that both microgels up to gel formation point and tightening reactive space cluster after it are self-organizing fractal structures, whose dimension is defined by curing system adaptability measure. The optimal temperature regime of curing process was established [72].

4.5   THE NANODIMENSIONAL EFFECTS IN THE CURING PROCESS IN FRACTAL SPACE

In paper [28], it has been shown that the epoxy polymers curing can occur in both Euclidean three-dimensional space and in the fractal one. In the latter case the space dimension is equal to fractal dimension D_f of microgels, formed in curing process. The main difference of conversion degree-reaction duration (Q–t) kinetic curves in the latter case is practically linear dependence $Q_{(t)}$ almost up to gelation point and variation (increase) of D_f value on this part of curve Q(t). The authors of papers [74–77] carried out further study of epoxy polymers curing in fractal space, in particular the reaction rate constant k_r and microgels self-diffusivity D_{sd} changing character on the example of the system EPS-4/DDM curing [39].

For the system EPS–4/DDM the average value $k_r = 0.97 \times 10^{-3}$ mole l/s was determined by IGC method [39]. From the Eq. (18) of Chapter 2 the value c can be determined at the average magnitudes of the parameters included into it: $t = 1.5 \times 10^3$ s, $D_f = 1.99$ and Q = 0.35. In this case c = 0.0244 mole l/s. As the calculation showed, kr reduction from 4.16 up to 0.76×10^{-3} mole l/s was observed within the range of t = 500–2500 s. The range of the obtained values D_f (1.61–2.38) assumes that the microgels formation occurs according to the cluster-cluster mechanism [78]. In this case the microgels molecular mass MM value is determined according to the scaling relationship (1.53). The microgels gyration radius R_g is connected with MM by the following relationship [13]:

$$R_g \sim MM^{1/D_f} \sim Q^{2/D_f\left(3-D_f\right)} . \qquad (40)$$

The obtained results allow to carry out the system EPS-4/DDM curing kinetics analysis within the frameworks of irreversible aggregation models [79]. In general case the relation between k_r and R_g can be presented as follows [79]:

$$k_r \sim R_g^{2\omega} \ . \tag{41}$$

In its turn, the exponent ω is defined by the parameters describing clusters (microgels) motion in space and their structure. This interconnection has the following form [79]:

$$2\omega = -\gamma + d - d_w \ , \tag{42}$$

where γ characterizes the dependences of microgels self-diffusivity D_{sd} on their sizes ($D_{sd} \sim R_g^{-\gamma}$), d is space dimension, in which the curing reaction occurs, d_w is dimension of microgels random walk trajectory.

For the reactions in Euclidean space d = 3, d_w = 2 (Brownian motion of microgels), γ = 1 and then ω = 0. This means that in the given case the condition should be fulfilled [76]:

$$k_r = \text{const} \ . \tag{43}$$

The condition (43) is confirmed experimentally (k_r value does not change at R_g growth [39]) and is general for any macromolecular reactions in Euclidean spaces [80]. For the curing reaction proceeding in fractal space the situation differs completely from the described above. This aspect attains special meaning within the frameworks of nanochemistry [81], therefore deserves consideration in more detail.

As it is known [81], in nanochemistry there are two fundamental notions – nanoparticle and nanoreactor: the first characterizes dimensional parameter while the second one defines nanoobject function. Thus, iron cluster loses almost completely its specific properties (ionization energy, magnetism) and approaches to metallic iron with an atoms number in cluster n = 15. At n > 15 it remains nanoobject in dimensional sense, but loses "nanoreactor" qualities, for which properties become a size function. In Fig. 44, the dependence of curing rate constant k_r on microgels diameter $2R_g$, which has a very specific form, is adduced.

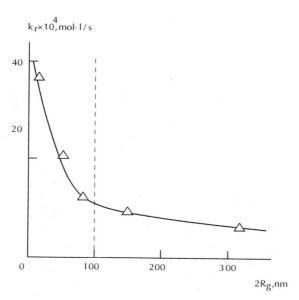

FIGURE 44   The dependence of reaction rate constant kr on microgels diameter $2R_g$ for the system EPS-4/DDM.

Within the range of microgels (although the term "nanogel" is more precise) there are fewer diameters than 100 nm, the value kr is a clearly expressed rapidly decreasing function of diameter $2R_g$ and at $2R_g \geq 100$ nm the indicated dependence is practically absent. Let us note that the size 100 nm is assumed as an upper limit (although conventionally enough) for nanoworld objects [81]. Hence, the data in Fig. 44 clearly demonstrate that microgel at $2R_g < 100$ nm is nanoreactor in which reaction (curing) rate is a strong function of its size, and at $2R_g \geq 100$ nm microgel loses this function and in essence becomes a chemically inert particle. Let us note that the indicated transition nanoreactor-nanoparticle is possible only in the fractal space. In Euclidean space these notions do not differ (kr=const).

In Fig. 45, the dependence $k_r(R_g)$ for the system EPS-4/DDM in double logarithmic coordinates is shown, which is well approximated by a straight line. From the slope of this straight line the value $2\omega = -0.58$ can be determined. As it was noted above, the space dimension, in which curing reaction occurs, is equal to D_f and the value d_w can be determined according to the Aarony–Stauffer rule (Eq. (93) of Chapter 1). Hence, the Eq. (42) for the considered case can be rewritten as follows (for any D_f value) [75]:

$$2\omega = -\gamma - 1. \tag{44}$$

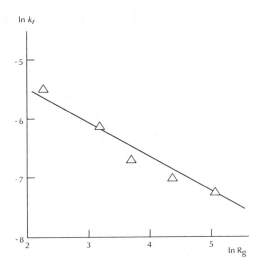

FIGURE 45   The dependence of reaction rate constant k_r on microgels gyration radius R_g in double logarithmic coordinates for the system EPS-4/DDM.

Then according to the Eq. (44) $\gamma = -0.42$ can be obtained. This means that self-diffusivity value D_{sd} decreases with R_g growth much slower ($D_{sd} \sim R_g^{-0.42}$) in comparison with the reaction in Euclidean space ($D_{sd} \sim R_g^{-1}$). The indicated difference is demonstrated in Fig. 46 in a diagram form.

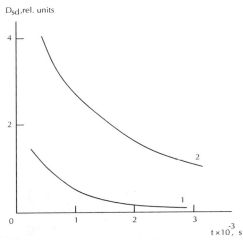

FIGURE 46   The dependences of microgels self-diffusivity D_{sd} on curing reaction duration t in Euclidean (1) and fractal (2) spaces for the system EPS-4/DDM.

The microgels molecular weight MM depends on curing duration t as follows [79]:

$$MM \sim t^{D_f/(D_f - 2\omega)}. \tag{45}$$

As it was noted above, in Euclidean space $2\omega = 0$ and the exponent in the relationship (45) is equal to 1. This assumes $MM \sim t$. For the reaction in fractal space $2\omega < 0$ and the exponent in the Eq. (45) is smaller than 1. This means that in fractal space the value MM grows slower than in Euclidean space. In diagram form this relation for the system EPS-4/DDM is shown in Fig. 47. Since Q value in the second case is larger than in the first then this means that the reaction in fractal space gives a larger number of small clusters (microgels).

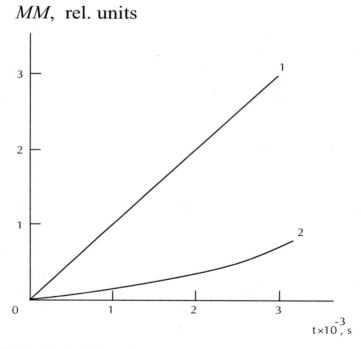

FIGURE 47 The dependence of microgels molecular weight MM on curing reaction duration t in Euclidean (1) and fractal (2) spaces for the system EPS-4/DDM.

The indicated above k_r change can be obtained immediately from Smoluchowski formula, which has the following form [79]:

$$k_r = 8\pi D_{sd} R_g \quad . \tag{46}$$

For the reaction in Euclidean space $D_{sd} \sim R_g^{-1}$ and k_r = const, for the reaction in fractal space for the system EPS–4/DDM $D_{sd} \sim R_g^{-0.42}$ and $k_r \sim R_g^{-0.58}$, i.e., k_r is supposed to be reduced at curing process proceeding (the growth of R_g or MM) [74].

Let us note in conclusion the strong dependence of k_r on the microgels structure, characterized by the fractal dimension D_f (Fig. 48). As it follows from Fig. 48, k_r sharp decay is observed for D_f growth at $D_f < 2$ and attainment of the values k_r on asymptotic branch at $D_f > 2$. As it is known [79], within the frameworks of irreversible aggregation models the following relationship is valid:

$$k_r \sim D_{sd} R_g^{d-2}. \tag{47}$$

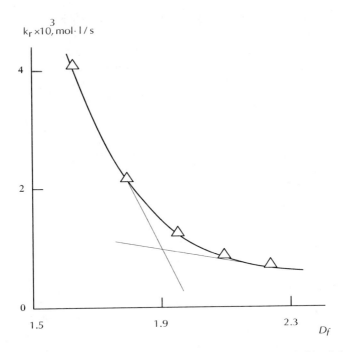

FIGURE 48   The dependence of reaction rate constant k_r on microgels fractal dimension D_f for the system EPS-4/DDM.

If for Euclidean space, for example, with d = 3, the exponent in the Eq. (47) is constant and equal to one, then for fractal space with variable value D_f the situation will be essentially different. For $D_f < 2$ the exponent in the Eq. (47) is less than zero and R_g growth results to k_r reduction under other equal conditions. At $D_f = 2$ k_r does not depended on R_g. And finally, at $D_f > 2$ kr value should increase at R_g growth. This is expressed by the sharp decay of k_r at $D_f < 2$, since both D_{sd} and $R_g^{D_f-2}$ reduce this parameter at microgels MM increasing. At $D_f > 2$ D_{sd} reduction is compensated to a certain extent by $R_g^{D_f-2}$ growth and kr decay at MM increasing is decelerated. It is easy to see that at $D_f=2.48$ for the system EPS-4/DDM the condition k_r=const is realized. Analytically the correlation $k_r(D_f)$ can be presented as follows [75]:

$$\lg k_r = -1.56 - 1.26\left(D_f - 1\right). \tag{48}$$

Therefore, the presented in this section results have shown that for curing reaction in fractal space the reaction rate constant reduction is typical at this reaction proceeding. The formation of a large number of microgels with smaller molecular mass in comparison with reaction in Euclidean space at the same conversion degree is also typical for such reaction. The dimensional border between nanoreactor and nanoparticle for the considered curing reaction has been obtained.

REFERENCES

1. Ligidov, M. Kh.; Kozlov ,G.V.; Bejev, A. A. Homogeneous and heterogeneous kinetics of haloid-contaiting epoxy polymers curing. Proceedings of Higher Schools. *Chemistry and Chemical Technology*, 2001, **44**(3), 27–30.
2. Dolbin, I. V.; Ligidov, M. Kh.; Bejev, A. A.; Kozlov, G. V. Homogeneous and non-homogeneous kinetics of curing of haloid-containing epoxy polymers. Theses of Reports of International conf. *"Vital Problems of Modern Science"*. Part 3. Samara, 12–14 September 2000, **26**.
3. Kokorevich, A. G.; Gravitis, Ya. A.; Ozol-Kalnin, V. G. A scaling approach development at lignin supramolecular structure study. *Chemistry of Wood*, 1989, 1, 3–24.
4. Hentschel, H. G. E.; Deutch, J. M. Flory-type approximation for the fractal dimension of cluster-cluster aggregates. *Phys. Rev. A*, 1984, **29**(3), 1609–1611.
5. Kozlov, G. V.; Bejev, A. A.; Lipatov, Yu. S. The fractal analysis of curing processes of epoxy polymers. *In book: Perspectives on Chem. and Biochemical Physics*. Ed. Zaikov, G. New York, Nova Science Publishers, Inc., 2002, 231–253.
6. Kozlov, G. V; Bejev, A. A.; Lipatov, Yu. S. The fractal analysis of curing processes of epoxy resins. In book: Fractal Ananlysis of Polymers: From Synthesis to Composites.

Ed. Kozlov, G.; Zaikov, G.; Novikov, V. New York, Nova Science Publishers, Inc., 2003, 201–223.

7. Kozlov, G. V.; Bejev, A. A.; Lipatov, Yu. S. The fractal analysis of curing processes of epoxy resins. In book: Polymer Yearbook 18. Ed. Pethrick, R.; Zaikov G. Shawbury, Rapra Technology Limited, 2003, 259–284.

8. Kozlov, G. V.; Zaikov, G. E. Fractal Analysis and Synergetics of Catalysis in Nanosystems, New York, *Nova Biomedical Books*, 2008, 163.

9. Naphadzokova, L. Kh.; Kozlov, G. V. Fractal Analysis and Synergetics of Catalysis in Nanosystems, Moscow, Publishers of Academy of Natural Sciences, 2009, 230.

10. Magamedov, G. M.; Kozlov, G. V.; Zaikov, G. E. Structure and Properties of Gross-Linked Polymers. Shawbury, A Smithers Group Company, 2011, 492.

11. De Gennes, P. The Scaling Concrpts in Physics of Polymers. Moscow, World, 1982, 368.

12. Kozlov, G. V.; Malkanduev, Yu. A.; Burya, A. I.; Sverdlikovskaya, O. S. Kinetics of initial polymerization of dimethyldiallylammoniumchloride. *Problems of Chemistry and Chem. Technology*, 2003, (2), 73–77.

13. Kozlov, G. V.; Shustov, G. B. The fractal physics of polycondensation processes. *In book: Achievements in Field of Polymers Physics-Chemistry*. Ed. Zaikov, G. a.a. Moscow, Chemistry, 2004, 341–411.

14. Kozlov, G. V.; Shustov, G. B.; Zaikov, G. E. Fractal analysis of copolymerization processes. J. Appl. *Polymer Sci.*, 2009, **111** (7), 3026–3030.

15. Botet, R.; Jullien, R.; Kolb, M. Gelation in kinetic growth models. Phys. Rev. A, 1984, **30**, 4, 2150–2152.

16. Kobayashi, M,; Yoshioka, T.; Imai, M.; Itoh, Y. Structural ordering on physical gelation of syndiotactic polystyrene dispersed in chloroform studied by time-resolved measurements and small angle neutron scattering (SANS) and infrared spectroscopy. Macromolecules, 1995, **28**(22), 7376–7385.

17. Djordjevič, Z. B. Observation of scaling in reaction with traps. *In book: Fractals in Physics*. Ed. Pietronero, L.; Tosatti, E.; Amsterdam, Oxford, New York, Tokyo, North-Holland, 1986, 581–585.

18. Vannimenus, J. Phase transitions for polymers on fractal lattices. *Physica D*, 1989, **38**, (2), 351–355.

19. Aharony, A.;Harris, A. B. Flory approximant for self-avoinding walks on fractals. *J. Stat. Phys.*, 1989, **54**, (3/4), 1091–1097.

20. Pfeifer, P.; Avnir, D.; Farin, D. Scaling behavior of surface irregularity in the molecular domain: from adsorption studies to fractal catalysts. *J. Stat. Phys.*, 1984, **36**, (5/6), 699–716.

21. Sahimi, M.; McKarnin, M.; Nordahl, T.; Tirrell, M. Transport and reaction on diffusion-limited aggregates. Phys. Rev. A, 1985, **32** (1), 590–595.

22. Alexander, S.; Orbach, R. Density of states on fractals: "fractons". J. Phys. Lett. (Paris), 1982, **43** 17, L625-L631.

23. Feder, E. Fractals. New York, Plenum Press, 1990, 242

24. Kozlov, G. V.; Temiraev, K. B.; Afaunov, V. V. Influence of reactionary mass stirring on main parameters of interfacial polycondensation. Plastics, 2000, (2), 23–24.

25. Hess, W.; Vilgis, T. A.; Winter, H. H. Dynamical critical behavior during chemical gelation and vulcanization. Macromolecules, 1988, **21** (8), 2536–2542.

26. Dolbin, I. V.; Kozlov, G. V. The features of curing process of epoxy polymers in fractal space. Mater. of All-Russian Conf. "Perspective–2005", *II*, Nal'chik, KBSU, 2005, 114–117.

27. Kozlov, G. V.; Bejev, A. A.; Zaikov, G. E. The physical reasons of homogeneous and nonhomogeneous reactions of haloid-containing epoxy polymers curing. *Oxidation Commun.*, 2002, **25** (4), 529–534.

28. Kozlov, G. V.; Bejev, A. A.; Zaikov, G. E. The physical reasons for the homogeneous and nonhomogeneous reactions of haloid-containing epoxy polymers curing. *J. Appl. Polymer Sci.*, 2003, **90** (5), 1202–1205.

29. Morgan, P. U. Polycondensation Methods of Polymers Synthesis, Moscow, Chemistry, 1970, 376 p.

30. Kozlov, G. V.; Bejev, A. A. Autoacceleration (autostopping) in reactions of curing of cross-linked polymers: fractal analysis. *In book: Fractals and Local Order in Polymeric Materials*. Ed. Kozlov, G.; Zaikov, G.; New York, Nova Science Publishers, Inc., 2001, 37–42.

31. Kozlov, G. V.; Beloshenko, V. A.; Kuznetsov, E. N.; Lipatov, Yu. S. The molecular parameters of epoxy polymers change in their curing process. *Reports of National Academy of Sciences of Ukraine*, 1994, **12**, 126–128.

32. Kozlov, G. V.; Zaikov, G. E.; Mikitaev, A. K. The Fractal Analysis of Gas Transport in Polymers: *The Theory and Practical Applications*. New York, Nova Science Publishers, Inc., 2009, 238.

33. Kozlov, G. V.; Zaikov, G. E.; Mikitaev, A. K. The Fractal Analysis of Gas Transport Process in Polymers, Moscow, Science, 2009, 199.

34. Kozlov, G. V.; Burya, A. I. Scaling and fractal analysis of haloid-containing epoxy polymer curing. *Composite Materials*, 2008, **2** (2), 31–35.

35. Shogenov, V. N.; Kozlov, G. V. Fractal Clusters in Physics-Chemistry of Polymers. Nal'chik, Polygraphservice and T, 2002, 268.

36. Kozlov, G. V.; Zaikov, G. E. Reaction and structure formation of polymers in fractal spaces. *J. Balkan Tribological Association*, 2004, **10** (1), 1–30.

37. Kozlov, G. V.; Bejev, A. A.; Zaikov, G. E. The physical reasons of homogeneous and nonhomogeneous reactions of haloid-containing epoxy polymers curing. *In book: New Perspectives in Chemistry and Biochemistry*. Ed. Zaikov, G. New York, Nova Science Publishers, Inc., 2002, 27–33.

38. Kozlov, G. V.; Bejev, A. A.; Dolbin, I. V. Change of microgel structure on curing of epoxy polymer in fractal space. *In book: Polymer Yearbook 18*. Ed. Pethrick, R.; Zaikov, G. Shawbury, Rapra Technology Limited, 2003, 373–378.

39. Kozlov, G. V.; Bejev, A. A.; Lipatov, Yu. S. The fractal analysis of curing processes of epoxy resins. *J. Appl. Polymer Sci.*, 2004, **92** (4), 2558–2568.

40. Meakin, P.; Stanley, H. E. Novel dimension-independent behavior for diffusive annihilation on percolation fractals. *J. Phys. A*, 1984, **17** (1), L173-L177.

41. Klymko, P. W.; Kopelman, R. Fractal reaction kinetics: exciton fusion on clusters. *J. Phys. Chem.*, 1983, **87** (23), 4565–4567.

42. Zelenyi, L. M.; Milovanov, A. V. Fractal topology and strange kinetics: from percolation theory to cosmic electrodynamics problems. *Achievements of Physical Sciences*, 2004, **174** (8), 809–852.

43. Kozlov, G. V.; Dolbin, I. V.; Zaikov, G. E. The theoretical estimation of effective spectral dimension for polymer melts. *J. Appl. Polymer Sci.*, 2004, **94** (4), 1353–1356.

44. Kozlov, G. V.; Malkanduev, Yu. A.; Burmistr, M. V.; Korenyako, V. A. Fractal model of reactionary medium for radical polymerization of dimethyldiallylammonium chloride. *Problems of Chemistry and Chem. Technology*, 2004, **4**, 101–105.

45. Matsushita, M.; Honda, K.; Toyoki, H.; Hayakawa, Y.; Kondo, H. Generalization and the fractal dimensionality of diffusion-limited aggregation. *J. Phys. Soc.* Japan, 1986, **55** (8), 2618–2626.

46. Kozlov, G. V.; Bejev, A. A.; Shustov, G. B.; Lipatov, Yu. S. The features of epoxy polymers curing in fractal space. *Theses of Reports of 7-th International Conf. by Chemistry and Physics-Chemistry of Oligomers "Oligomers–2000"*. Moscow-Perm'-Chernogolovka, 4–8 September 2000, 207.

47. Kozlov, G. V.; Bejev, A. A.; Shustov, G. B.; Ligidov, M. Kh.; Bejeva, D. A. A microgels formation in epoxy polymers curing process. Mater. of International Sci.-Techn. Conf. *"New Materials and Technologies on Centuries Boundary"*. Part I. Penza, PSU, 2000, 37–39.

48. Pakter, M. K.; Paramonov, Yu. M.; Belaya, E. S. Structure of Epoxy Polymers. Moscow, NIITEKhIM, 1984, 46.

49. Lipatova, T. E. Catalytic polymerization of oligomers and formation polymeric networks. *Kiev, Scientific Thought*, 1974, 298.

50. Bejev, A. A.; Kozlov, G. V.; Ligidov, M. Kh. The conditions of microgels formation in epoxy polymers on the basis of hexachlorethane curing process. Proceedings of Higher Schools. *Chemistry and Chem. Technology*, 2001, **44** (1), 47–49.

51. Kozlov, G. V., Bejev, A. A.; Dolbin, I. V. Change of microgel structure at curing of epoxy polymers in fractal space. *J. Balkan Tribological Association*, 2004, **10** (1), 31–35.

52. Shklowskii, B. I.; Efros, A. L. Percolation theory and strongly heterogeneous mediums conductivity. *Achievements of Physical Sciences*, 1975, **117** (3), 401–436.

53. Witten, T. A.; Meakin, P. Diffusion-limited aggregation at multiple growth sites. *Phys. Rev. B*, 1983, **28** (10), 5632–5642.

54. Bobryshev, A. N.; Kozomazov, V. N.; Babin, L. O.; Solomatov, V. I. Synergetics of Composite Materials, Lipetsk, NPO ORIUS, 1994, 154.

55. Kozlov, G. V.; Bejev, A. A.; Dolbin, I. V. Change of microgels structure at curing of epoxy polymers in fractal space. *Russian Polymer News*, 2003, **8** (2), 65–68.

56. Meakin, P.; Majid, I.; Havlin, S.; Stanley, H. E. Topological properties of diffusion limited aggregation and cluster-cluster aggregation. *J. Phys. A*, 1984, **17** (8), L975-L981.

57. Kozlov, G. V.; Zaikov, G. E. A reactive medium viscosity change in cross-linking process of epoxy polymers. *Polymer Research J.*, 2008, 2 3, 315–322.

58. Muthukumar, M. Dynamics of polymeric fractals. *J. Chem. Phys.*, 1985, **83** (6), 3161–3168.

59. Balankin, A. S.; Ivanova, V. S.; Kolesnikov, A. A.; Savitskaya, E. E. Fractal kinetics of dissipative structures self-organization in mechanical alloying process in attritors. *Letters in Journal of Technical Physics*, 1991, **17** (14), 27–30.

60. Kozlov, G. V.; Burmistr, M. V.; Korenyako, V. A.; Zaikov, G. E. Kinetics of dissipative macrostructures formation in cross-linked polymers curing process. *Problems of Chemistry and Chem. Technology*, 2002, **6**, 77–81.

61. Barns, F. S. Influence of electromagnetic fields on chemical reactions rate. *Biophysics*, 1996, **41** (4), 790–802.

62. Jullien, R.; Kolb, M. Hierarchical method for chemically limited cluster-cluster aggregation. *J. Phys. A*, 1984, **17** (12), L639-L643.

63. Brown, W. D.; Ball, R. C. Computer simulation of chemically limited aggregation. *J. Phys. A*, 1985, **18** (9), L517-L521.

64. Kozlov, G. V.; Zaikov, G. E.; Artsis, M. I. A reactive medium viscosity change in cross-linking process of epoxy polymers. In book: *Chemistry and Biochemistry. From Pure to Applied Science*. New Horizons. Ed. Pearce E. M.; Zaikov G.; Kirshenbaum G.; New York, Nova Science Publishers, Inc, 2009, 71–78.

65. Baranov, V. G.; Frenkel, S. Ya.; Brestkin, Yu. V. Dimensionality of different states of linear macromolecule. *Reports of Academy of Sciences of SSSR*, 1986, **290** (2), 369–372.

66. Adolf, D.; Hance, B.; Martin, J. E. Remnant percolative disorder in highly-cured networks. *Macromolecules*, 1993, **26** (11), 2754–2758.

67. Kozlov G.V.; Novikov V.U.; Mikitaev A.K. Fractal analysis of structure elements connectivity with elasticity modulus for cross-linked polymers. Materialovedenie, 1997 4, 2–5.

68. Family, F. Fractal dimension and grand universality of critical phenomena. *J. Stat. Phys*, 1984, **36** (5/6), 881–896.

69. Kozlov, G. V.; Beloshenko, V. A.; Varyukhin, V. N. Simulation of cross-linked polymers structure as diffusion-limited aggregate. *Ukrainian Phys. Journal*, 1998, **43** (3), 322–323.

70. Kolb, M.; Jullien, R. Chemically limited versus diffusion limited aggregation. *J. Phys. Lett. (Paris)*, 1984, **45** (10), L977-L981.

71. Ivanova, V. S.; Kuzeev, I. R.; Zakirnichaya, M. M. Synergetics and Fractals. *Universality of Materials Mechanical Behavior*, Ufa, Publishers USSTU, 1998, 366.

72. Kozlov, G. V.; Bejev, A. A.; Dolbin, I. V. Change of microgels structure at curing of epoxy polymers in fractal space. In book: *Perspectives on Chem. and Biochemical-Phys.*. Ed. Zaikov G. New York, Nova Science Publishers, Inc,2002, 225–230.

73. Kozlov, G. V.; Yanovskii, Yu. G.; Zaikov, G. E. Synergetics and Fractal Analysis of Polymer Composites Filled with Short Fibers. New York, Nova Science Publishers, Inc,2011, 223 p.

74. Kozlov, G. V.; Bashorov, M. T.; Mikitaev, A. K.; Zaikov, G. E. Transition nanoreactor-nanoparticle in curing process of epoxy polymers. *J. Balkan Tribological Association*, 2008, **14** (2), 215–220.

75. Kozlov, G. V.; Bashorov, M. T.; Mikitaev, A. K.; Zaikov, G. E. Transition nanoreactor-nanoparticle in epoxy polymers curing process. *Chem. and Chemical Technology*, 2008, **2** (4), 281–284.

76. Kozlov, G. V.; Bashorov, M. T.; Mikitaev, A. K.; Zaikov, G. E. The transition nanoreactor-nanoparticle in curing process of epoxy polymers. *Polymer Research J.*, 2009, **3** (1), 95–102.

77. Kozlov, G. V.; Bashorov, M. T.; Mikitaev, A. K.; Zaikov, G. E. The transition nano-reactor-nanoparticle in curing process of epoxy polymers. In book: *Chem. and Biochemistry. From Pure to Applied Science*. New Horizons. Ed. Pearce, E.; Zaikov, G.; Kirshenbaum, G.; New York, Nova Science Publishers, Inc., 2009, 345–352.

78. Kozlov, G. V.; Shustov, G. B.; Zaikov, G. E. The fractal and scaling analysis of chemical reactions. *J. Appl. Polymer Sci*, 2004, **93**(5), 2343–2347.

79. Smirnov, B. M. Fractal clusters. *Achievements of Phys. Sciences*, 1986, **149**(2), 177–219.

80. Kozlov, G. V.; Zaikov, G. E. The physical significance of reaction rate constant in Euclidean and fractal spaces at consideration of polymers thermooxidative degradation. *Theoretical Grounds of Chemical Technology*, 2003, **37**(5), 555–557.

81. Buchachenko, A. L. Nanochemistry – the direct way to new century high technologies. *Achievements of Chemistry*, 2003, **72**(5), 419–437.

CHAPTER 5

FRACTAL ANALYSIS AND SYNERGETICS OF CATALYTIC SYSTEMS

CONTENTS

5.1   FRACTAL MODELS OF CATALYSIS PROCESS

At present it is known [1] that the majority of catalytic systems are nanosystems. At the heterogeneous catalysis active substance one tries to deposit on
the bearers in a nanoparticles form in order to increase their specific surface.
At homogeneous catalysis active substance molecules by themselves have
often nanometer sizes. It is known too [2] that the operating properties of
heterogeneous catalyst systems depend on their geometry and structure of the
surface, which can influence strongly on catalytic properties, particularly, on
catalysis selectivity. It has been shown experimentally [3] that the montmorillonite surface is a fractal object. Proceeding from this, the authors [4] studied
the montmorillonite fractal surface effect on its catalytic properties in isomerization reaction.

In paper [5] two types of montmorillonite were used: Na– montmorillonite
(SW) and Ca– montmorillonite (ST). The indicated types of the layered silicate
were applied as a catalyst at isomerization of 1-butene (B) by obtaining *cis*-2-bu-
tene (C) and *trans*-2-butene (T).

Meakin [6] considered the simplest catalysis scheme, which was used later for
the estimation of catalyst selectivity S_c. It demonstrates general features, inherent
to all catalysis models, and is expressed by a simple reaction scheme, which can
be written as follows [4]:

$$A + P \rightarrow A_a,   \qquad (1)$$

$$A \xrightarrow[(cat)]{k_1} B,   \qquad (2)$$

$$A + A_a \xrightarrow[(cat)]{k_2} C,   \qquad (3)$$

where the Eq. (1) represents molecule A adsorption on catalyst surface P. The
Eq. (2) represents unimolecular process, transformating an adsorbed molecule
$A(A_a)$ in a new molecule B, which is assumed as fastly leaving catalyst surface. In real systems this can be an isomerization reaction (as in the consid-

ered case) or a reaction of secondary products decay. The Eq. (3) represents the molecule A addition to catalyst surface in the site already, occupied by the adsorbed molecule $A(A_a)$ with subsequent reaction of molecule C formation, which is also assumed by fastly leaving this surface in order to make the model maximum simple. The selectivity S_c is determined as a molecules C number, divided by a molecules B number [6].

Within the framework of this model, using computer simulation, the following expression for S_c was obtained [6]:

$$S_c = \frac{k_i \Sigma_i P_i^2 / (k_1 + 2P_i k_f)}{k_i \Sigma_i P_i / (k_1 + 2P_i k_f)},$$ (4)

where k_f is molecules A receiving rate on catalyst surface, P_i is contact probability for i-th site of the indicated surface.

The Eq. (4) is simplified essentially for two limiting cases. In the large k_1 limit it accepts the form [6]:

$$S_c = \frac{k_f}{k_1} \Sigma_i P_i^2,$$ (5)

and in the small k_1 limit let's obtain the equation [6]:

$$S_c = \frac{k_f}{Nk_1},$$ (6)

where N is a total sites number of catalyst surface, which can be estimated according to the following general fractal relationship [7]:

$$N \sim L^{d_{surf}},$$ (7)

where L is characteristic size of nanofiller particle, accepted equal to 100 nm [3], d_{surf} is catalyst surface fractal dimension.

In its turn, d_{surf} value can be estimated with the help of the equation [8]:

$$S_u = 410\left(\frac{L}{2}\right)^{d_{surf}-d}, \qquad (8)$$

where S_u is specific surface of nanofiller particles, d is dimension of Euclidean space, in which fractal is considered (it is obvious, that in our case $d=3$). The value S_u is given in m²/g and L – in nm. In Fig. 1, the dependence $S_c(N)$ is adduced for two types of the studied catalyst – montmorillonite. In case of 1–butene isomerization reaction the Eqs. (1)–(3) can be written as follows [4]:

$$B+P \rightarrow B_a, \qquad (9)$$

$$B_a \underset{(cat)}{\overset{k_1}{\rightarrow}} C, \qquad (10)$$

$$B+B_a \underset{(cat)}{\overset{k_2}{\rightarrow}} T, \qquad (11)$$

and then the value S_c is defined as the ratio T/C.

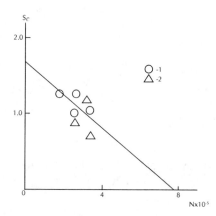

FIGURE 1   The dependence of the catalyst selectivity S_c on its surface sites total number N for Na– montmorillonite (1) and Ca– montmorillonite (2).

From Fig. 1 data S_c linear decrease at N growth follows. According to the Eq. (6) this assumes isomerization reaction course at small k_1 and at fulfillment of the conditions k_1=const, k_f=const or k_f/k_1=const. In case of larger k_1 (Eq. (5)) $\Sigma_i P_i^2$ >1/N [6]. For Witten–Sander clusters the value $\Sigma_i P_i^2$ is scaled as follows [6]:

$$\Sigma_i P_i^2 \sim N^{-\gamma}, \qquad (12)$$

where the exponent γ varies within the limits of 0.5–0.8 [6].

One of the fractal analysis merits is a clear definition of limiting values of its main characteristics – fractal dimensions. So, the value d_{surf} changes within the limits of $2 \leq d_{surf} < 3$ [9]. At d_{surf}=2.0 the value N=0.1×10^5 relative units and according to the Fig. 1 plot maximal value S_c=1.65. At the greatest for real solids dimension d_{surf}=2.95 [10] N=7.94×10^5 of relative units and according to Fig. 1 the plot $S_c \to 0$. This means that at such conditions trans-2-butene conversion degree goes to zero.

It was found out that d_{surf} increase resulted to the decrease of trans-2-butene and cis-2-butene general conversion degree. This is explained by the fact that the formed in synthesis process a polymeric chain has finite rigidity and consists of statistical segments of finite length. In virtue of this circumstance it cannot "repeat" the growing catalyst surface roughness at d_{surf} increase and "perceives" it as still smoother surface. In this case the effective fractal dimension of montmorillonite surface d_{surf}^{ef} is determined as follows [11]:

$$d_{surf}^{ef} = d_{surf} \qquad (13)$$

within the range of d_{surf}=2.0–2.5 and according to the Eq. (10) (at d_{surf}^{ef}=d_{surf}

and d_{surf}=d_{surf}^0) – within the range of d_{surf}=2.5–3.0.

Since for the studied catalysts the values d_{surf}=2.637–2.776 (let's note their

closeness to the experimental value d_{surf}=2.78 [3]) then for d_{surf}^{ef} estimation the Eq.

(10) was used. In Fig. 2, the dependence $Q(d_{surf}^{ef})$ is adduced, which turns out to

be linear and extrapolates to Q=0 at d_{surf}^{ef}=2.0 (or d_{surf}=3.0) and to Q=1.0 at d_{surf}^{ef} =d_{surf}=2.5. Thus, the combined consideration of Figs. 1 and 2 allows to assume the catalyst optimal value d_{surf}, which is equal to 2.5. At this d_{surf} magnitude Q=1.0 and S_c=1.54, i.e., close to maximum value S_c for montmorillonite in the considered reaction.

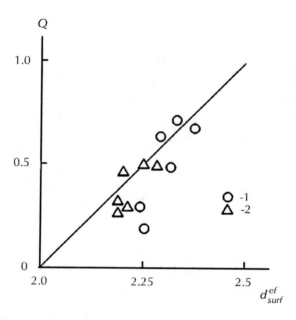

FIGURE 2   The dependence of total conversion degree Q on effective fractal dimension d_{surf}^{ef} of catalyst surface for Na– montmorillonite (1) and Ca– montmorillonite (2).

Hence, the results stated above demonstrated the important role of catalyst (montmorillonite) surface fractal geometry in its catalytic properties definition [4].

In work [12] it has been shown that in the liquid-phase epoxidizing of ethyl-allylethylacrylate (EAEA) by hydroperoxide of tert-butyl (HPTB), catalyzed by molybdenum boronide Mo_2B, the nonlinear dependence of the initial epoxidizing rate k_1 on the catalyst contents [cat] is observed. This effect has been explained by the decrease of redispersion degree of Mo_2B aggregates at the increase of [cat] in reactionary medium and the corresponding decrease of accessible for catalysis surface area of aggregates Mo_2B, which is accessible for catalysis [12]. Neverthe-less, the authors [12] have admitted that in the case considered by them only the redispersion effect cannot adequately explain the received experimental data.

It is known [13-15], that the catalysts used at present have a rough surface, which is, as a rule, fractal one. Besides, as it is shown in paper [16], the disperse powder has in general the fractal surface of particles, whose dimension d_{surf} changes within the broad range ($2<d_{surf}\leq3$). These circumstances allow to assume that Mo_2B surface also has fractal structure. One more argument in favor of such assumption is nonlinearity of the dependence $k_1([cat])$, that is a typical sign of the reaction fractality, i.e., either the reactions of fractal objects or the reactions occurring in fractal spaces [17]. Therefore the authors [18] have shown the possibility of the found in paper [12] effects qualitative description within the framework of fractal analysis.

As it is known [19], one of the most outstanding surface characteristics is its capability to control the physical and chemical processes concerning both distance and rate. For diffusion-controlled catalytic reactions the authors [19] obtained the following relationship:

$$Q(t) \sim t^{(d-d_{surf})/2} , \tag{14}$$

where Q is a number of particles, reacted on catalyst surface or a conversion degree, t is a reaction duration, d is dimension of Euclidean space, in which one reaction occurs (it is obvious, that in the considered case $d = 3$), d_{surf} is a dimension of catalyst surface.

As it has been shown in paper [19], the relationship is general enough and can be used for the description of any diffusion-controlled reaction. So, the authors [20] used this relationship for the description of radical polymerization reaction in the form of the Eq. (61) of Chapter 2. Differentiating the Eq. (14) by time t and using the analogue with the Eq. (16) of Chapter 2, we can obtain [18]:

$$k_1([cat],t) \sim [cat]t^{(1-d_{surf})/2} , \tag{15}$$

since at the change $[cat]$ because of its small absolute value a considerable change of reactionary medium initial viscosity η_0 is not expected.

For calculation of the value k_1 according to the Eq. (15), it is necessary to estimate the fractal dimension d_{surf} of the aggregates surface of the initial particles of catalyst Mo_2B by the independent method. For this purpose the relationship was used [16, 19]:

$$S_0 \sim r^{d_{surf}-3} , \tag{16}$$

where S_0 is a specific area of the surface of the initial powder Mo_2B, which is equal to 0.31 m²/g [12], r is a radius of particle aggregates Mo_2B, adopted according to the data [12] within the interval 1.4–2.4 mcm at [cat] variation, equal to 0.5–0.6 g/dm³.

Assuming for [cat] = 0.5 g/dm³ and for value r = 1.4 mcm d_{surf}= 2.2, corresponding to it [13], the constant of the proportionality in the Eq. (16) can be obtained, that allows to calculate d_{surf} as a function [cat]. This dependence is shown in Fig. 3 (solid line), from which the fast growth of d_{surf} at small [cat] and the achievement of plateau – for [cat]>4 g/dm³ follows.

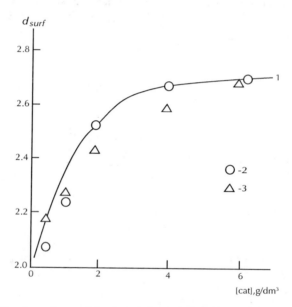

FIGURE 3 The dependence of the fractal dimension d_{surf} of particle aggregates surface of catalyst Mo_2B on its contents [cat], calculated according to the Eqs. (16) (1), (17) (2) and (15) (3).

This shape of the dependence d_{surf} on the concentration of the initial disperse material is typical and is observed for other condition of aggregation, namely, the aggregation of graphite powder with the size of the initial particles ~ 10 mcm in polymer composites, which is described by the empirical equation [21]:

$$d_{surf} = d_0 + \left(1 - \frac{\ln r_0}{\ln r}\right),$$

(17)

where d_0 and r_0 are a fractal dimension of surface and a radius of the initial powder particles, respectively.

In Fig. 3, the calculation d_{surf} according to the Eq. (17) at d_0=2.08 and r_0=1.4 mcm is also shown. As one can see, a good enough correspondence of the dependences d_{surf}([cat]), calculated according to the Eqs. (16) and (17) is received, though the values d_0 for them differ a little (d_0 = 2.20 and 2.08, respectively). This points out that the increase of d_{surf} at [cat] growth is only due to the aggregation process of the initial particles of catalyst.

Then the value k_1 can be calculated theoretically, using the Eq. (15). In Fig. 4, the comparison of the experimental and calculated according to the Eq. (15) dependences k_1([cat]) is adduced, where the values d_{surf} were calculated according to the Eqs. (16) and (17). As one can see, at both methods of estimation d_{surf} the fractal model of catalysis [19] gives the exact enough both qualitative, and quantitative description of the change k_1 at variation of the concentration of catalyst Mo_2B. In other words, the model [19] gives the correct explanation to the effect described in paper [12]. As it follows from the Eq. (15), at d_{surf}=2 (the smooth surface of a catalyst) the dependence k_1 on [cat] will be linear at t=const (the curve 4 in Fig. 4), but k_1 at [cat]=const is a function of the time, decreasing at t increase. At d_{surf}=1.0 the dependence k_1 on [cat] will be linear too and will not depend on t.

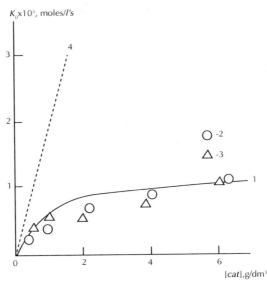

FIGURE 4   The dependence of the epoxidation initial rate k_0 on the contents of catalyst Mo_2B [cat]: experimental data (1); calculation according to the Eq. (15) at d_{surf}, determined according to the Eqs. (16) (2) and (17) (3) and d_{surf}=2.0 (4). ([EAEA]$_0$=2.2 mole/l, [TBHP]$_0$=0.52 mole/l, T=383 K).

Using the Eq. (15), the opposite procedure can be made, i.e., to calculate d_{surf} on k_1, assuming again d_{surf}=2.2 at [cat]=0.5 g/dm³. The comparison of the dependences d_{surf}([cat]), calculated according to the Eqs. (15) and (16), in Fig. 3 is shown. In spite of the definite disagreements, both the shape of these dependences, and the absolute values d_{surf} agree rather well (the disagreement is smaller than 10 %). This circumstance points out, that on the initial stage of catalysis "poisoning" of catalyst does not happen [19]. If such an effect happened, then the values d_{surf} calculated according to the Eq. (15) would be smaller than the corresponding magnitudes calculated according to the Eq. (16) [19].

Then let's consider the nature of aggregates particles of catalyst. As it is known [12], the density of these aggregates ρ_{fr} can be determined according to the equation:

$$\rho_{fr} = \frac{3}{rS_0}. \tag{18}$$

The calculation ρ_{fr} according to the Eq. (18) at S_0=0.31 m²/g=const and r=1.4–2.4 mcm shows the decrease ρ_{fr} at the increase r within the limits of 6910–4030 kg/m³. As it is known [12], the density of the initial particles of catalyst Mo_2B ρ_{Euc} is equal to 9260 kg/m³. Therefore, for particle aggregates the condition $\rho_{fr}<\rho_{Euc}$ is true, that is typical for the loose fractal aggregates, which density is determined according to the Eq. (73) of Chapter 1, the calculation according to this equation has shown, that the value d_f of aggregates Mo_2B structure within the range of r=1.4–2.4 mcm changes insignificantly, namely, within the range of d_f=2.13–2.05. Let's note, that such value d_f is typical for aggregates, restructuring during the growth process, i.e., including the processes of restructuring [22]. This value d_f also well agrees with the dimension of the aggregates of nickel particles, obtained experimentally [23].

At last, let's find the physical significance of the catalyst structural characteristics, controlling the epoxdizing rate. The catalysis process can be considered as an analogue of fractal object growth process with the difference, that instead of particles adding to a fractal object we will take into account the reacting molecules. According to [24], we assume that the catalyst particles aggregate with a radius r has three types of sites:
- Visited by molecules (infectious or poisoning ones) nodes, belonging to a cluster;
- The nodes already visited, but blocked (not susceptible ones);
- The nodes still non visited, which are the neighbors of the visited ones (nodes of growth).

The nodes of growth form an "open boundary" of a fractal [25] and their multitude has a fractal dimension d_g:

$$N_g \sim r^{d_g} . \tag{19}$$

Between the mass of cluster (aggregate) N and N_g there is the following relationship [24]:

$$N_g \sim N^x , \tag{20}$$

where $x = d_g / d_f = 1/2$. Since $N \sim r^{d_f}$, then from the Eqs. (19) and (20) we obtain [18]:

$$N_g \sim N^{d_g / d_f} \sim r^{d_f d_g / d_f} \sim r^{d_g} \sim r^{d_f / 2} . \tag{21}$$

As it was shown above, $d_f \approx 2$ and then $k_1 \sim N_g \sim r$. This conclusion is confirmed by the data of Fig. 5, where the dependence $k_1(r)$ is adduced, plotted according to the experimental data of paper [12].

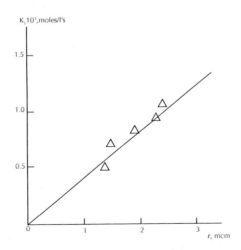

FIGURE 5 The dependence of the epoxidation initial rate k_1 on radius of particles aggregates of the catalyst Mo_2Br.

Thus, the stated above results have shown that the methods of the fractal analysis explain qualitatively and quantitatively the nonlinear character of the dependence of the epoxidation initial rate on the catalyst concentration, obtained by the authors [12]. Let's note that the continual models which use the representations of Euclidean geometry, do not allow to obtain the adequate explanation.

5.2   THE SYNERGETICS OF CATALYSIS BY METAL OXIDES

For studying filler catalytic properties the preliminary carried out transesterification model reaction allows to determine the catalytic activity of 14 metals oxides [26]. As these studies have shown, the oxides catalytic activity, estimated according to the first order reaction rate constant the value k_1, varies widely – from 0.004×10^{-4} s^{-1} for BeO up to 8.91×10^{-4} s^{-1} for CaO, i.e., by more than three orders. For explanation of this difference a number of factors was considered: the electronic configuration of metals atoms, its solubility in reactionary medium and so on [26]. Nevertheless, to obtain general explanation of the observed effect wasn't successful. Therefore the authors [27, 28] undertake the attempt to give such explanation within the framework of the nanoparticles synergetic [29].

The sizes interval of the used metal oxides particles (220–700 nm) allows to attribute them to a nanoparticles type (at any rate, formally) and to use for their description nanoparticles synergetics laws. Ivanova [29] introduced atom structural stability measure Δ_p and showed, that this parameter was in periodical dependence on the atom mass M while adaptability threshold of atom structure A_m with M increase corresponds to the condition [29]:

$$A_m = \frac{M_n^*}{M_{n+1}^*} = \Delta_p^{1/m} \,, \tag{22}$$

where M_n^* and M_{n+1}^* are threshold masses of atoms, within the limits of which the stable connection remains, $\Delta_p = f(M)$, m is an exponent of structure periodicity at M change, equal to $m=1$ for linear and to $m = 2$ for nonlinear feedback.

In Table 1 the relationships for Δ_p calculation as M function for stable connection six intervals are listed. The dependence of k_1 on the values A_m calculated according to the Eq. (22) at the condition of linear feedback ($m = 1$) in logarithmic coordinates is adduced in Fig. 5. As one can see, the approximately linear correlation $k_1(A_m)$ is obtained, analytically described as follows [27]:

$$\ln k_1 = -17 + 26.2 A_m \ .$$ (23)

TABLE 1 The formula for the calculation of structural stability of metal atoms Δ_p.

The threshold element and its precursors	The form of stable connection between Δ_p and M
Ca: 1–20	$\Delta_p = 1.50 \times 10^{-2} M$
Kr: 20–36	$\Delta_p = 0.55 \times 10^{-2} M$
La: 36–57	$\Delta_p = 0.232 + (M-84) \times 0.423 \times 10^{-2}$
Lu: 57–71	$\Delta_p = 0.232 + (M-139) \times 0.423 \times 10^{-2}$
Ac: 71–89	$\Delta_p = 0.232 + (M-175) \times 0.287 \times 10^{-2}$
Db: 89–105	$\Delta_p = 0.232 + (M-227) \times 0.287 \times 10^{-2}$

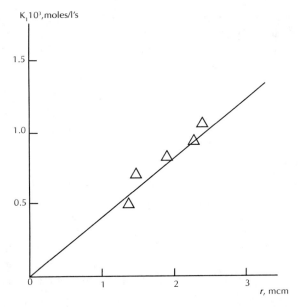

FIGURE 5 The dependence of the transesterification reaction rate constant k_1 on metal oxides nanoparticles adaptability A_m in logarithmic coordinates.

Therefore, the higher metal atom structure adaptability, included in oxide composition, characterized by value A_m is, the higher its catalytic activity, characterized by value k_1 is. At minimum value $\Delta_p = 0.232$ ($A_m = 0.232$) according to the law of "gold proportion" for the considered metals [29] from the Eq. (23) we obtain $k_1 \approx 0.18 \times 10^{-4}$ s^{-1}, i.e., the value k_1 for the transesterification reaction without filler [26]. Therefore, at any value $\Delta_p < 0.232$ ($A_m < 0.232$ at $m = 1$) the transesterification reaction will be slowed down [28].

Hence, the stability measure of metal oxides particles and their adaptivity are in periodical dependence on metal atom mass. The higher adaptivity of the metal atom, including in oxide composition, characterized by the value A_m is, the higher its catalytic activity, characterized by the first order reaction rate constant k_1 is. This dependence is general for metal oxides of II–VII groups of periodic system [28].

The structural analysis shows [30] that the fractal dimension D_f of transesterification reaction product (heptylbenzoate) changes in case of the indicated metal oxides presence from 1.52 up to 2.29. Therefore the authors [30] studied the factors, controlling heptylbenzoate molecule structure change and, hence, transesterification reaction rate. The calculation of fractal dimension D_f of a fractal-like heptylbenzoate molecule was made according to the Eq. (18) of Chapter 2.

As it has been noted above, the decrease of reaction rate, characterized by the value k_1, is due to the dimension D_f increase within the interval 1.52–2.29. Such D_f change has not only quantitative character, but qualitative as well. The value D_f depends on the mechanism of aggregation (synthesis) and indicated above the change of this dimension means the transition from the diffusion-limited aggregation to the chemically limited one [31]. D_f increase was noted at electronegativeness raising of metals included in oxides composition [26]. As it is known [32], relative electronegativeness obeys a general law: in period it increases with the element number raising while in group it decreases. Its values serve as elements of nonmetallicity measure. It is obvious that the more relative electronegativeness is the stronger the element reveals its nonmetallic properties [32].

Using general tendencies indicated above it is possible to construct the dependence of D_f on the most general relative electronegativeness characteristic: the ratio of a period number N_{per} to a group number N_{gr}. Such dependence is shown in Fig. 7. As one can see, a good enough linear correlation is obtained, showing D_f increase at ratio N_{per}/N_{gr} raising, i.e., the raising of electronegativeness degree of the element included in the oxide composition. The greatest D_f value for fractal-like molecules equal to 2.28 [33] was also shown by a horizontal shaded line in Fig. 7.

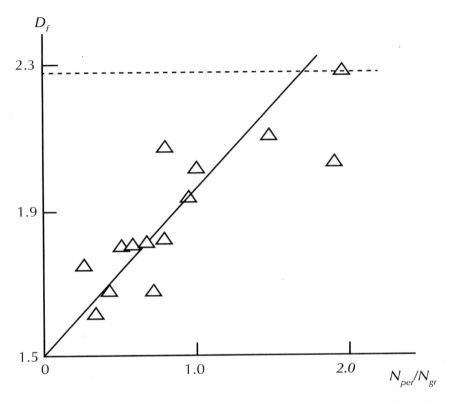

FIGURE 7 The dependence of the heptylbenzoate molecule fractal dimension D_f on the ratio N_{per}/N_{gr} for metal oxides. The horizontal shaded line shows the greatest value D_f of fractal-like molecule.

This D_f value corresponds well to the heptylbenzoate molecule fractal dimension obtained in transesterification reaction process in BeO presence. Let's note, that the obtained within the synergetic approach structural stability measure of metal atom Δ_p is also the function of relative electronegativeness expressed by ratio N_{per}/N_{gr} (Fig. 8). One can see, that Δ_p magnitude decreases with relative electronegativeness raising from maximum value $\Delta_p = 0.618$ and at $N_{per}/N_{gr} \approx 1.25$ reaches the minimum asymptotic value $\Delta_p = 0.232$ [29]. The last value means minimum oxide adaptivity, i.e., its minimum catalytic activity that results to values $k_1 = (0.004\text{–}0.13) \times 10^{-4}$ s^{-1} (let's remind, that in case of CaO presence $k_1 = 8.91 \times 10^{-4}$ s^{-1} [26]).

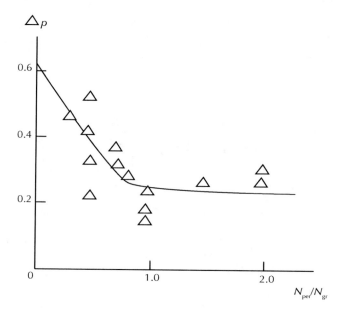

FIGURE 8   The dependence of structural stability measure Δ_p on value of the ratio N_{per}/N_{gr} for metal oxides.

In its turn, as it has been shown in paper [34], Δ_p increase results to the reduction of metal oxide solubility C_m in methylbenzoate. This dependence is expressed analytically as follows [34]:

$$C_m^{1/2} = 2.5\Delta_p \ , \tag{24}$$

where C_m is expressed in mol. %.

Let's consider metal oxides solubility synergetics in more detail. For solubility degree estimation of metal compounds they were heated with methylbenzoate or heptanol-1 at the same temperature and temporal conditions, as in kinetic studies. After heating the liquid was filtered from any undissolved oxide and in the filtrate the contents of metal was determined by -ray fluorescent method. It was shown [26], that at 443 K in methylbenzoate only those oxides were dissolved, which were better catalysts and in heptanol-1 all the studied metal compounds had not been dissolved practically. Therefore in Fig. 9 the dependence of metal contents (in mol. % from the reagent – methylbenzoate) C_m as the function of A_m for the six studied oxides is adduced. As it follows from the adduced plot, the value $C_m^{1/2}$ (the quadratic form of dependence was chosen for its linearization) is the increasing function of A_m, as well as k_1 (see Fig. 6). The relation between C_m and A_m is given analytically by the following empirical equation [34]:

$$C_m^{1/2} = 5.61(A_m - 0.213) ,$$ (25)

where C_m is given again in mol. % from a reagent.

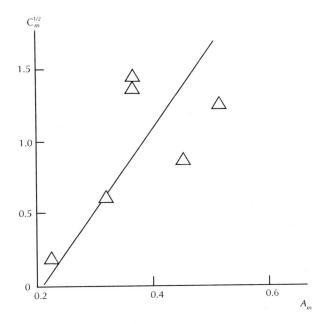

FIGURE 9   The dependence of metal contents in methylbenzoate C_m on its adaptability A_m.

Therefore, the metal adaptability A_m degree raising results to its solubility in methylbenzoate growth. This has a definite influence on the value k_1 of transesterification reaction. However, it should be noted, that this influence is not decisive, since C_m increase at A_m raising happens slower, than the function $k_1(A_m)$ growth (see the Eq. (23)). So, A_m increase from 0.3 up to 0.5 results to k_1 growth approximately in 190 times, whereas C_m raising makes up only 11 times. At minimum (according to the law of gold proportion [4]) the value $A_m = 0.213$ C_m=0, i.e., metal in reactionary medium does not dissolve [34].

Further the authors [30] carried out theoretical calculation of fractal dimen-

sion $D_f(D_f^T)$, using the methods offered in paper [35]. In the mentioned paper aggregation analysis of gold colloidal particles was accomplished and it was shown
that the added pyridine concentration increase results to D_f decrease and aggregation mechanism change from chemically-limited to diffusion-limited. To explain
this effect the interaction energy nature of short-range was considered. The statistical probability (steric factor) p of two particles (in our case – methylbenzoate
and heptanol–1) molecules sticking is determined as follows [35]:

$$p \sim \exp\left(-\frac{E_r}{RT}\right) , \qquad (26)$$

where E_r is repulsion energetic barrier of Coulomb's origin, R is universal gas
constant, T is reaction temperature ($T = 443$ K).

The energy of bonds, formed at the reaction, $\gg RT$ since heptylbenzoate molecule formation is irreversible. Metal ions dissolved in methylbenzoate decrease
E_r value and therefore influence directly on p magnitude. Consequently, E_r value
can be considered as the value [30] Eq. (27):

$$E_r = \frac{I + E}{C_m} , \qquad (27)$$

where I is atom ionization energy, E is the energy of electron affinity.

The sum of energies ($I + E$) was calculated according to the method cited in
paper [32]. As the standard value the total energy ($I + E$) for lithium was accepted,
which is equal to 5.61 Ev, and which has relative electronegativeness $X_{Li} \approx 1.0$.
Further according to the known literary data [32] value X can be obtained for arbitrary metal and calculated for it ($I + E$) magnitude as product 5.61X in Ev [32].
Then according to the Eq. (26) p value is calculated [30]. It is significant that at
transition from CaO to BeO p decrease from 0.857 up to 0.057 is observed, i.e., as
it was noted above, the transition from diffusion-limited aggregation (synthesis)
mechanism to chemically-limited one was realized [36]. And lastly, theoretical

value of fractal dimension D_f^T of heptylbenzoate molecule can be calculated now
using the Eq. (76) of Chapter 1.

In Fig. 10, the comparison of values D_f and D_f^T calculated according to the Eqs. (18) of Chapter 2 and (76) of Chapter 1, respectively, is shown. As one can see, between these parameters a good correspondence (the average discrepancy of D_f and D_f^T is equal to 6%) is obtained confirming the treatment of metal oxides catalytic activity offered above.

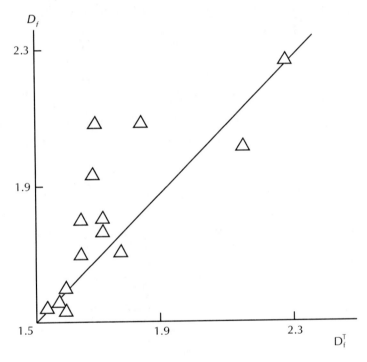

FIGURE 10 The comparison of the heptylbenzoate molecule fractal dimension, calculated according to the Eq. (18) of Chapter 2 D_f and Eq. (76) of Chapter 1 D_f^T.

Therefore, the results considered above have shown that the catalytic activity of metal oxides nanoparticles is defined by their relative electronegativeness. The higher this parameter is the smaller metal solubility in methylbenzoate is. This results to the increase of the energetic barrier of methylbenzoate and heptanol-1 molecules repulsion, that decreases steric factor and results to raising the fractal dimension of heptylbenzoate molecule. The last factor decelerates sharply the transesterification reaction.

It is important to note that the synergetic effect of metal oxides is also preserved at the formation of high-molecular macromolecules in the polytransesterification reaction. At poly(butylenes terephthalate) (PBT) synthesis study by polytransesterification reaction in six metal oxides (CaO, PbO, BaO, ZnO, CoO and MgO) presence substantial increase of intrinsic viscosity [η] was found out in comparison with the reaction without nanoparticles [26]. So, the last reaction gives PBT with [η] = 0.06 dl/g, whereas the presence of 10 mass % metal oxides allows to obtain the [η] values within the interval 0.15–0.68 dl/g. Therefore, in case of using metal oxides as nanofillers higher-molecular polymers can be obtained. The authors [26] have assumed that one of the reasons of this effect can be an increase of the system viscosity in polytransesterification reaction course and the reaction transition from kinetic region in the diffusive one, where the prevailing factor becomes not the catalyst activity, but its active centers mobility. However, with regard to such explanation one should note, that in virtue of the same contents of oxides in reactionary medium its viscosity should be higher than at their absence, but approximately the same for the indicated oxides, whereas the value [η] for them differs approximately in five times. The authors [37] offered the explanation of the observed effect within the framework of nanoparticles synergetics [29].

As it is shown above, the more is the adaptivity of the metal atom structure, included in oxide composition, characterized by the value A_m the higher is its catalytic activity, characterized by the value k_1. Let's note, that in paper [26] [η] increase at k_1 growth was obtained. Therefore in Fig. 11 the correlation between the intrinsic viscosity [η] for PBT and the value A_m in case of using the six metal oxides, indicated above, is adduced. As one can see, that approximately linear correlation is obtained demonstrating [η] growth at A_m increase. This correlation can be described analytically by the following empirical equation [37]:

$$[\eta] = 2.03 (A_m - 0.213) , \tag{28}$$

where [η] is given in dl/g.

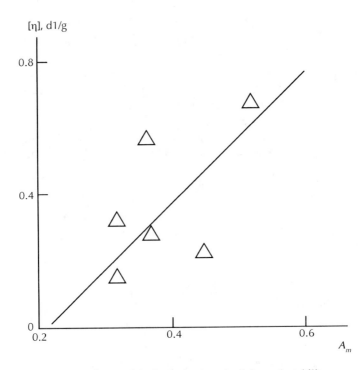

FIGURE 11 The dependence of the intrinsic viscosity [η] on adaptability measure A_m of metal oxides for polytransesterification reaction PBT.

It is significant that the extrapolation of the dependence $[\eta](A_m)$ to $[\eta]=0$ gives the value $A_m = 0.213$, i.e., minimum possible metal adaptability measure according to the law of gold proportion [29]. The Eq. (28) also allows to estimate maximum reached in polytransesterification reaction process [η] magnitude for PBT: at $A_m \approx 1.0$ this [η] value is equal approximately to 1.6 dl/g. However, the greatest real value A_m at $m = 1$ is equal to 0.618 [29] and corresponding to it [η] value is equal to 0.822 dl/g, that is close to value [η] = 0.68 dl/g obtained in CaO presence. Let's note, that the nonlinear feedback, i.e., $m = 2$, allows to increase [η]: so, in this case for CaO will be obtained $A_m = 0.718$ and [η] = 1.025 dl/g. Therefore, the offered treatment allows to predict [η] value for PBT polytransesterification reaction.

5.3 FEATURES OF THE KINETICS OF TRANSESTERIFICATION MODEL REACTION, CATALYZED BY METAL OXIDES

Very strong influence of metal inorganic compounds nanoparticles on transesterification model reaction process, both quantitative and qualitative was shown above. This influence is expressed quantitatively in essential change of the first order reaction rate constant k_1: from 8.91×10^{-4} s^{-1} for transesterification reaction in CaO presence up to 0.004×10^{-4} s^{-1} – in BeO presence, i.e., more than on three orders [26]. The qualitative difference consists of a type of kinetic curves conversion degree – reaction duration $Q(t)$: for the reaction with larger k_1 ($>2.80 \times 10^{-4}$ s^{-1}) they have autoaccelerated character and for the reaction with smaller k_1 – autodecelerated one. The authors [38] made analytical inference of the curves $Q_{(t)}$ transition criterion from autoaccelerated regime to autodecelerated one.

In Fig. 12, the kinetic curves $Q_{(t)}$ for transesterification reaction without catalyst and in the presence of metal inorganic compounds BeO, Fe$_2$O$_3$, CeO$_2$ and BaO are adduced.

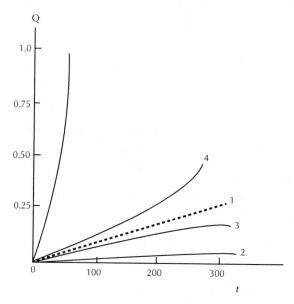

FIGURE 12 The kinetic curves conversion degree – reaction duration $Q_{(t)}$ for transesterification reaction without catalyst (1) and in BeO (2), Fe$_2$O$_3$ (3), CeO$_2$ (4) and BaO (5) presence.

As one can see, the reaction without catalyst has approximately linear kinetics, the reaction in BeO and Fe_2O_3 is autodecelerated one and in CeO_2 and BaO is autoaccelerated one, while the kinetic curve $Q_{(t)}$ for reaction without catalyst represents the boundary, dividing both indicated regimes.

In papers [39, 40] it was shown, that the cause of transition from autodecelerated regime of polymers thermooxidative degradation to autoaccelerated one is a structural factor, namely, the increase of fractal dimension of macromolecular coil in melt. We have all reasons to assume, that this factor will be decisive in the considered case as well. The fractal dimension D_f of the heptylbenzoate fractal-like molecule can be determined with the help of the Eq. (18) of Chapter 2. In Table 2 the values k_1 according to the data of paper [26], the values D_f, calculated according to the Eq. (18) of Chapter 2 and the type of the curve $Q_{(t)}$ for transesterification reaction in 16 indicated inorganic compounds with metal presence and also for this reaction without catalyst are listed. From the data of Table 2 clearly enough the expressed tendency follows: for small D_f, about <1.80, the large k_1 and autoaccelerated kinetics are observed while for $D_f > 1.80$ – small values k_1 and autodecelerated kinetics of transesterification reaction are also observed. This allows to make the conclusion about structural origin of transesterification reaction kinetic regime change. In case of small D_f fractal-like molecules of heptylbenzoate and its components are less compact and the reaction can occur in internal regions of their structure. At D_f increase the indicated molecules are compacted and the reaction occurs only on their surface. This assumption is placed in the basis of analytical inference of kinetic regime change criterion or, more exactly, of determination of the critical value $D_f(D_f^{cr})$, at which the indicated transition is realized. Further, by analogy with papers [39, 40] the authors [38] have assumed that the observed change of kinetic curve $Q_{(t)}$ type occurs in the case, when molecule reactive sites number on its surface N_{surf} and in its volume N_{vol} becomes equal [38]:

$$N_{surf} = N_{vol} . \tag{29}$$

TABLE 2 The characteristics of transesterification model reaction in 30 mass % inorganic compounds of metal presence.

The metal compound	$k_1 \times 10^4$, s^{-1} [26]	D_f	The kinetic curve $Q_{(t)}$ type
–	0.18 ± 0.02	1.85	linear
CaO	8.91 ± 0.52	1.54	autoaccelerated
PbO	5.38 ± 0.33	1.54	autoaccelerated
BaO	4.40 ± 0.21	1.62	autoaccelerated

TABLE 2   *(Continued)*

ZnO	3.63 ± 0.20	1.67	autoaccelerated
CoO	2.80 ± 0.20	1.70	autoaccelerated
MgO	1.96 ± 0.14	1.80	linear
MoS_2	1.85 ± 0.11	1.80	linear
CuO	0.26 ± 0.02	1.81	autodecelerated
Al_2O_3	0.25 ± 0.04	2.03	autodecelerated
CeO_2	0.20 ± 0.01	1.78	autoaccelerated
La_2O_3	0.19 ± 0.01	1.80	autodecelerated
ZnS	0.16 ± 0.01	1.86	autodecelerated
Fe_2O_3	0.13 ± 0.01	1.95	autodecelerated
TiO_2	0.11 ± 0.008	2.10	autodecelerated
ZrO_2	0.08 ± 0.004	2.06	autodecelerated
BeO	0.004 ± 0.0002	2.29	autodecelerated

Let's consider methods of estimation of parameters in the Eq. (29). As it has been shown above (see section 1.2) for fractal objects in Euclidean spaces with $d>1$ as a fractional exponent v_f fractional part of D_f is accepted, that allows one to determine v_f value according to the Eq. (43 of Chapter 1). In this case the value v_f characterizes reactive molecules fraction, not participating in evolution (reaction) process. Then the molecule fraction β_f, capable of chemical transformations in reaction course, is defined as follows [39]:

$$\beta_f = 1 - v_f = d - D_f. \tag{30}$$

Hence, the real value of reactive sites number N_{vol} in molecule volume can be written as follows [38]:

$$N_{vol} = N_m \left(d - D_f \right) - N_{surf}, \tag{31}$$

where N_m is a reactive sites number per molecule at the condition of their full accessibility. It is obvious that the value N_m is defined by reagent chemical composition.

The relation of parameters N_{vol} and N_{surf} is controlled by relation of volume V_{fr} and surface S_{fr} of a fractal-like object, which is equal to [10]:

$$\frac{V_{fr}}{S_{fr}} = \frac{R}{D_f} , \qquad (32)$$

where R is fractal-like object radius, in our case equal to gyration radius of heptylbenzoate molecule R_g.

The exact estimation of the last parameter is difficult enough. Therefore in paper [38] the formula R_g estimation in case of polycarbonate was used [41]:

$$R_g = 0.988 \times 10^{-10} M_w^{0.5} , \text{ m}, \qquad (33)$$

where M_w is average weight molecular mass of the heptylbenzoate molecule, with the replacement of exponent 0.5 by the value 0.65 typical for good solvent [41]. For the heptylbenzoate molecule $M_w = 215$, that gives $R_g = 32.4$ Å according to the Eq. (33). Since the mentioned estimation has approximate character then further the value $R_g = 3$ nm will be used [38].

The combination of the Eqs. (29)–(33) allows to obtain the structural criterion of kinetic curves $Q(t)$ transition from autodecelerated regime to autoaccelerated one and vice versa [38]:

$$\frac{D_f^{cr}}{d - D_f^{cr}} = \frac{R_g}{2} , \qquad (34)$$

where D_f^{cr} is critical value D_f at the indicated transition and the value R_g is given in nm.

Let's note a number of the Eq. (34) important features. Firstly, in this equation the parameter N_m is absent, i.e., the value D_f^{cr} is independent on polymer chemical composition. Secondly, R_g increase results to D_f^{cr} raising. This fully corresponds to the assumption, served as a basis for the Eq. (34) inference: the value N_{vol} increases proportionally to cube R_g and N_{surf} to quadrate R_g, i.e., R_g increase results to the ratio V_{fr}/S_{fr} raising. Thirdly, the Eq. (34) explains autoaccelerated regime absence for the majority of reactions of polymers synthesis: in these reactions fast growth of R_g increases the value D_f^{cr} unlike the transesterification model reaction, where R_g=const,. So, already at R_g= 5 nm D_f^{cr} = 2.14, i.e., equal to fractal dimension of macromolecular coil, at which reaction of polymers synthesis is practically stopped [42]. At typical sizes of macromolecular coil at synthesis end of the order R_g=15 nm D_f^{cr}=2.65, i.e., higher than this coil dimension in gelation process, where $D_f \approx 2.5$ [43, 44]. Fourthly, radical polymerization reaction autoacceleration is observed very clearly on its initial part because of D_f sharp decrease [45, 46] and, hence, at the condition $D_f < D_f^{cr}$ reaching.

Let's consider the boundary conditions for the Eq. (34). For D_f= 0, i.e., for a dotted object, R_g = 0 and Eq. (34) is true for a zero-dimensional object. For D_f= 3 $R_g \to \infty$ and, as one can expect, for Euclidean object measure scale has no physical significance.

In Fig. 13, the dependence k_1 on D_f in logarithmic coordinates is adduced. As one can expect, the linear decrease ln k_1 at D_f growth is observed. The critical value D_f^{cr} obtained according to the Eq. (34) is pointed in Fig. 13 by a vertical shaded line. As one can see, it serves actually as a boundary for autoaccelerated and autodecelerated kinetics of transesterification reaction.

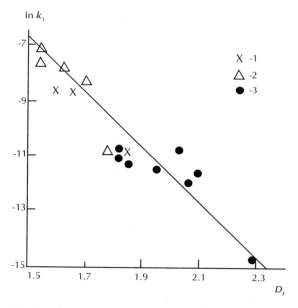

FIGURE 13   The dependence of the first order rate constant k_1 on reaction product fractal dimension D_f for transesterification reaction in logarithmic coordinates. Type of kinetic curves $Q(t)$: 1 – linear, 2 – autoaccelerated, 3 – autodecelerated. Vertical shaded line indicates the value D_f^{cr}.

The authors [38] also made the estimation of methylbenzoate and heptanol–1 molecules interpenetration depth l for reactions in CaO ($D_f = 1.54$) and BeO ($D_f = 2.29$) presence. The value l can be estimated according to the equation [47] Eq. (35):

$$ l = \frac{1}{2} a \left(\frac{R_g}{a} \right)^{2(d - D_f)/d} , \qquad (35) $$

where a is a lower linear scale of molecule fractal behavior, d is the dimension of Euclidean space, in which fractal is considered (it is obvious, that in our case $d = 3$).

Assuming $a = 3$ Å, $M_w = 110$ for the molecules indicated above and estimating the value R_g for them as 21 Å according to the Eq. (33), we obtain $l = 10$ Å for the molecule with $D_f = 1.54$ (CaO) and $l = 3.7$ Å for the molecule with $D_f = 2.29$ (BeO). In other words, if in the first case molecules penetrate one into another almost completely ($2l \approx R_g$), then in the second case they can react only in a surface layer with thickness 3.7 Å, i.e., about 18 % from R_g.

Hence, the results quoted above demonstrated clearly, that the transition of transesterification reaction kinetic curves from an autodecelerated regime to an autoaccelerated one was defined by a structural factor, namely, by reaching of fractal-like molecule (reaction product) fractal dimension critical value, at which the number of reactive sites in volume and on the surface of molecule became equal. Within the framework of the fractal analysis the analytic relationship is obtained, confirming this hypothesis.

In paper [48] the autoacceleration effect in transesterification model reaction kinetics in a number of metal oxides nanoparticles was studied within the framework of chemical reactions fractal kinetics [49–51]. As it is known [50], the average number of cluster $\langle S$, visited by random walk by which methylbenzoate and heptanol–1 molecules are simulated depends on time t as follows:

$$ \langle S \rangle \sim t^{d_s'/2} , \qquad (36) $$

where d_s' is an effective spectral dimension of reactionary medium in which random walks diffuse. The value d_s' is connected with reactionary medium heterogeneity exponent h ($0 \leq h \leq 1$) according to the Eq. (1). At $h=0$ the system (reactionary medium) becomes homogeneous (Euclidean) and at $h>0$ it is heterogeneous (fractal).

In Fig. 14 experimental kinetic curves $Q_{(t)}$ (points) are shown for transesterification reaction catalyzed by the three metal oxides (CaO, BaO and ZnO). As one can see, these curves $Q_{(t)}$ have approximately quadratic form that allows to write the Eq. (36) as follows [48]:

$$Q \sim t^{d_s'}.\qquad(37)$$

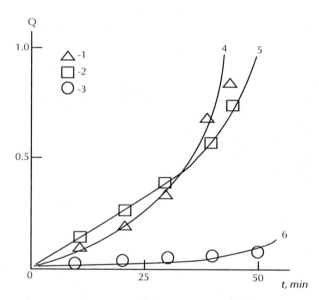

FIGURE 14 The experimental (1–3) and calculated according to the Eq. (38) (4–6) kinetic curves $Q_{(t)}$ for transesterification reaction catalyzed by CaO (1, 4), BaO (2, 5) and ZnO (3, 6) nanoparticles.

In Fig. 15, the dependences $Q_{(t)}$ are shown in double logarithmic coordinates for transesterification reaction catalyzed by CaO an BaO nanoparticles, which correspond to the Eq. (37). As one can see, in both cases the linear dependences breaking down into two sections were obtained. At small durations ($t \leq 27$ min) the value d_s' determined according to the slope of these linear sections is equal to 1.1 and according to the Eq. (1) $h = 0.45$ will be obtained. Therefore, at small durations the transesterification reaction occurs in homogeneous medium with small values d_s', that defines its relative small rate (see Fig. 15). At relatively large durations ($t > 27$ min) the reaction rate increases essentially and from the slope of Fig.

15 linear plots the value d_s' can be determined for this part, which is equal to 2. According to the Eq. (1) this means $h = 0$, i.e., at $t > 27$ min the transesterification reaction proceeds in homogeneous medium (Euclidean space). In other words, at $t = 27$ min the time-dependent transesterification reaction transition from heterogeneous (fractal) kinetics to homogeneous (Euclidean) one is observed. The indicated transition defines the sharp increase of reaction rate and corresponding growth of conversion degree Q.

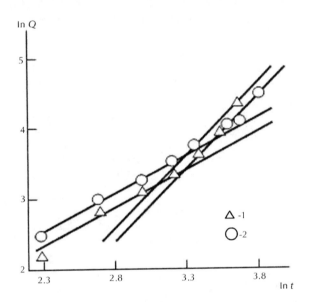

FIGURE 15 The dependences of conversion degree Q on reaction duration t in double logarithmic coordinates for transesterification reaction catalyzed by CaO (1) and BaO (2) nanoparticles.

Therefore, the curves $Q_{(t)}$ for the considered transesterification reaction can be simulated by the following simple equation [48]:

$$Q = c_1 t^{d_s'}, \qquad (38)$$

where the coefficient c_1 is the function of heptylbenzoate molecule structure, which can be characterized quantitatively by its fractal dimension D_f, determined with the help of the Eq. (18) of Chapter 2.

For the considered metal oxides D_f the increase from 1.54 for CaO up to 1.81 for ZnO results to c_1 decrease from 0.042 up to 0.002, respectively, i.e., approximately in twenty times. Hence, the sections of kinetic curves $Q_{(t)}$ at $t \le 27$ min can be simulated by the linear correlation and at $t > 27$ min by the quadratic one according to the Eq. (38). The results of such simulation are shown in Fig. 14 by solid lines. As one can see, a good correspondence of theory and experiment is obtained.

Let's note an interesting aspect following from the indicated simulation. Transesterification reaction rate which is very different for catalyst cases by CaO and ZnO nanoparticles in the considered case is independent from reactionary medium connectivity, but is defined by the structure of reaction product (hep-

tylbenzoate molecule), i.e., the value D_p since d_s' value in the three considered examples is the same [48].

Reaction order n is connected with heterogeneity exponent h as follows [51]:

$$n = \frac{2-h}{1-h} .$$
(39)

As it follows from the data, cited above, at $t27$ min $n=2.82$ and at $t > 27$ min transesterification reaction becomes classical reaction of the second order.

Let's consider the physical reasons for dimension d_s' increase at t raising. As it is known [52], the value $\langle S \rangle$ is connected with steps number N (or time t) of a particle up to the contact with cluster according to the following relationship:

$$\langle S \rangle \sim N^{d_s'/2} .$$
(40)

It is obvious, that as the transesterification reaction occurs a number of reactive molecules of methylbenzoate and heptanol-1 is decreased and they need to make more steps (the Eq. (40)) or to spend more time (Eq. (36)) in order to "find" reactive molecule-partner for realization heptylbenzoate molecule formation reaction. In its turn, in paper [49] it is shown, that N (or t) increase results to

d_s' growth, the value of which tries to attain asymptotically to the corresponding value for Euclidean space. This is a physical base of the observed time-dependent transition from fractal kinetics to Euclidean one for the transesterification model reaction [48].

Therefore, the results quoted above have shown that for the transesterification reaction catalyzed by nanoparticles of metal inorganic compound autoaccelerated regime of kinetics is defined by the time-dependent transition from the fractal behavior to Euclidean one. This transition is due to the increase of effective spectral dimension of reactionary medium owing to time interval raising between collisions of reacting molecules.

The permanent object of the study in organic and inorganic chemistry is the question of chemical reactions (for example, polymers synthesis) rate increase that allows to reduce their course duration and, hence, to raise the productivity of the process. For this purpose many sufficient methods are applied, for example, reaction temperature raising, catalysis, intensive stirring of reactionary medium and so on [53]. However, all these methods should be supposed external factors of influence on chemical reaction rate. The problem of intensification of internal factors has been studied much less. In this connection one can note the appearance of a number of new tendencies lately.

One of such tendencies is polymers synthesis in the presence of all kinds of fillers, which serve simultaneously as reaction catalyst [26, 54]. The second tendency is the chemical reactions study within the framework of physical approaches [55–59], from which the fractal analysis obtained the largest application [36]. Within the framework of the last approach in synthesis process consideration such fundamental conceptions as the reaction products structure, characterized by their fractal (Hausdorff) dimension D_f [60] and the reactionary medium connectivity, characterized by spectral (fracton) dimension d_s [61], were introduced. In its turn, diffusion processes for fractal reactions (strange or anomalous) differ principally from those occurring in Euclidean spaces and described by diffusion classical laws [62]. Therefore the authors [63] give transesterification model reaction kinetics description in 14 metal oxides presence within the framework of strange (anomalous) diffusion conception.

The authors [62] formulated the fractional equation of transport processes, namely, the Eq. (23), where depending on concrete value α one distinguishes persistent (superdiffusive, $1<\alpha\leq2$) and antipersistent (subdiffusive, $0\leq\alpha<1$) processes. In its turn, in paper [64] the equation was obtained, which connects the values α and D_f:

$$\alpha = 0.5\left(2 - D_f\right) .$$
(41)

In the Eq. (41) the condition $=$ const $= 0.25$ was accepted, where β is an exponent in the Eq. (23).

As it has been shown above, the kinetic curves $Q_{(t)}$ for transesterification reaction in different metal oxides presence are distinguished both qualitatively and quantitatively (Fig. 12). This allows the authors [63] to plot the dependence of conversion degree Q on active time t^α for 14 studied transesterification reactions in metal oxides presence (see Table 2) and analogous reaction without filler-catalyst for $t = 50$ min, shown in Fig. 15. As one can see, this dependence is approximated well by linear correlation, and described analytically as follows [63]:

$$Q = 0.138 t^\alpha .$$

(42)

In Fig. 16, the dependence $Q(t^\alpha)$, obtained for transesterification reaction in the presence of mica with different surface treatment is also adduced, which is described according to the empirical equation [24]:

$$Q = 0.108 t^\alpha .$$

(43)

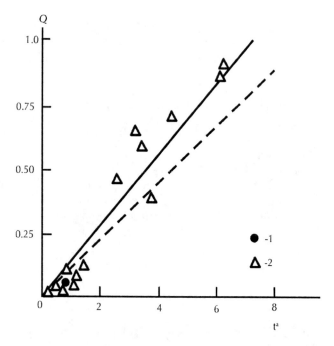

FIGURE 16   The dependence of conversion degree Q at $t = 50$ min on active time t^α for transesterification reaction without filler (1) and in metal oxides presence (2). The dependence $Q_{(t\alpha)}$ for transesterification reaction in mica presence is shown by a shaded line.

It is significant that the Eq. (42) describes kinetic curves better with autoacceleration and the Eq. (43) does linear and autodecelerated kinetic curves. Therefore, a kinetic curve type also affects Q value: for the same value of active time t^α the value Q is approximately by 22 % higher for the autoaccelerated kinetics than for the linear or autodecelerated one [63].

Let's consider physical aspects of heptylbenzoate molecules formation, which, as it has been shown above, are fractal-like objects. The increase D_f means the decrease of steric factor p ($p \le 1$), which defines reaction probability at reagents first contact. Values D_f and p are connected between themselves by the Eq. (76) of Chapter 1. This equation for the reaction in CaO presence ($D_f = 1.54$) gives $p = 0.852$ and for reaction in BeO presence ($D_f = 2.29$) gives $p = 0.042$. Therefore, small D_f correspond to diffusion-limited aggregation mechanism ($p \to 1$) and large D_f to chemically-limited one ($p \to 0$) [31]. It is important to note the decisive role of the fractal-like heptylbenzoate molecule structure, characterized by dimension D_f, in both aggregation type definition (the Eq. (41 of Chapter 2)) and strange (anomalous) diffusion type definition (the Eq. (41)). The introduction of either nanofiller-catalyst is only a method of structure regulation, moreover not the only one. Other methods are solvent variation, reaction temperature change or reagent relation and so on [36].

Hence, the results considered above have shown the possibility of product of transesterification reaction (heptylbenzoate molecule) structure regulation by filler-catalyst change which are used as metal oxides nanoparticles. The indicated structure change results to active time t^α essential variation, i.e., real chemical reaction duration, namely, for the considered conditions of reaction approximately in 20 times. In its turn, this tells on transesterification reaction main characteristics variation. The dependence of conversion degree on active time for all the used metal oxides is described by general linear correlation [63].

REFERENCES

1. Melikhov, I. V. Tendecies of nanochemistry development. *Russian Chemical Journal*, 2002, *46(5)*, 7–14.
2. Tyan, H. –L.; Liu, Y. -C.; Wei, K. –H. Enhancement of imidization of poly(amic acid) through forming poly(amic acid)/organoclay nanocomposites. *Polymer*, 1999, *40(11)*, 4877–4886.
3. Pernyeszi, T.; Dekani, I. Surface fractal and structural properties of layered clay minerals monitored by small-angle X-ray scattering and low-temperature nitrogen adsorption experiments. *Colloid Polymer Sci.*, 2003, *281(1)*, 73–78.

4. Naphadzokova, L. Kh.; Shustov, G. B.; Kozlov, G. V. Fractal analysis of nanoparticles catalytic properties. Proceedings of 9-th International symposium "Order, Disorder and Oxides Properties". *Rostov-on–Don, Loo*, 19–23 September 2006, 49–51.

5. Moronta, A.; Ferrer, V.; Quero, J.; Arteaga, G.; Choren, E. Influence of preparation method on the catalytic properties of acid-activated tetramethylammonium-exchanged clays. *Appl/ Catalysis A*, 2002, *230(1)*, 127–135.

6. Meakin, P. Simulation of the effects of fractal geometry on the selectivity of heterogeneous catalysts. *Chem. Phys. Lett.*, 1986, *123(5)*, 428–432.

7. Stanley, H. E. Fractal surfaces and "termite" model for two-component random materials. In book: *Fractals in Physics*. Ed. Pietronero L., Tosatti E. Amsterdam, Oxford, New York, Tokyo, North–Holland, 1986, 463–477.

8. Mikitaev, A. K.; Kozlov, G. V.; Zaikov, G. E. *Polymer Nanocomposites: Variety of Structural Forms and Applications*. New York, Nova Science Publishers, Inc., 2008, 319.

9. Avnir, D.; Farin, D.; Pfeifer, P. Molecular fractal surfaces. *Nature*, 1984, *308(5959)*, 261–263.

10. Balankin, A. S. *Synergetics of Deformable Body*. Moscow, Publishers of Ministry of Defence of SSSR, 1991, 404.

11. Van Damme, H.; Levitz, P.; Bergaya, F.; Alcover, J. F.; Gatineau, L.; Fripiat, J. J. Monolayer adsorption on fractal surfaces: a simple two-dimensional simulation. *J. Chem. Phys.*, 1986, **85**(1), 616–625.

12. Komarenskaya, Z. M.; Nikipanchuk, M. V.; Trach, Yu. B.; Yaremko, Z. M.; Fedushinskaya, L. B. Redispersion of Mo_2B powder in reactionary medium of ethylallylmethacrylate epoxidizing by tret-butyl hydroperoxide. *Problems of Chemistry and Chemical Technology*, 2002, (1), 38–41.

13. Farin, D.; Peleg, S.; Yavin, D.; Avnir, D. Applications and limitations of boundary-line fractal analysis of irregular surfaces: proteins, aggregates and porous materials. *Langmuir*, 1985, 1(4), 399–407.

14. Kozlov, G. V.; Zaikov, G. E. *Fractal Analysis and Synergetics of Catalysis in Nanosystems*. New York, Nova Biomedical Books, 2008, 163.

15. Wellner, E.; Rojanski, D.; Ottolenghi, M.; Huppert, D.; Avnir, D. Catalytic structure sensitivity: the effect of pore size on the oxygen quenching of excited aromatic molecules adsorbed on silica surfaces. *J. Amer. Chem. Soc.*, 1987, **109**(3), 575–576.

16. Avnir, D.; Farin, D.; Pfeifer, P. Chemistry in noninteger dimensions between two and three. II. Fractal surfaces of adsorbents. *J. Chem. Phys.*, 1983, **79**(7), 3566–3571.

17. Shogenov, V. N.; Kozlov, G. V. *Fractal Clusters in Ohysics–Chemistry of Polymers*. Nal'chik, Polygraphservice and T, 2002, 270.

18. Kozlov, G. V.; Afaunova, Z. I.; Zaikov, G. E. Fractal treatment of catalysis in the process of liquid-phase epoxidation of the ethylallylethylacrylate by tetrabutyl hydroperoxide. *Oxidation Commun.*, 2005, *28*(4), 863–868.

19. Pfeifer, P.; Avnir, D.; Farin, D. Scaling behavior of surface irregularity in the molecular domain: from adsorption studies to fractal catalysts. *J. Stat. Phys.*, 1984, *36*(5/6), 699–716.

20. Novikov, V. U.; Kozlov, G. V. The fractal analysis of macromolecules. *Achievements of Chemistry*, 2000, **69**(4), 378–399.

21. Kozlov, G. V.; Lipatov, Yu. S. Fractal and structural aspects of adhesion in particulate-filled polymer composites. *Composite Interfaces*, 2002, **9**(6), 509–527.
22. Kolb, M. Reversible diffusion-limited aggregation. *J. Phys. A*, 1986, **19**(5), L263–L268.
23. Shamurina, M. V.; Roldugin, V. I.; Pryamova, T. D.; Visotskii, V. V. Aggregation of colloidal particles in curing systems. *Colloidal Journal*, 1994, **56**(3), 451–454.
24. Naphadzokova, L. Kh.; Kozlov, G. V. *Fractal Analysis and Synergetics of Catalysis in Nanosystems*. Moscow, Publishers of Academy of Natural Sciences, 2009, 230.
25. Rammal, R.; Toulouse, G. Random walks on fractal structures and percolation clusters. *J. Phys. Lett.* (Paris), 1983, **44**(1), L13–L22.
26. Vasnev, V. A.; Naphadzokova, L. Kh.; Tarasov, A. I.; Vinogradova, S. V.; Lependina, O. L. Influence of metal inorganic compounds on poly(butylenes terephthalate) synthesis and properties of forming in situ compositions. *High–Molecular Compounds. A*, 2000, **42**(12), 2065–2071.
27. Burya, A. I.; Naphadzokova, L. Kh.; Kozlov, G. V. Description of metal oxides catalytic activity within the frameworks of synergetics. *Proceedings of IV-th Ukranian–Polish Scientific Conf. "The Polymers of Special Application"*. Dnepropetrovsk, DSU, 2006, 88.
28. Naphadzokova, L. Kh.; Kozlov, G. V.; Shustov, G. B. Catalytic activity of metal oxides in transesterification reaction of methylbenzoate by heptanol–1. *Catalysis in Industry*, 2007, (5), 61–63.
29. Ivanova, V. S. The periodical dependence of atoms and nanoparticles structure stability on their mass as nanoworld properties reflection. *Proceedings of International interdisciplinary symposium "Fractals and Applied Synergetics, FaAS–03"*. Moscow, Publishers of MSOU, 2003, 271–274.
30. Naphadzokova, L. Kh.; Kozlov, G. V.; Zaikov, G. E. Structure of the product of re-esterification reaction catalyzed by metal oxides nanoparticles. *Oxidation Commun.*, 2007, **30**(4), 788–792.
31. Kokorevich, A. G.; Gravitis, Ya. A.; Ozol–Kalnin, V. G. The scaling approach development at lignin supramolecular structure study. *Chemistry of Wood*, 1989, (1), 3–24.
32. Fomchenko, G. P. *Chemistry*. Moscow, Publishers of "Higher School", 1993, 368.
33. Family, F. Fractal dimension and grand universality of critical phenomena. *J. Stat. Phys.*, 1984, **36**(5/6), 881–896.
34. Naphadzokova, L. Kh.; Kozlov, G. V. Synergetics of metal oxides solubility in trans-esterification reaction process. *Proceedings of I-th International Sci.–Techn. Conf. "Analytical and Numerical Methods of Natural Scientific and Social Problems Simulation"*. Penza, PSU, 2006, 190–193.
35. Weitz, D. A.; Huang, J. S.; Lin, M. Y.; Sung, J. Limits of the fractal dimension for irreversible kinetic aggregation of gold colloids. *Phys. Rev. Lett.*, 1985, **54**(13), 1416–1419.
36. Kozlov, G. V.; Shustov, G. B.; Zaikov, G. E. Fractal physics of polycondensation processes. In book: *Essential Results in Chem. Phys.* Ed. Goloshchapov A., Zaikov G., Ivanov V. New York, Nova Science Publishers, Inc., 2004, 193–241.
37. Naphadzokova, L. Kh.; Kozlov, G. V. Synergistic effect of metal oxides in polytrans-esterification reaction. *Proceedings of VIII-th International Sci.–Pract. Conf. "New Chemical Technologies: Production and Application"*. Penza, PSU, 2006, 63–66.

38. Naphadzokova, L. Kh.; Kozlov, G. V.; Ligidov, M. Kh.; Pakhomov, S. I. Strange (anomalous) diffusion as a method of chemical reactions kinetics operation. *Proceedings of Higher Educational Institutions*. Chemistry and Chemical Technology, **2008**, *51(9)*, 79–81.

39. Kozlov, G. V.; Dolbin, I. V.; Zaikov, G. E. Structural criterion of change of a kinetic curves type in the process of a thermooxidative degradation. *Oxidation Commun.*, **2005**, *28(1)*, 143–152.

40. Kozlov, G. V.; Zaikov, G. E. *The Structural Stabilization of Polymers: Fractal Models*. Leiden, Boston, Brill Academic Publishers, **2006**, 345.

41. Shnell, G. *Chem. and Phys. of Polycarbonates*. Moscow, Chemistry, **1967**, 229.

42. Kozlov, G. V.; Temiraev, K. B.; Kaloev, N. I. A solvent nature influence on polyarylate structure and formation mechanism in a conditions of low-temperature polycondensation. *Reports of Academy of Sciences*, **1998**, *362 4*, 489–492.

43. Botet, R.; Jullien, R.; Kolb, M. Gelation in kinetic growth models. *Phys. Rev. A*, **1984**, *30(4)*, 2150–2152.

44. Kobayashi, M.; Yoshioka, T.; Imai, M.; Itoh, Y. Structural ordering on physiccal gelation of syndiotactic polystyrene dispersed in chloroform studied by time-resolved measurements of small angle neutron scattering (SANS) and infrared spectroscopy. *Macromolecules*, **1995**, *28(22)*, 7376–7385.

45. Kozlov, G. V.; Ozden, S.; Malkanduev, Yu. A.; Zaikov, G. E. Autoacceleration in the process of radical polymerization: fractal analysis. In book: *Fractals and Local Order in Polymeric Materials*. Ed. Kozlov G., Zaikov G. New York, Nova Science Publishers, Inc., **2001**, 11–19.

46. Kozlov, G. V.; Ozden, S.; Malkanduev, Yu. A.; Zaikov, G. E. Autoacceleration in the process of radical polymerization: fractal analysis. *Russian Polymer News*, **2002**, *7(3)*, 38–44.

47. Hentschel, H. G. E.; Deutch, J. M. Flory-type approximation for the fractal dimension of cluster-cluster aggregates. *Phys. Rev. A*, **1984**, *29(3)*, 1609–1611.

48. Naphadzokova, L. Kh.; Kozlov, G. V.; Zaikov, G. E. Time-dependent transition from fractal to Euclidean kinetics for model reaction of re-esterification. *J. Balkan Tribological Association*, **2007**, *13(3)*, 329 333.

49. Ardyrakis, P.; Kopelman, R. Random walk on percolation clusters. *Phys. Rev. B*, **1984**, *29(1)*, 511–514.

50. Kopelman, R.; Klymko, P. W.; Newhouse, J. S.; Anacker, L. W. Reaction kinetics on fractals: random-walker simulations and exciton experiments. *Phys. Rev. B*, **1984**, *29(6)*, 3747–3748.

51. Anacker, L. W.; Kopelman, R. Fractal chemical kinetics: simulations and experiments., **1984**, *81(12)*, 6402–6403.

52. Sahimi, M.; McKarnin, M.; Nordahl, T.; Tirrell, M. Transport and reaction on diffusion-limited aggregates. *Phys. Rev. A*, **1985**, *32(1)*, 590–595.

53. Korshak, V. V.; Vinogradova, S. V. *Nonequilibrium Polycondensation*. Moscow, Science, **1972**, 695.

54. Naphadzokova, L. Kh.; Vasnev, V. A.; Tarasov, A. I. Study of transesterification reaction kinetics in inorganic catalyst presence. *Plastics*, **2001**, *(3)*, 39–41.

55. Grassberger, P.; Procaccia, I. The long time properties of diffusion in a medium with static traps. *J. Chem. Phys.*, **1982**, *77(12)*, 6281–6284.

56. Meakin, P.; Stanley, H. E. Novel dimension – independent behavior for diffusive annihilation on percolation fractals. *J. Phys. A*, **1984**, *17(1)*, L173–L177.
57. Redner, S.; Kang, K. Kinetics of the "scavenger" reaction. *J. Phys. A*, **1984**, *17*(2), L451–L455.
58. Kang, K.; Redner, S. *Scaling approach for the kinetics of recombination processes*, **1984**, *(12)*, 955–958.
59. Kozlov, G. V.; Shustov, G. B.; Zaikov, G. E. The fractal and scaling analysis of chemical reactions. *J. Appl. Polymer Sci.*, **2004**, *93*(5), 2343–2347.
60. Vilgis, T. A. Flory theory of polymeric fractals – intersection, saturation and condensation. *Physica A*, **1988**, *153*(2), 341–354.
61. Alexander, S.; Orbach, R. Density of states on fractals: "fractons". *J. Phys. Lett* (Paris), **1982**, *43*(*17*), L625–L631.
62. Zelenyi, L. M.; Milovanov, A. V. Fractal topology and strange kinetics: from percolation theory to cosmic electrodynamics problems. *Achievements of Physical Sciences*, **2004**, *174*(8), 809–852.
63. Naphadzokova, L. Kh.; Kozlov, G. V.; Zaikov, G. E. Description of the model transesterification reaction within the framework of a strange diffusion concept. *J. Appl. Polymer Sci.*, **2008**, *109*(5), 2791–2794.
64. Naphadzokova, L. Kh.; Kozlov, G. V. Description of the model transesterification reaction within the framework of an anomalous diffusion concept. *Chemical Technology*, **2006**, *10*, 36–39. reaction within the framework of an anomalous diffusion concept. *Chemical Technology*, **2006**, *10*, 36–39.

THE STRUCTURAL MODEL OF TRANSESTERIFICATION REACTION IN MELT

At present it has been established that the polymers synthesis in solution is controlled by macromolecular coil structure, which is a fractal object [1–3]. In addition the synthesis process regardless of the used method (radical polymerization, polycondensation and so on) proceeds according to the mechanism of irreversible aggregation cluster–cluster, i.e., it presents itself small macromolecular coils joining in larger ones. In this case fractal dimension of macromolecular coils in solution varies within the limits of approx. 1.50–2.12 [4], which corresponds to the indicated aggregation mechanism. However, at polymers synthesis in melt macromolecular coil environment change occurs instead of solvent molecules it is surrounded now by the same coils. As it has been shown in paper [5] this results to fractal dimension enhancement of macromolecular coil in melt up to about 2.5 (compare the Eqs. (39) and (100) of Chapter 1). The macromolecular structure change should tell on the synthesis process course. Therefore the authors [6] offered the quantitative treatment of the assumed changes within the framework of fractal analysis and irreversible aggregation models on the example of polyester polyols polyesterification reaction [7].

In Fig. A1, the kinetic curves conversion degree — reaction duration Q_{-t} for two polyols on the basis of ethylenglycole (PO-1) and propyleneglycole (PO-2) are adduced. As it was to be expected, these curves have an autodecelerated character, i.e., the reaction rate ϑ_r decreases with time. Such type of kinetic curves is characteristic for fractal reactions, to which either fractal objects reactions or reactions in fractal spaces belong [8]. In case of Euclidean reactions linear kinetics is observed. For the fractal reactions the general Eq. (26) of Chapter 1 was obtained, from which it follows that the construction of the plot $Q_{(t)}$ in double logarithmic coordinates allows one to determine the exponent value in this relationship and, hence, the value of macromolecular coil in melt fractal dimension Δ_r.

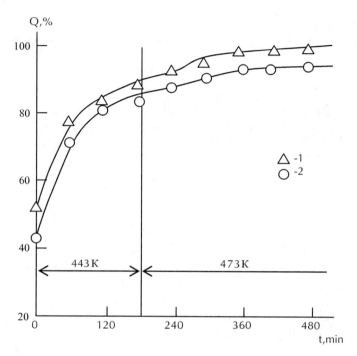

FIGURE A1   The kinetic curves conversion degree-reaction duration $Q_{(t)}$ for polyols on the basis of ethyleneglycole (1) and propyleneglycole (2).

In Fig. A2 such dependence for PO-1 is adduced, from which it follows that it consists of two linear parts, allowing to perform the mentioned above estimation. For small t (t ≤ 50 min) the slope of linear part is larger one and Δ_f=2.648 and t > 50 min Δ_f=2.693. Such Δ_f increase or macromolecular coil density enhancement in reaction course is predicted by irreversible aggregation model for multiple nucleation sites, the most fully corresponding to the conditions of real polymerization reactions [9]. The indicated dimensions correspond to irreversible aggregation mechanism particle–cluster, i.e., unlike synthesis in solutions a reaction in melt proceeds by particles (oligomer molecules) addition to the growing macromolecular coil. It is obvious, that Δ_f enhancement in the Eq. (26) of Chapter 1 in comparison with synthesis in solution should essentially increase reaction duration in virtue of this relationship power form that was observed in practice [10]. However, this reduction is compensated to a certain extent by higher c_0 values and much higher η_0 values for melt in comparison with solution as a result of that synthesis duration in melt remains on the acceptable level [7].

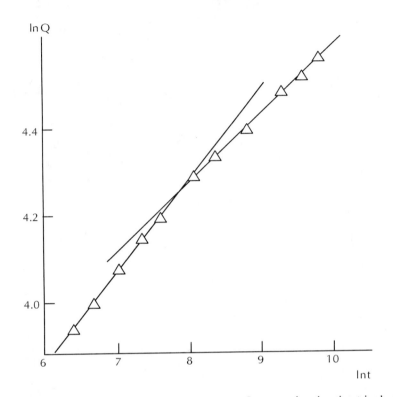

FIGURE A2   The dependence of conversion degree Q on reaction duration t in double logarithmic coordinates for polyol on the basis of ethyleneglycole.

The authors [7] have shown the polyesterification reaction ϑ_r decrease on about an order within the range of Q = 0.5–0.9. Within the framework of fractal analysis Δ_r-value can be received by the Eq. (26) of Chapter 1 differentiation by time t, and as a result the Eq. (69) of Chapter 1 was obtained. In Fig. A3 the comparison of the dependences $\Delta_r(Q)$, received experimentally [7] and calculated according to the Eq. (69) of Chapter 1, where the product $c_0\eta_0$ was supposed as constant and proportionality coefficient was obtained by method of the best correspondence of theory and experiment, is adduced. As it follows from the data of Fig. A3, between the indicated dependences a good correspondence is obtained. Let us note that according to the Eq. (69) of Chapter 1 ϑ_r decrease at t growth is due only to reaction fractal character.

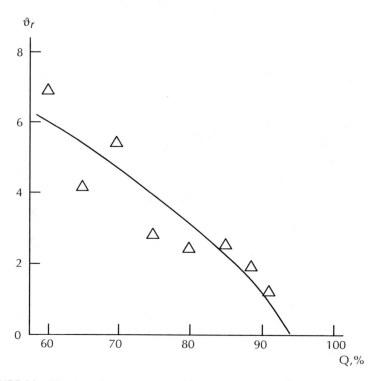

FIGURE A3   The dependences of polyeterification reaction rate ϑ_r on conversion degree Q, obtained experimentally (1) and calculated according to the Eq. (69) of Chapter 1 for polyol on the basis of ethyleneglycole.

ϑ_r estimation another variant is given in paper [11]. In this case for fractal reactions the Eq. (14) of Chapter 5 is valid. In Fig. A4 the dependence $\vartheta_r(t)$ in double logarithmic coordinates is adduced. As it was to be expected, this dependence turns out to be linear, that allows to determine from its slope heterogeneity exponent h according to the Eq. (14) of Chapter 5, which is equal to 0.56. Thus, this result confirms the polyesterification reaction proceeding in heterogeneous (fractal) medium. For such mediums (especially at elevated temperatures) connectivity is characterized not by a spectral dimension ds, but by its effective value , which takes into account temporal (energetic) disorder availability in the system [11]. The value   is linked with the exponent h by the Eq. (1) of Chapter 5, which for h = 0.56 gives = 0.88.

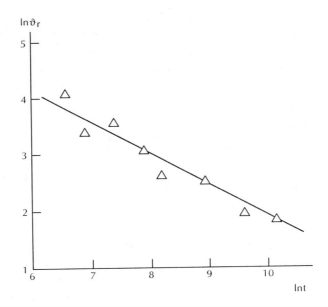

FIGURE A4 The dependence of reaction rate ϑ_r on its duration t in double logarithmic coordinates for polyol on the basis of ethyleneglycole.

Ardyrakis [12] showed that at the study of chemical reactions on fractal ob-jects the cor-rections on small clusters availability in the system were necessary. Just such corrections require the usage in theoretical estimations not generally accepted spectral (fracton) dimension ds [13], but its effective value . For perco-lation system two cases are possible [12]:

a) A random walk is placed on the largest percolation cluster of the system, characterized by the dimension ds;

b) A random walk can get on any cluster, including the cluster of a small size. Such clusters multitude is characterized by the dimension .

It is obvious, that molecular weight distribution availability in real polymers requires the usage of just effective spectral dimension value. The authors [14] of-fered the following formula for the value determination in case of polymer melts:

$$d_s' = \frac{2(2\Delta_f - d)}{d + 2}$$

(A.1)

Calculation according to the Eq. (1) gives for $\Delta_f = 2.648$ (T = 443 K) the value $= 0.918$ and for $\Delta_f = 2.693$ (T = 473 K) $= 0.954$. These both values correspond well to the estimation according to the Eq. (1) of Chapter 5 (the discrepancy makes up less than 8%). This circumstance confirms, that in melts of synthesized polyester polyols macromolecular coils of different sizes are formed, i.e., certain molecular weight distribution takes place.

Let us note that the synthesis temperature enhancement results to some h reduction (from 0.541 up to 0.523). This corresponds to the subordination theorem (the Eq. (15) of Chapter 5) [15].

And lastly, as it has been shown in paper [16], the medium character defines the reaction order n according to the Eq. (33) of Chapter 6. For the obtained above h magnitudes the value $n \approx 3$. Let us note, that for Euclidean spaces (objects) at h $= 0$ only one value of the reaction order is possible, namely, $n = 2$.

Hence, the results stated above have shown that the structural analysis of polyesterification reaction in melt, using ideas of fractal analysis and irreversible aggregation models, allows give precise enough description of this reaction even without handling of purely chemical aspects. Let us note, that fractal analysis is a much stricter mathematical method than the one often used for synthesis kinetics description empirical equations.

REFERENCES

1. Kozlov, G. V.; Shustov, G. B.; Zaikov, G. E. The fractal physics of the polycondensation processes. *J. Balkan Tribological Association*, 2003, **9**(4), 467–514.
2. Kozlov, G. V.; Shustov, G. B.; Zaikov, G. E. The fractal and scaling analysis of chemicalreactions. *J. Appl. Polymer Sci.*, 2004, **93**(5), 2343–2347.
3. Kozlov, G. V.; Dolbin, I. V.; Zaikov, G. E. Fractal physical chemistry of polymer solutions. *Polymer Research J.*, 2007, **1**(1/2), 167–210.
4. Kozlov, G. V.; Temiraev, K. B.; Kaloev, N. I. Solvent nature influence on polyarylate struc-ture and formation mechanism in the low-temperature polycondensation conditions. *Reports of Academy of Sciences*, 1998, **362**(4), 489–492.
5. Vilgis, T. A. Flory theory of polymeric fractals — intersection, saturation and condensation. *Physica A*, 1988, **153**(2), 341–354.
6. Naphadzokova, L. Kh.; Kozlov, G. V.; Zaikov, G.E. The structural model of polyesterifica-tion reaction in melt. *Chem. Phys., and Mesoscopy*, 2008, **10**(1), 72–76.
7. Vaidya, U. R.; Nadkarni, V. M. Polyester polyols from glycolyzed PET waste: effect of glycol type on kinetics of polyesterification. *J. Appl. Polymer Sci.*, 1989, **38**(6), 1179–1190.
8. Kozlov, G. V.; Zaikov, G. E. The physical significance of reaction rate constant in Euclidean and fractal spaces at consideration of polymers thermooxidative degradation. *Theoretical Grounds of Chem. Technology*, 2003, **37**(5), 555–557.

9. Witten, T. A.; Meakin, P. Diffusion-limited aggregation at multiple growth sites. *Phys. Rev. A*, 1983, **28**(10), 5632–5642.

10. Korshak, V. V.; Vinogradova, S. V. Nonequilibrium Polycondensation. Moscow, *Science*, 1972, 695

11. Klymko, P. W.; Kopelman, R. Fractal reaction kinetics: exciton fusion on clusters. *J. Phys. Chem.*, 1983, **87**(23), 4565–4567.

12. Ardyrakis, P. Percolation and fractal behavior on disordered lattices. In book: *Fractals in Phys*. Ed. Pietronero, L.; Tosatti, E. Amsterdam, Oxford, New York, Tokyo, North-Holland, 1986, 513–518.

13. Alexander, S.; Orbach, R. Density of states on fractals: "fractons." *J. Phys. Lett.* (Paris), 1982, **42**(17), L625–L631.

14. Kozlov, G. V.; Dolbin, I. V.; Zaikov, G. E. The theoretical estimation of effective spectral dimension for polymer melts. *J. Appl. Polymer Sci.*, 2004, **94**(4), 1353–1356.

15. Kopelman, R. Excitons dynamics resembling fractal one: geometrical and energetical disorder. In book: *Fractals in Phys*. Ed. Pietronero, L.; Tosatti, E. Amsterdam, Oxford, New York, Tokyo, North–Holland, 1986, 524–527.

16. Anacker, L. W.; Kopelman, R. Fractal chemical kinetics: simulations and experiments. *J. Chem. Phys.*, 1984, **81**(12), 6402–6403.

INDEX

M